D1326035

PANAMA FEVER

By the same author

The Battle of Britain
Monte Cassino

PANAMA FEVER

The Battle to Build the Canal

MATTHEW PARKER

HUTCHINSON
London

First published by Hutchinson in 2007

1 3 5 7 9 10 8 6 4 2

Hutchinson
The Random House Group Limited
20 Vauxhall Bridge Road, London SW1V 2SA

www.randomhouse.co.uk

Addresses for companies within The Random House Group Limited can be found at:
www.randomhouse.co.uk/offices.htm

The Random House Group Limited Reg. No. 954009

A CIP catalogue record for this book is available from the British Library

ISBN 9780091797041

The Random House Group Limited makes every effort to ensure that the papers used in its books are made from trees that have been legally sourced from well-managed and credibly certified forests. Our paper procurement policy can be found at:
www.randomhouse.co.uk/paper.htm

Typeset by Palimpsest Book Production Limited,
Grangemouth, Stirlingshire

Printed and bound in Great Britain by William Clowes Ltd, Beccles, Suffolk

In loving memory of Roger Durman

CONTENTS

ILLUSTRATIONS

Section Two

Maps

Text illustrations

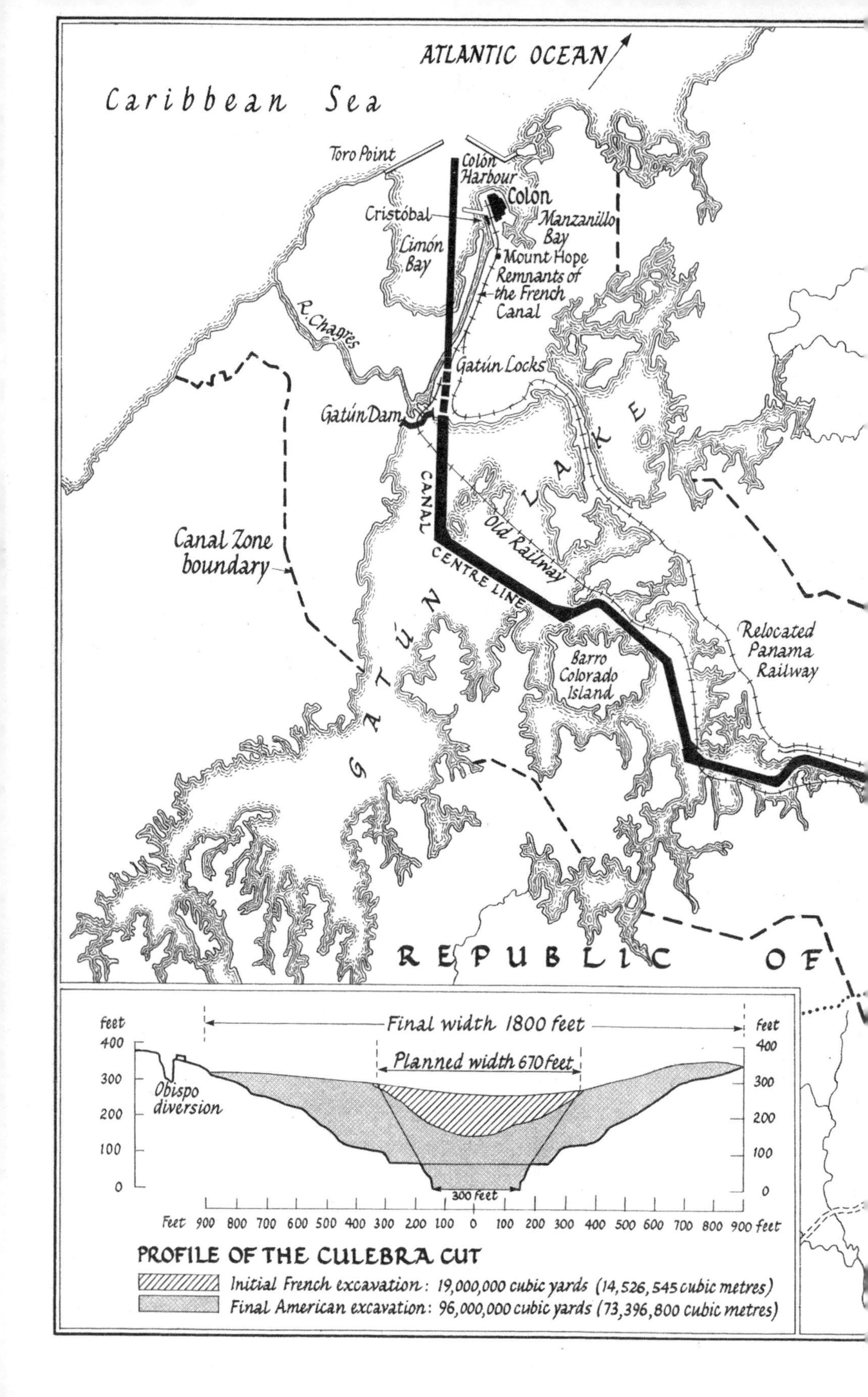
ATLANTIC OCEAN
Caribbean Sea
Toro Point
Colón Harbour
Colón
Cristóbal
Manzanillo Bay
Limón Bay
Mount Hope
Remnants of the French Canal
R. Chagres
Gatún Locks
Gatún Dam
CANAL CENTRE LINE
Old Railway
GATÚN LAKE
Canal Zone boundary
Barro Colorado Island
Relocated Panama Railway
REPUBLIC OF
feet
400
300
200
100
0
Final width 1800 feet
Planned width 670 feet
Obispo diversion
300 feet
Feet 900 800 700 600 500 400 300 200 100 0 100 200 300 400 500 600 700 800 900 feet
PROFILE OF THE CULEBRA CUT
Initial French excavation: 19,000,000 cubic yards (14,526,545 cubic metres)
Final American excavation: 96,000,000 cubic yards (73,396,800 cubic metres)

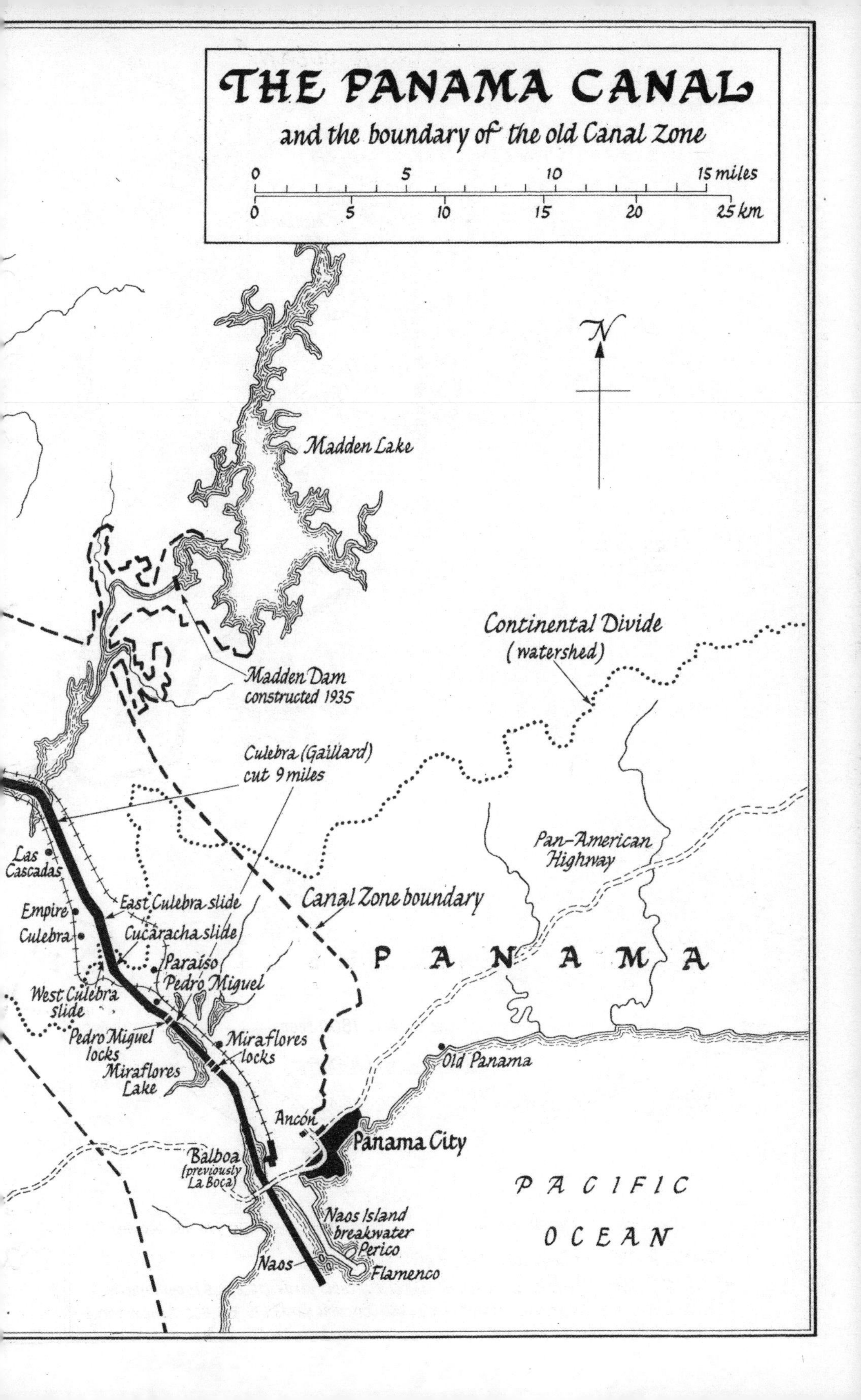
THE PANAMA CANAL
and the boundary of the old Canal Zone
0 5 10 15 miles
0 5 10 15 20 25 km
N
Madden Lake
Continental Divide
(watershed)
Madden Dam
constructed 1935
Culebra (Gaillard)
cut 9 miles
Pan-American
Highway
Las
Cascadas
Empire
Culebra
East Culebra slide
Cucaracha slide
Canal Zone boundary
PANAMA
Paraíso
Pedro Miguel
West Culebra
slide
Pedro Miguel
locks
Miraflores
locks
Miraflores
Lake
Old Panama
Ancón
Panama City
Balboa
(previously
La Boca)
PACIFIC
OCEAN
Naos Island
breakwater
Perico
Naos
Flamenco

PREFACE

The Battle to Build the Canal

'You here who are doing your work well in bringing to completion this great enterprise are standing exactly as a soldier of the few great wars of the world's history. This is one of the great works of the world.'

– President Theodore Roosevelt to the American canal workforce, 1906

Every year, on the anniversary of the opening of the Panama Canal on 14 August 1914, there is a special celebration. A small motor launch takes a group of about twenty-five on a short boat ride from a dock near the top of the Pedro Miguel lock. In August 2004, almost everyone is of Antillean descent, mainly from Jamaica or Barbados. The commemorative trip has been organised by Panama's British West Indian community. All ages are represented, and the vast majority are Panama-born. The others are from the islands, and one visitor with her young daughter has come all the way from Boston.

As the boat sets off from the dock, huge container ships loom past. One is a giant German carrier on the way to the US East Coast crewed by Philippinos. Another is transporting car parts from China to Brazil. We are floating in one of the world's key commercial arteries. Since the opening of the canal, over a million ships have used it to travel between the great oceans, some fourteen thousand a year, now passing through twenty-four hours a day. In places the sides of the canal consist of sheer rock, in others there are elaborate terraces of crumbly, reddish soil mixed with dark boulders. Where the ground is flat lie neat, clipped lawns, the rich tropical jungle kept at a respectable distance. As it is before noon, the container-carriers are heading from the Pacific to the Atlantic (travelling in a northerly direction). They have sailed past Panama City and ascended the double locks at Miraflores, crossed the small lake and climbed the single lock just behind us at Pedro Miguel. They are now at the top of the 'bridge of water', 85 feet above sea level. Beyond the narrow and windy stretch ahead, the canal opens up into the huge man-made Lake Gatún, created by the damming of the River Chagres with a massive earthen structure. At the other end of the lake, the vessels will descend through the spectacular triple lock at Gatún, and thence out into the Caribbean Sea and away.

Within a few minutes we are approaching the infamous Culebra Cut, the site of the maximum excavation for the canal. Here, for up to twenty years, thousands of men, the vast majority British West Indians, laboured in torrential rain or in burning heat to break the back of the Continental Divide, the rocky spine that links the great mountain ranges of North and South America.

The occasion is a celebration of the contribution made to the canal by the Antillean workers, but also a solemn memorial of their sufferings and losses. Between the two highest mountains of the Cut, where the sides of the canal rise almost vertically, the boat stops for petals and flowers to be thrown on the water. With all the passengers crowded out on the small deck, prayers are said by several of the rectors present, and hymns are sung as the warm tropical rain starts pouring down.

Bahamas-born Albert Peters was twenty-years old when, in 1906, he and two friends decided to head for the Isthmus. 'We were all eager for some adventure and experience,' he wrote. 'My parents were against the idea. They told me about the Yellow Fever, Malaria and Small Pox that infested the place but I told them that I and my pals are just going to see for ourselves.'

He was in for a shock. Seeing the working conditions, and the 'heavy rain and mud', he wanted to be back at home, but having spent all his money on the trip, he had no choice but to stick it out. Within a month he had malaria, and was hospitalised. 'The first night in there the man next to me died,' he writes, 'and that's when I remembered my parents' plea and wished I had taken their advice.'

'Death was our constant companion,' remembered another West Indian worker, Alfred Dottin. 'I shall never forget the trainloads of dead men being carted away daily, as if they were just so much lumber. Malaria with all its horrible meaning those days was just a household word. I saw mosquitoes, I say this without fear of exaggerating, by the thousands attack one man. There were days that we could only work a few hours because of the high fever racking our bodies – it was a living hell. Finally typhoid fever got me . . .'

Although the hardships of the construction period were shared in part by all the numerous nationalities who built the canal – French, American, Spanish, Italian, Greek and many others – the West Indian workers were three times as likely as the others to die from disease or accidents, and their startling accounts are dominated by stories of appalling conditions and dangerous work. The loss of life was astronomical. With safety precautions incredibly primitive by modern standards, accidents 'were numberless'. Constantine Parkinson, a Jamaican born in Panama, lost his right leg and left heel when the spoil-carrying train he was working on crashed off its rails. He was taken to hospital and operated on. 'After coming out of the operation in the ward,' he writes, 'I noticed all kinds of cripples around my bed without arms foot one eye telling me to cheer up not to fret we all good soldiers.'

'Some of the costs of the canal are here,' wrote a sympathetic American policeman in his account of visiting the main hospital's black wards, which were always situated on the least favourable side of the building. 'Sturdy black men in pyjamas sitting on the verandas or in wheel chairs, some with one leg gone, some with both. One could not help but wonder how it feels to be hopelessly ruined in body early in life for helping to dig a ditch for a foreign power that, however well it may treat you materially, cares not a whistle blast more for you than for its old worn-out locomotives rusting away in the jungle.'

Certainly the West Indians were treated as cheap and expendable by both the French and the Americans. The working conditions were described by one as 'some sort of semi-slavery', and, particularly under the Americans, there was a rigid apartheid system in place throughout the Canal Zone. Nonetheless, in spite of obvious resentments, the West Indian accounts are full of pride in knowing they were part of a great, heroic and civilising achievement. 'Many times I met death at the door,' wrote one worker fifty years after the completion of the canal, 'but thank God I am alive to see the great improvement the Canal had made and the wonderful fame it has around the world.' Another commented: 'We worked in rain, sun, fire, gunpowder, explosions from dynamite . . . but our interest was to see the Canal finished because we came here to build it . . . most of us came here with the same spirit as a soldier going to war, don't dodge from work or we will never finish it.'

The men who built the canal did indeed go to Panama as soldiers to a great battle and the fight to build the canal can be compared with an armed conflict. It has been estimated that three out of four of the French engineers who set out to be part of Ferdinand de Lesseps' heroic dream were dead within three months of arriving on Panama's 'Fever Coast'. Yellow fever and a mysterious and particularly vicious form of malaria known as 'Chagres Fever' accounted for thousands. Many others were carried off by accidents, pneumonia or sheer exhaustion. The most conservative estimate of the death toll is 25,000, five hundred lives for every mile of the canal. Many, many more were maimed or permanently debilitated by disease. Even in 1914, when the Isthmus was supposedly 'sanitised' by the Americans, over half the work force was hospitalised at some point during the year.

Apart from actual wars, it was the costliest project ever yet attempted in history, as ambitious a construction as the Great Pyramids. Hundreds of millions of francs were invested - and lost - during the ten-year struggle by the French in the 1880s, and the Americans spent nearly four hundred million dollars between 1904 and 1914, in the days when a couple of dollars a day was a good working wage. Although much shorter than the canal at Suez, it cost four times as much and required three times the amount of excavation. Mountains, literally, had to be moved. One observer called it the 'greatest liberty ever taken with nature.'

And 'nature' was not going to be conquered easily. In the way of the path between the seas were huge, geologically complex mountains, thick jungle teeming with deadly creatures and seemingly bottomless swamp. For eight months of the year there was almost continual rain. Because of the peculiar geographical configurations of the Isthmus it is one of the wettest places on earth. Two inches of rain have been known to fall in one hour. This made the Chagres River, which lies along the path of the waterway, a powerful and unstable force to have as a bedfellow to the canal. Rising up to thirty yards in an hour, it would regularly burst its banks, sweeping away men and materials. Vast mudslides buried workers, supplies, and machines. And in the swamps and puddles, fever-carrying mosquitoes bred in their millions to launch themselves on the toiling labourers.

What impresses now about the story of the canal is not just the extraordinary number of 'firsts' its achievement entailed – financial, technical and medical – but the astonishing, almost arrogant ambition of it all. Nothing like it had ever been attempted in the tropics before. The leaders of the project, be they French or American, simply believed they could do anything, that innovation and technology – the forces of Progress, of the Industrial Revolution and the great Victorian Age – were able to conquer any challenge.

The French effort, in particular, powered as it was by private capital and a sublime belief in emerging technology, sees this Age overreach itself, with tragic consequences. The Americans were driven less by idealism than by national, racial and military ambition, but they too would be humbled by the challenges that the jungles of Panama presented. The US construction project succeeded because of state funding, local political control and access to scientific and technical expertise beyond the reach of the French. But it also opens the door to a new era where the efforts of individuals would be controlled and channelled by the state for its own purposes, the machine age that was ushered in by the industrial slaughter of the Western Front.

In both cases, and throughout the history of the canal dream, almost everyone involved with the project, from the humblest pick-and-shovel man to the most venal Wall Street speculator, became gripped by the 'great idea' of the canal, by 'Panama Fever'. But it is striking, too, how many enemies the canal project attracted through its history. Vested interests feared the change in the status quo that such a radical altering of the geography of the world would usher in. The Americans, in particular, were fiercely opposed to a foreign power controlling any transcontinental waterway. The French attempt would bring heavy criticism of the 'over-optimism' of its promoter, Ferdinand de Lesseps, and its failure would see his ruin and disgrace, as well as financial and political disaster for France. The American project was even more controversial, entailing at its inception the murky activities of political lobbyists and a vivid demonstration of a new kind of United States, casting off its historical aversion to imperialism and aggression on the international stage.

The successful opening of the canal in August 1914, at almost precisely the moment when Old Europe was embarking on a ruinous war, was the climax of the United States' spectacular rise to world power. The Isthmus was the key to the struggle for mastery of the Western Hemisphere as well as to wider international commercial and naval strength. With the successful completion of the Washington-funded and dominated canal, the United States emerged as a truly global power and the 'American Century' could begin.

In Panama itself, the canal was realisation of a dream that went back four hundred years. It had been the destiny of the Isthmus ever since 1513, when the Spanish conquistador Vasco Núñez de Balboa ventured inland from Panama's Caribbean coast and, 'silent, upon a peak in Darien', discovered a previously unknown great ocean separated from the Atlantic by only a narrow bridge of land forty miles wide. Balboa's discovery immediately engendered a belief that a waterway could be built linking the two oceans. Thereafter, the Isthmus became of crucial strategic importance and the focus of fierce international rivalry between Spain, France, Great Britain and, as it emerged, the United States.

Panama was not only a magnet for empire-builders; its transit route – first a paved road, then a railway – had brought the world to the Isthmus even before a canal was started. An international crossroads a hundred years before the *Mayflower* landing, Panama played host at times to traders, bullion carriers, pirates, missionaries, soldiers, and then a California-bound gold rush. The canal dream would bring explorers, doctors, engineers, more soldiers (this time to stay) and, at one time, a workforce of fifty thousand from twenty-seven different countries. Many of the canal builders believed that fever was the result of vice, but that did not prevent Panama's two cities from becoming roaring dens of gambling, drinking and prostitution. All this descended on an unstable region still struggling to find political solutions to its problems. At times the canal has been an awkward destiny for Panama.

In the wider world, the great dream of the canal attracted idealists, dreamers and scoundrels from the very outset. The four hundred years after Balboa's discovery saw Panama's unique geography inspire grandiose canal schemes from each age's greatest engineers, promoters and visionaries. It was the great unfulfilled engineering challenge. But for those four centuries all efforts had ended in failure or disaster.

PART ONE

The Golden Isthmus

'I am not capable by writing to give you ane account of the miserable condition we have undergone, first befor we came offe from Caledonia, being starved and abandoned by the world'

– Letter from Robert Drummond, 11 August 1699

I

'THE KEYS TO THE UNIVERSE'

The 'Darien Disaster', the calamitous effort at the beginning of the eighteenth century to establish a Scottish colony in Panama, has many parallels with de Lesseps' French adventure nearly two hundred years later. Each was financed by a host of small investors in their own countries and motivated by idealism, patriotism and naivety, as well as by the chance to make a fast buck. Both had leaders with more front than particular expertise. William Paterson was born in Dumfriesshire, Scotland, in 1658, and as a young man had travelled, as part missionary, part buccaneer, to the West Indies. Returning to England, he had made his fortune in business and had become a 'projector', a promoter of speculative moneymaking schemes.

But ever since his sojourn in the Caribbean, Paterson had been in the grip of a 'Great Idea', the venture to cap everything. He was not the first, nor would he be the last, to fall for the 'lure of the Isthmus'. It was so obvious. If ports could be established on both coasts, cargoes could be transferred over the narrow strip of land, saving ships the long and dangerous voyage around Cape Horn. News from British coastal raiders had identified a spot where there was 'no mountain range at all' and where 'broad, low valleys' extended from coast to coast. It was perfect enough to envisage not just a road, but, in time, a waterway. Paterson, with more than a few ideas before his time, intended to welcome traders from all over the world to the new settlements, regardless of race of creed. It would be a truly global entrepôt, to rival any in the world, and whoever controlled it, proclaimed the Scot, would possess 'the Gates to the Pacific and the keys to the Universe'.

Having unsuccessfully touted the project around England and Europe, Paterson had a stroke of luck when the Scottish Parliament, jealous of the new riches flowing into England from trade, passed an Act to encourage new settlements and commerce. Paterson rushed to Edinburgh, parading his knowledge of

America and hinting that he possessed a great secret. In fact, he said, he knew of a place ideally situated for trade with all parts of the world where there were no Spaniards, and in which there were rich gold mines a plenty. It was the Isthmus. 'Do but open these doors,' wrote Paterson in his proposal to Parliament, 'and trade will increase trade, and money will beget money.' There were warnings that the area was jealously guarded by Spain, and that Paterson 'talks too much and raises people's expectations', but the government were sold. In June 1695 an Act of the Scottish Parliament established the 'Company of Scotland trading to Africa and the Indies'.

But then, startled by the extensive privileges granted to the new company and under pressure from the East India Company, the English Parliament turned against the project. William III, who had given royal assent to the Act, in part to assuage the bad feeling after the Massacre of Glencoe, now backtracked. The substantial English subscriptions of £300,000 were withdrawn, and similar amounts raised in Hamburg also taken away under English pressure. But a wave of patriotic indignation in Scotland saw money pouring in from all quarters and all levels of society. Like the French public in the 1880s, Scottish investors saw the scheme as a means of re-establishing national pride after recent disappointments. From 1400 individuals, including craftsmen and servants, £400,000 was quickly raised, about half the country's available capital. It was a colossal risk for so much of the national silver. Now the future of Scotland itself was at stake.

Five large vessels were loaded with merchandise, including large quantities of cloth, Scotland's principal export, as well as shoes, slippers, stockings, hats, wigs and 1500 bibles in English. Aristocracy sent their younger sons, and the ranks were swelled by army officers who had been kicking their heels since the end of fighting in Europe in 1697. Twelve hundred people embarked. Hundreds of others had to be restrained from stowing away in the ships. Almost the entire population of Edinburgh came out to see off the fleet on 26 July 1698.

More than forty of the colonists died on the three and a half month voyage, and many others were ill when the small fleet anchored of the coast of the Isthmus on 3 November. But their first meeting with the local inhabitants was a success. They found the Indians 'very free and not at all shy . . . They got drunk and lay on board all night.' The Scots went ashore at a deserted beach in a sheltered cove a hundred miles north of the end of the Isthmus, opposite the supposed 'valleys leading from coast to coast'. The inlet was named Caledonia Bay, and a Fort St Andrew, equipped with sixteen guns, was rapidly established on a thumb of rock overlooking the anchorage.

At first, all went well. Friendships were formed with the most powerful of the local tribes, who saw the newcomers as possible allies against their brutal Spanish

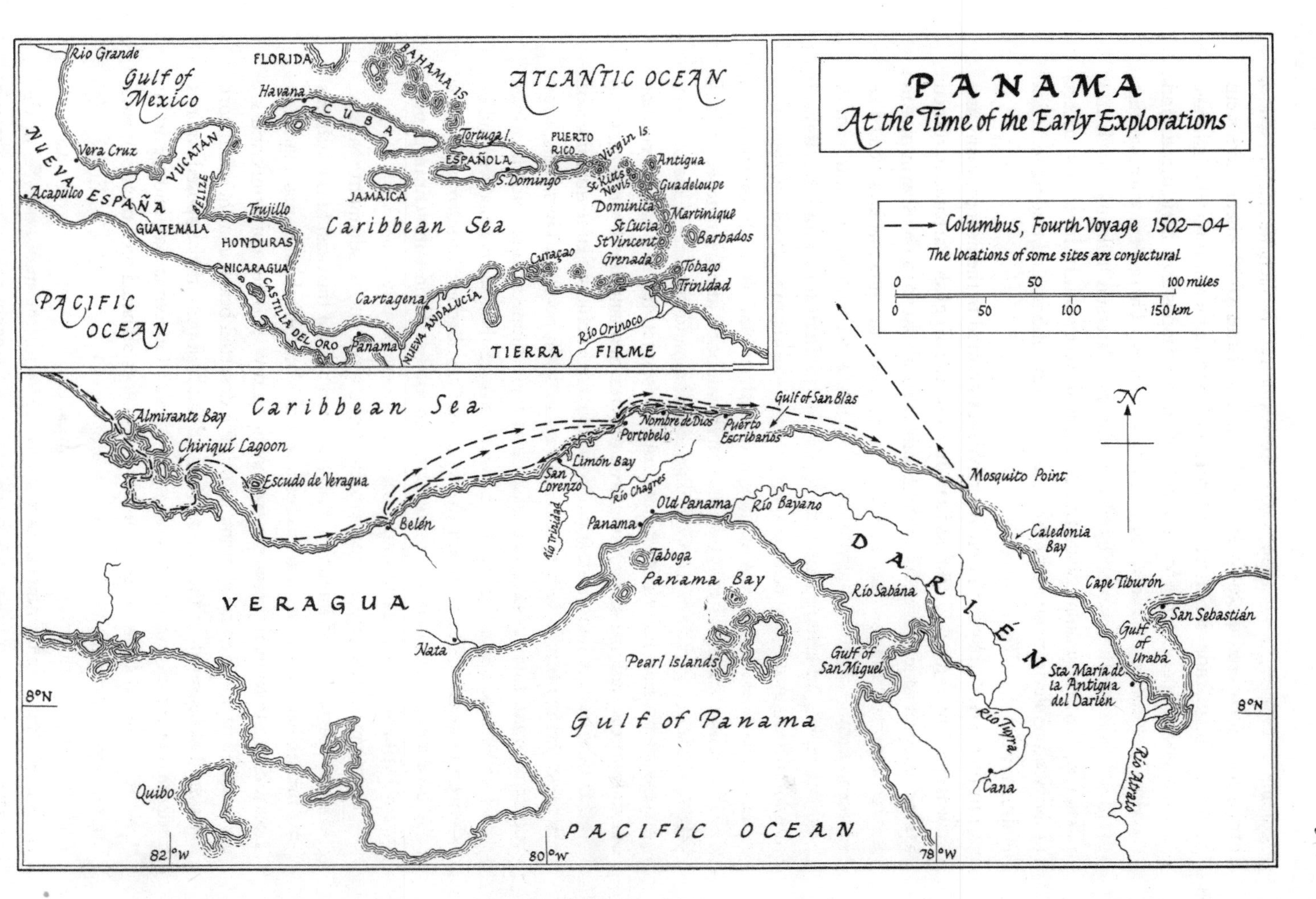

PANAMA
At the Time of the Early Explorations
Columbus, Fourth Voyage 1502–04
The locations of some sites are conjectural
0 50 100 miles
0 50 100 150 km
N
Rio Grande
Gulf of Mexico
FLORIDA
Havana
CUBA
BAHAMA IS.
ATLANTIC OCEAN
Tortuga I.
ESPAÑOLA
S. Domingo
PUERTO RICO
Virgin Is.
St Kitts
Nevis
Antigua
Guadeloupe
Dominica
Martinique
St Lucia
St Vincent
Barbados
Grenada
Tobago
Trinidad
JAMAICA
Caribbean Sea
Curaçao
NUEVA ESPAÑA
Vera Cruz
Acapulco
YUCATÁN
BELIZE
Trujillo
GUATEMALA
HONDURAS
NICARAGUA
CASTILLA DEL ORO
Cartagena
NUEVA ANDALUCÍA
Panama
Río Orinoco
TIERRA FIRME
PACIFIC OCEAN
Caribbean Sea
Almirante Bay
Chiriquí Lagoon
Escudo de Veragua
Belén
VERAGUA
Nata
Quibo
San Lorenzo
Río Trinidad
Limón Bay
Río Chagres
Portobelo
Nombre de Dios
Puerto Escribanos
Gulf of San Blas
Mosquito Point
Panama
Old Panama
Río Bayano
Taboga
Panama Bay
Pearl Islands
Gulf of Panama
PACIFIC OCEAN
DARIÉN
Río Sabána
Gulf of San Miguel
Río Tuyra
Cana
Caledonia Bay
Cape Tiburón
San Sebastián
Gulf of Urabá
Sta María de la Antigua del Darién
Río Atrato
8°N
8°N
82°W
80°W
78°W

colonial overlords. Land was cleared and planted; those who had been sick on the voyage recovered quickly. One of the first official reports, written at the end of December, although requesting provisions to replace those that had perished on the voyage, spoke of the 'rich earth' and proclaimed 'here is food hunting and fowling and excellent fishing'. 'The Country is Healthfull to a wonder,' wrote a settler to a friend in Boston. 'The Country is exceeding Fertile, and the Weather Temperate. If Merchants should Erect factories here, this place will soon become that best and surest Mart in all America . . .' There was one caveat, however: 'We have had but little Trade as yet; most of our Goods Unsold.'

And there was the rub. Far from establishing a custom-free entrepôt welcoming the traders of the world, the settlers had only made enemies, keen to protect their own privileges. As soon as the Scotsmen landed in the New World, there were fierce protests in London from the Spanish ambassador as well as from English merchants. In response, William III issued orders to the Governors of Virginia, New York, New England, Jamaica and Barbados forbidding them to trade with or supply food or other provisions to the Darién colonists. For a settlement established as a trading station, it was a fatal blow.

Everything started to unravel. The death rate from fever was now rising. In the absence of word from the Company directors in Scotland, the leadership of the colony became divided and ineffectual. It soon emerged that far too many of the settlers were 'gentlemen', with neither the inclination nor the strength for the hard labour required in starting the settlement. The 'valleys' 'extending coast to coast' turned out to be a fiction, and no realistic attempt was made to open up an overland route to the Pacific as planned. Relations with the Indians cooled when it became apparent that the new arrivals were preparing no blow against the common enemy from Spain. Now the Indians brought alarming news that the Spaniards, determined to protect an area vital to their shipment of treasure from Peru, were massing to destroy the settlement.

With the international embargo, the only trading partners were the local Indians. The settlers could not eat the unwanted piles of bibles and heavy Scottish cloth, and many regretted that the space had not been given to more useful provisions. Scarcity of food brought increasing weakness, disease and demoralisation; among the first to die was Paterson's wife. Within six months, nearly four hundred settlers had perished of fever or starvation. The onset of the rainy season in May, and the concurrent further worsening of living conditions, was the final straw.

Utterly discouraged, on 20 June 1699, after just seven months on the Isthmus, the Scots abandoned Panama and sailed for New York, en route to Europe. The last man to embark was William Paterson; suffering from a fever, he had to be carried on board, protesting all the time at the too-hasty evacuation. For a while on board ship he lost his reason. Only half of the weakened settlers survived the journey. One of the lucky ones wrote from New York: 'In our passage from Calidonia hither our sickness being so universal aboard, and Mortality so great, that I have hove over board 105 Corps.' The survivors, described by an eyewitness as looking 'rather like Skelets than men, being starved', barely numbered enough to fill one ship on the cross-Atlantic voyage back home.

In fact, help had been on the way. During the month before the abandonment of the colony, two ships carrying three hundred new recruits and plentiful supplies had left Scotland heading for Darién. The previous December, the first, optimistic reports had arrived at last in Scotland, and an anxious nation had learnt that the colony was established and running well. Euphoria had ensued. Thanksgiving services were held in the churches, bells rang out and bonfires were lit across the country.

The second expedition arrived in Caledonia Bay in August to the find the settlement deserted, with 'nothing left but a howling wilderness'. Despite their shortage of supplies, they determined to stay and restart the colony, in the expectation of another relief expedition in the process of being prepared at home. But soon after their arrival one of the two ships, containing most of the expedition's supplies and provisions, was accidentally burnt, and that settled the matter. 'We are a poor, graceless shiftless and heartless company labouring under all discouragements,' wrote a settler home. All but a dozen set sail for Jamaica, with many dying on the way or soon afterwards from sickness.

Shortly after the arrival of the second expedition on the Isthmus, a third set sail from Scotland in late September 1699, this time the fleet comprising four ships with 1300 people and copious supplies. More than one in ten were lost on the voyage, but the remainder reached Caledonia Bay on 30 November. Again a ghostly scene of an abandoned settlement was encountered, although this time the newcomers found twelve of their countrymen living peaceably among the Indians. New huts were built and the ground cleared once more. But, as before, the Scottish leadership was divided in the face of the Spanish threat. Paralysed by their strong religious faith that declared warfare illegal, the Scots allowed their enemies to assemble strong enough forces to end their occupation of the Isthmus. At last convinced that William III really did mean to abandon his subjects and not send the English Navy to the rescue, the Spanish moved against the isolated settlement. In January 1700 the Scots finally roused themselves to take the fight to the Spanish and inflicted on them and their allies a sharp reversal, capturing their camp and dispersing their mulatto and Negro allies. But the following month powerful Spanish ships appeared off Caledonia Bay and the colony was effectively blockaded. Skirmishes were fought around the perimeter as troops from Panama City and Santa María closed in.

On 30 March the Spanish governor of Panama and Cartagena offered honourable capitulation with fourteen days allowed for the Scots to leave. The settlers had no choice but to accept the face-saving deal. On 11 April 1700 they departed New Caledonia still in possession of their arms and with colours flying and drums beating. Many of the settlers were so weakened that the Spaniards had to help them hoist the sails of their getaway ships. Few were strong enough to cope with the difficult voyage. Most of the vessels were wrecked on the journey north and only 360 of the original 1300 survived, most to end up scattered around the British communities of the West Indies.

In all, Paterson's 'Great Idea' had cost over two thousand lives and the precious savings of an entire nation. As de Lesseps and many others would discover, the Isthmus could be a graveyard of men, dreams and reputations.

The 'Darien Disaster' hastened the coming of the Act of Union that dissolved the Scottish Parliament. Seeing the futility of trying to compete with England, and stripped of capital from the disaster, Scotland was merged into Great Britain in 1707, an early but spectacular casualty of Panama Fever.

What had motivated the voyages that led to the discovery of the New World was exactly what the Panama Canal would eventually deliver – a through passage to the East. On his fourth voyage in 1502 Columbus had sailed all along Panama's northern coast, obsessively searching every tiny cove for a 'hidden strait'. At one point he anchored in Limón, or 'Navy', Bay, now the Atlantic terminus of the canal. Even after Columbus's failure to find an open passage to the East, the idea died hard. In 1507, the first map ever printed of the New World optimistically showed an open strait about where the Isthmus of Panama is located.

But Columbus did report back that the *Tierra Firma* he had discovered was rich in gold and pearls. Having set out to discover a route to the wealth of the East, the Spaniards had effectively found far greater riches on the way. A settlement was established, Santa María de La Antigua del Darién, some sixty miles south-east of what would later be named Caledonia Bay. Then, in 1513, the colony's leader, Vasco Núñez de Balboa, his curiosity aroused by Indian stories of a Great Ocean across the mountains, put together an expedition of 190 Spaniards, accompanied by a number of bloodhounds, which the natives found particularly terrifying. On 6 September, they set off across the mountains on a route about a hundred miles east of the modern canal, their heavy loads of supplies carried by a mixture of press-ganged local Cuna Indians and black slaves. The expedition's rate of advance through the Darién jungle was at times only a mile a day. The rivers were in spate and numerous bridges had to be improvised from tree trunks. Even in the sweltering jungle, the Spaniards wore helmets and breastplates of polished steel, thick leather breeches, woollen stockings and thigh boots.

Heatstroke, hostile Indians and disease began to thin their number. On 25 September, with only a third of his men left, Balboa reached a small hill. From its summit, promised the guides, you could see the Great Ocean. Balboa set off alone at midday. At the top, he turned one way and then the other; he could see both oceans quite clearly. He fell to his knees in prayer and then called up his men, 'shewing them the great maine sea heretofore vnknowne to the inhabitants of Europe, Aphrike, and Asia'.

They struggled down to the shore, on the way defeating and then befriending Indians who had barred their route to the Ocean. On the afternoon of 29 September they reached the sea. That evening Balboa, in full armour, waded into the muddy water and laid claim in the name of Ferdinand of Castile to what he called the 'South Sea'.

The party remained on the Pacific coast for over three months, exploring the bay and trading trinkets with the local Indians. Balboa heard stories of a rich land away to the south, but wrongly deduced that he must be close to Asia. He at last returned, heavily laden with pearls and gold, to Santa María and a hero's welcome. Along with the fifth of his treasure, Balboa sent the King of Spain a report, which included, rather as an afterthought, the musing of a Castilian engineer, Alvaro de Saavedra – a suggestion that although the search for a strait between the two oceans should continue, if it was not found, 'yet it might not be impossible to make one'.

Five years after Balboa's discovery, a paved road had been established linking Nombre de Dios, a port on the Caribbean, with a new Spanish settlement at Panama, a prosperous Indian village on the Pacific coast. The transit route opened up the Pacific. Although Magellan found a way around the southern tip of the continent in 1519–21, the voyage was so remote and hazardous that it did nothing to discourage the quest for a way through the Isthmus to the newly found ocean. In 1522 explorers sailing north from Panama discovered Lake Nicaragua. The following year Hernando Cortés, the conqueror of Mexico, was ordered by Charles V to continue the search for an open strait. By 1530 it was clear that no such waterway existed in the tropics, and in 1534 Charles ordered that the Chagres River be mapped and cleared as far as possible in the direction of Panama City, and that the intervening land be studied with a view to excavation. This was the first survey for a proposed ship canal through Panama, and it more or less followed the course of the current Panama Canal. At the same time, the San Juan River, which runs from Lake Nicaragua to the Caribbean coast, was also to be surveyed as part of a possible canal. The great rivalry between the two routes was thus started.

Detailed, reliable information on these very early surveys has not survived, although Charles seems to have received mixed messages. Some reported that the project was totally unfeasible; others, like the Spanish priest Francisco López de Gómara, writing to the King in 1551, thought anything was possible. In an early example of the hubris that the canal dream attracted throughout its history, the priest wrote: 'If there are mountains there are also hands . . . To a King of Spain with the wealth of the Indies at his command, when the object to be attained is the spice trade, what is possible is easy.'

Then Spanish priorities in Panama changed. Philip II, Charles V's successor and a religious fanatic, shared little of his enthusiasm for a canal, seeing it, among other evils, as 'unnatural', as meddling with God's creation. More importantly, the conquest of Peru led to concerns that an Isthmian canal could be a strategic liability. As early as 1534, the Governor of Panama had warned against the construction of a canal as it 'would open the door to the Portuguese and even the French'. By the 1560s most believed that it was safer to have an unbroken wall of land between the gold and silver of Peru and Spain's maritime enemies in the Atlantic. Similar strategic concerns would arise three hundred years later when the United States debated building a canal.

With the conquest of the Incas, the Panama Isthmus became the overland route for the treasure pouring back to Europe, whose value dwarfed anything that could have come from the Indies. Once a year a grand fleet would arrive at the Pacific terminus of the trail, unload the bullion, which would be transferred across the Isthmus to waiting ships at Nombre de Dios. One witness recounted that he saw 1200 muleloads of precious metal leave Panama City in 1550. The 'Royal Road' was now the most important thoroughfare in the Spanish empire, and the Isthmus the key to the Spanish commercial and defence system in the New World. Panama City quickly became one of the three richest centres in the Americas, outshone only by Lima and Mexico City. At the other end of the trail, Nombre de Dios grew into an important port, and the site of an annual trade fair of dazzling opulence, where European goods were bought for transhipment throughout Spanish America. The experience of visiting the fair was described by a travelling Englishman, Thomas Gage, as highly risky: it was 'an unhealthy place . . . subject to breed fevers . . . an open grave'. But, he wrote, 'I dare boldly say and avouch, that in the world there is no greater fare.'

The great wealth and strategic importance of Panama led to numerous attacks from Spain's enemies. In 1572, Francis Drake carried back to Plymouth an enormous pile of looted silver; he returned twenty years later to attempt to capture the Isthmus for England, only to die of dysentery off Nombre de Dios. The infamous Sir Henry Morgan sacked Panama City in 1671, causing a new city to be built in a more secure location nearby.

The new Panama City was never overrun, and Spanish power was still sufficient to see off the Scottish settlers in Darién at the beginning of the eighteenth century. But the English Navy continued to flex its muscles in the region, and frequent plans were laid to seize the Isthmus for the English crown. To take Panama, it was believed, would end Spanish rule in the Americas, and open up the Pacific to English trade. In 1739, during a period of official war with Spain, the English Admiral Edward Vernon, leading six ships of the line and nearly three thousand men, took the Caribbean port of Portobelo and destroyed its defences, although he was unable to cross the Isthmus to seize Panama City itself.

Confronted by growing threats on the Caribbean side of the Isthmus, in 1748 the bullion ships abandoned the Panama route and started sailing around Cape Horn. Thus Panama City lost her place as the treasure house of the New World. Soon afterwards, the famous fair had declined and ceased. Panama was attached to the Vice-Royalty of New Granada based in Bogotá, beginning a century and a half of struggle on the part of the Panamanians to regain their autonomy.

During the rest of the eighteenth century Panama shared Spain's steep decline. Weakened by incessant European warfare, falling birth rates and intermittent bankruptcy, Spain gave way to the new aggressive mercantile and maritime powers of northern Europe. However, economic decline on the Isthmus was not matched by a falling off of geopolitical or military importance. Spain's new rivals in the Caribbean, now a key arena of international conflict, were more interested than ever in the strategic value of the Isthmus.

In 1735 the French government had sent an astronomer through Central America on a scientific expedition to investigate the possibility of a trans-Isthmian canal. He had reported back in 1740 to the French Academy of Science advocating a canal at Nicaragua, making use of the San Juan River that flowed from Lake Nicaragua to the Caribbean coast. In the same year, however, the British were establishing control over a section of the Nicaraguan coast through an alliance with the Mosquito Indians, who refused to recognise Spanish sovereignty. It was not a coincidence that this gave the British control over the mouth of the San Juan River, and therefore the Atlantic terminus of any future Nicaraguan canal.

Portobelo was attacked again by the British in 1745, and when in 1779 Spain joined France in support of the rebellious American colonies, the British decided to launch a naval expedition to seize Lake Nicaragua and thus divide Spain's American empire in two. A force was assembled by 1780 and in command of the naval section was a young Horatio Nelson, who was well aware of the importance of the potential transit route. 'I intend,' he wrote, 'to possess the Lake of Nicaragua which, for the present, may be looked upon as the inland Gibraltar of Spanish America. As it commands the only water pass between the oceans, its situation must ever render it a principal post to insure passage to the Southern Ocean.'

The expeditionary force succeeded in overcoming the Spanish defenders, but was decimated by disease and was forced to withdraw. The idea of the canal, however, retained its wide appeal, and over the next twenty years no fewer than four French proposals were made.

With the independence of Gran Colombia*, officially declared on 28 November 1821, the dead hand of Spanish rule was at last removed, and a major barrier to the construction of a canal disappeared at the same time. Furthermore, there was now a new emerging power to the north beginning to take a keen interest in Central American affairs.

* Gran Colombia consisted of modern-day Panama, Colombia, Venezuela and Ecuador. This federation dissolved in 1830 with the later two becoming independent and the remainder renamed the Republic of New Grenada, which became 'Colombia' in 1863.

2

RIVALRY AND STALEMATE

Even in the midst of the American Revolution, Benjamin Franklin, then the United Colonies' representative in Paris, became gripped by the idea of a trans-Isthmian canal. In 1781 he printed on his own press a pamphlet written by a French peasant called Pierre-André Gargaz, which advocated the cutting of canals at Panama and Suez, which Gargaz believed would bring about world peace through enhanced commerce and communication. When Thomas Jefferson became the United States minister to France, he, too, became interested in a canal at Panama. Jefferson saw expansion southwards as the natural destiny of the United States and was intrigued by rumours that there had been recent Spanish surveys of a canal route at Panama. 'I am assured . . . a canal appeared very practicable,' he wrote in 1788 to a fellow US diplomat in Madrid, 'and that the idea was suppressed for political reasons.'

Before independence from Spain, revolutionaries in Latin America had looked to the United States and Britain as their natural allies. When freedom from Spanish rule, having been achieved, was subsequently threatened by the French-led Holy Alliance, the British Foreign Secretary George Canning contacted the leadership of the United States to ask them to make a joint declaration warning of their shared opposition to the reconquest plans. But President Monroe was persuaded to make a statement purely on behalf of the United States. His famous Doctrine, delivered in a message to Congress on 2 December, 1823, is of huge importance to the Panama canal story: 'The American continents,' he announced, 'are henceforth not to be considered as subjects for future colonization by any European Powers.'

In the meantime, impetus for a trans-Isthmian canal had received a major boost with the publication of 'A Political Essay on the Kingdom of New Spain' in 1811. Its author, the German Alexander von Humboldt, had recently explored

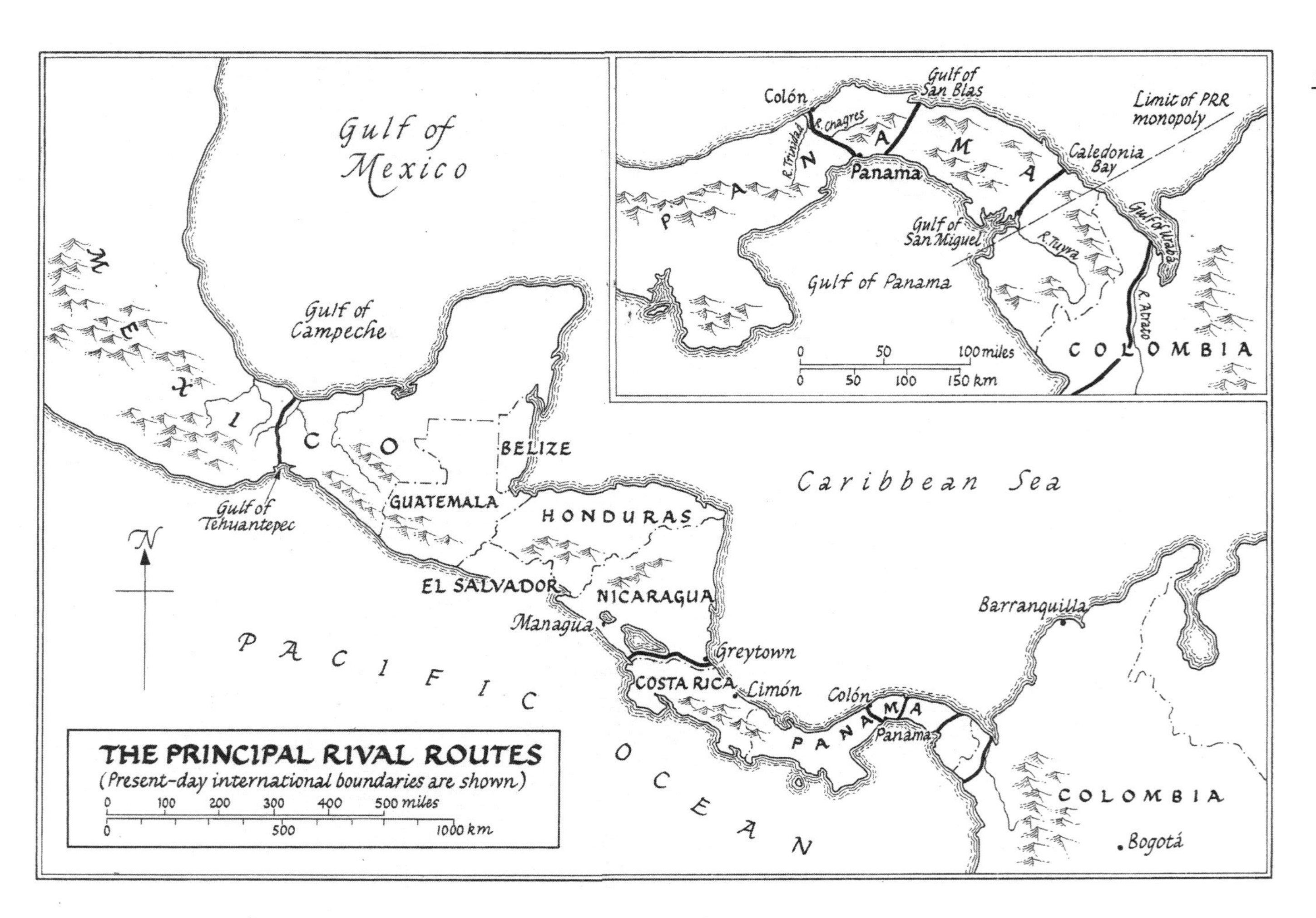
THE PRINCIPAL RIVAL ROUTES
(Present-day international boundaries are shown)
0 100 200 300 400 500 miles
0 500 1000 km
Gulf of Mexico
Gulf of Campeche
MEXICO
Gulf of Tehuantepec
BELIZE
GUATEMALA
HONDURAS
EL SALVADOR
NICARAGUA
Managua
Greytown
COSTA RICA
Limón
Colón
PANAMA
Panama
Caribbean Sea
Barranquilla
COLOMBIA
Bogotá
PACIFIC OCEAN
Colón
Gulf of San Blas
Limit of PRR monopoly
R. Chagres
R. Trinidad
PANAMA
Panama
Caledonia Bay
Gulf of San Miguel
R. Tuyra
Gulf of Urabá
Gulf of Panama
R. Atrato
COLOMBIA
0 50 100 miles
0 50 100 150 km

the regions of South and Central America on an epic journey that thrilled readers all over the world. Although he never actually visited any of the sites, he identified five possible Central American routes for a trans-Isthmian canal. Going from north to south, they were: the narrowing of Mexico at Tehuantepec; in Nicaragua, using the giant inland lake; at Panama; and two using the Atrato River in modern-day northern Colombia. Humboldt's book was widely read and admired, and can be seen as the inspiration for all the many subsequent surveys of the Isthmus looking for the best route for a canal. The account was full of errors – he calculated the height of the Continental Divide in Panama at three times its correct elevation – but it was a hugely exciting work nonetheless. His personal favourite was the Nicaragua route, where he envisaged something along the lines of Thomas Telford's Caledonian Canal, then the most ambitious construction of its kind. Humboldt reckoned a canal was possible and, furthermore, 'would immortalise a government occupied with the interests of humanity'.

One of those inspired by the book was the German poet Heinrich Goethe, who in 1827 predicted the 'incalculable results' of a 'crosscut' through the South American Isthmus. With remarkable prescience he also foretold how intimately interwoven would be the destinies of Panama, the United States and the canal. 'I, however, would be surprised if the United States would miss the chance to get such a work into her hands,' he wrote. 'It is entirely indispensable for the United States to make a passage from the Gulf of Mexico to the Pacific Ocean, and I am certain that she will accomplish it.'

It was also the climax of the 'canal age' in Europe and the United States. The 1820s saw the opening of the Erie Canal, joining the Hudson River and the Great Lakes, as well as Telford's monumental achievement in linking the Atlantic and the North Sea across Scotland, which included twenty-eight locks of a size big enough for most of the ocean-going vessels of the time. All across Europe and North America, canals transformed communication, slashing journey times and transport costs for raw materials and finished goods. Crucially, steam power had arrived, and with it the possibility of vessels transiting canals without the need for towpaths. A steam tug was first used on the Forth and Clyde Canal in Scotland as early as 1802, and by the time of the independence of Gran Colombia in 1821, the prospect of a trans-Isthmian canal seemed much closer.

Simón Bolívar chose Panama City as the site for his Latin-American Congress, which finally met on 22 June 1826. For Bolívar, Panama was the 'veritable capital of the world, the centre of the globe, with one face turned toward Asia and the other toward Africa and Europe'. Britain and the United States were both invited to send delegates.

Bolívar was in favour of a canal, and soon after independence a stream of proposals were put to him, first by an American army officer (a relation of

Benjamin Franklin), then a British naval captain, then a Jamaican merchant, all focused on the narrow Isthmus at Panama. Each was rejected by his Congress, quite rightly, as unrealistic, and the Bogotá government tried unsuccessfully to get funding from London to build the canal itself. But in 1827 Bolívar at last granted permission for a survey to be carried out by a British army captain, John Augustus Lloyd, and Captain Maurice Falmarc, a Swedish officer in the Colombian military. Hampered by the weather, they nonetheless produced the most reliable survey yet of the Isthmus, even though construction never started. Their plan was to use the Chagres River up to its junction with the Trinidad, then build a railway to the coast. In due course, this would be replaced by a ship canal. Labour would be supplied by British convicts, who would be accompanied by colonising settlers. This body combined would, Lloyd suggested, 'present a human barrier of such formidable power, as to limit . . . any attempts of the United States towards aggrandizement and increase of territory . . .'. Lloyd had little time for the Panamanians, whom he described as 'superstitious . . . Billiards, cockpits, gambling and smoking in low company, are their exclusive amusements . . . Their best quality is great liberality to the poor, and especially to the aged and infirm.'

In fact the Panamanians were almost all desperately poor. The long economic decline started in the mid-eighteenth century had continued unabated since independence. A visitor from Bogotá in the 1830s was shocked by what he found at Panama City: buildings in ruins, vagrants everywhere and prices fallen to unbelievable levels. The population of the Isthmus had shrunk, with many who had the will or the means leaving to find prosperity elsewhere. Those remaining knew they had to re-establish the transit route that had always been the reason for Panama's existence. In 1829 Bolívar was petitioned by the Panamanians to do what he could to facilitate the construction of a 'clear route or canal'.

There was also a real fear that Panama would lose out to Nicaragua or some other site on the Isthmus as the beneficiary of a future canal. Like Panama, Nicaragua played host to a multitude of optimistic surveyors and explorers from the United States, Britain, France, Italy, Denmark and Holland. Their backers were sometimes private companies, sometimes kings or emperors. The King of the Netherlands and Louis-Philippe of France were at various times interested in trans-Isthmian canals. Telford himself considered a 'grand scheme' for a canal at Darién, starting at the ill-fated Caledonia Bay. The Great Idea of the canal now attracted not only proven engineers such as Telford, but almost all the millionaires, dreamers, amateur engineers and crackpots of the nineteenth century. Wildly unrealistic schemes abounded. One of the greatest businessmen of English fiction, Sir Roger Scatchered, the irascible, drunken construction tycoon in Anthony Trollope's 1858 *Dr Thorne*, is called upon to mastermind a project to 'cut a canal from sea to sea through the Isthmus of Panama'. It was the great unfulfilled engineering dream of the nineteenth century.

In Panama itself concessions to build a canal or railroad were handed out freely after 1834. An eccentric, and probably insane, Frenchman was the first concession-holder, but his scheme came to nothing. In 1835, US President Andrew Jackson, alarmed by Dutch efforts to secure a monopoly in Nicaragua, ordered Charles Biddle to visit Nicaragua and Panama and carefully to document the possibilities of building a canal or railroad. Biddle's effort, too, ended in failure when he ignored Nicaragua and negotiated a concession for Panama on his own behalf, but from then on American policy towards an interoceanic canal was established: if such a waterway could be built, it would not be allowed to be under the sole control of any foreign power.

At the same time, New Granada still hoped to have any future canal under its control, and petitioned the governments of the United States, France and Britain to act as funders and international guarantors of their sovereignty over the Isthmus. In response to this, in 1843 the French government sent a senior civil engineer, Napoléon Garella to map the Panama route. His survey was the most complete yet, although he failed to find the lowest pass over the Continental Divide, and his plan for a canal included nearly forty locks as well as a huge tunnel over three miles long. Humboldt himself was among those denouncing the scheme as 'an absurdity'. The next concession was more modest – for a French syndicate to build a railroad over the Isthmus.

All this activity by the French had not gone unnoticed in London. In fact, British steamers had started operating out of Panama in the 1830s, and several reports were sent back detailing the possibilities of a canal or railway there. In 1836, however, an incident occurred that was to sour relations between New Granada and Great Britain. After a personal disagreement, there was a fight between Joseph Russell, Her Majesty's vice-consul in Panama City, and a local resident, in which both were injured. Russell was given a stiff sentence, which led, in true gunboat diplomacy style, to British warships blockading the mouth of the Chagres River and the harbour of Cartagena. War was averted when the local authorities agreed to release Russell and pay an indemnity. It was all part and parcel, however, of an increasingly high-handed and interventionist approach on behalf of the British. Everyone in Panama and Washington feared that the British, using another excuse, might one day seize the Isthmus. There was also American concern about the railway concession given to the French syndicate, which might be operated as a monopoly. Thus, at the end of 1845, the United States sent a new chargé, Benjamin A. Bidlack, to Bogotá to ensure that 'no other nation should obtain either an exclusive privilege or an advantage'. He did not have high hopes, such was Britain's regional dominance, but found that the new President, Tomás Cipriano de Mosquera, who had been minister to London, had also come to fear British aggression and presented a sympathetic ear.

Much of the subsequent treaty negotiated by Bidlack with the New Grenadan Foreign Minister, Manuel María Mallarino, was humdrum, concerning the removal of discriminatory tariffs on American products. The key article, which would shape relations between the two countries and decisively affect the story of the Panama Canal, was number thirty-five: this guaranteed 'that the right of way or transit across the isthmus of Panama upon any modes of communication that now exist, or that may be hereafter constructed, shall be open and free to the Government and citizens of the United States . . .'. In return, the US guaranteed 'the perfect neutrality of the before-mentioned Isthmus, with the view that the free transit from the one to the other sea may not be interrupted'. The New Granadan Foreign Minister, Mallarino, declared that 'to surround the isthmus with the vigorous will of a powerful and benign democracy . . . was to save the isthmus'.

It was clear that the treaty, by which the US would protect the Isthmus from British seizure in return for transit rights and favourable customs rates, would come perilously close to an 'entangling alliance', something the United States Senate had set its face against. So the New Granadans emphasised British aggression, warning that British ambitions would soon stretch from the Cape to California. The treaty was ratified by the US Senate in 1848, and, at a stroke, Mosquera had reversed Bolívar's traditional policy of using Britain as a balance against the potentially more powerful United States. More than that, he had, in fact, effectively handed control of the Isthmus to a foreign power, as French diplomats and others in Bogotá warned at the time. From now on, the United States had the right to land troops on the Isthmus if 'free transit' was threatened. For the United States it was a unique treaty, the only foreign alliance ratified throughout the nineteenth century, and for almost the first time the country had strategic interest outside its continental borders.

The treaty also further escalated tensions in the region between Britain and the United States. Seeing the Panama route now under US control, the British moved to strengthen their hold over the Caribbean terminus of any potential Nicaraguan canal. They were also alarmed (as was the whole of South America) by the acquisition by the United States at the end of the war with Mexico of vast territories including California. The southward march of American power seemed unstoppable, even to the extent of taking over all of Central America and threatening British colonies and investments in the entire hemisphere. Nicaragua protested about British incursions on their territory and asked for – and received – the backing of the United States, which then negotiated a treaty with Nicaragua that gave it exclusive control over a canal there.

On both sides the sabre-rattling increased ominously, and there was talk of war. In December 1849 the British Foreign Secretary Lord Palmerston sent Sir Henry Lytton Bulwer to Washington to try to find a solution to the standoff. The result was the Clayton–Bulwer Treaty, at the heart of which lay the desire by both sides to avoid having a canal exclusively controlled by the other. In effect, the fear of a foreign-controlled canal – which would hand to its owner total dominance of the region – was greater than the benefits that such a waterway might provide. The treaty stipulated that neither the United States nor Great Britain would take exclusive control over any waterway in Central America or try to fortify any that might be built. The canal was now on hold, with the hands of the US firmly tied.

But there were still other ways of improving the Isthmus transit route. Because of the upheavals in Europe in 1848, the French syndicate which had been granted a railway concession failed to raise the initial surety, and the contract was taken over by a syndicate headed by William Aspinwall, a New York businessman, who had been lobbying for permission to build a railway for some years, and who was running a new steamship service between Panama and San Francisco. His concession gave him exclusive transit rights across the Isthmus at Panama to a point ten miles south of Caledonia Bay, including the first option to build a canal within this zone.

The timing was, to put it mildly, fortunate. On 24 January, gold was discovered near Colonel Sutter's sawmill in the vicinity of Sacramento in California. 'The Eldorado of the Spaniards is discovered at last,' announced the New York *Herald* after President Polk himself confirmed that the rumours of huge deposits were true. With the transcontinental railroad still twenty years away, the quickest and safest way to get from the East Coast to California was by travelling to the Chagres River, taking a boat upriver to Gorgona, crossing overland to Panama City, then sailing up to San Francisco. Once more, in a different way, Panama was to become 'the Golden Isthmus'.

3

GOLDRUSH

The small steamer the *Falcon* left New York on the first of December 1848 with sacks of mail and twenty-nine passengers bound for California via Panama. The majority were government officials and missionaries. On the way the boat docked at New Orleans. But in the meantime the news of the discovery of gold in California had swept the nation, and by the time it left New Orleans on 18 December, the *Falcon* carried 193 passengers. The new arrivals were Southern backwoodsmen bearing pans, picks and axes and all the paraphenalia of the gold hunter. The small ship was swamped; efforts by the captain to reduce the number boarding had been met with brandished revolvers.

The impatient party arrived off the mouth of Chagres River on 27 December 1848. From that moment on, the Isthmus was transformed forever. After a decade of hard times on the East Coast of the United States, there was a flood of young men eager to try their luck in the California goldfields. The Panama route was the most expensive, but the quickest, and speed was vital if the best claims in Sacramento were to be staked out. Following the *Falcon*, one ship after another arrived, and within a few weeks more than five hundred prospectors had crossed the Isthmus to Panama City, which had not known such commotion and wealth since the early days of the Spaniards.

Because of a sand bar, arriving vessels had to anchor about a mile from the landing place, the small village of Chagres, on the south side of the river, a 'low, miserable town, of thirty thatched huts', as an early traveller reported. Native canoes carried them ashore across the swirling muddy water. Chagres did not make a good impression, and reaffirmed the Americans' sense of superiority. 'The houses are only hovels,' wrote one to his mother back in Alabama, 'that in the States, would not even do for Negro quarters or even for a respectable cow house.' The people, too, were seen by the Americans as little better than savages: 'Half are full-blooded negroes . . . The dress for the men is little short of indecent.'

The locals were not slow, however, to realise that they, too, had struck gold. The prices charged to the new arrivals were exorbitant. One traveller, a seventeen-year-old New Englander called Stephen Davis, reported, 'Early in the morning went ashore in a skiff, for which we paid $2 each . . . We engaged passage in a large flat-bottomed boat, with 15 others, for $10 each to Gorgona.' For him, it was worth the money just to get away from Chagres near the river mouth, 'one of the filthiest places we ever saw'. The town also quickly acquired a reputation as 'the birthplace of a malignant fever', as an English traveller, Frank Marryat, noted. Within the town and in a new shanty across the river, known as 'American Chagres', bars and brothels soon sprang up. At the Silver Dollar Saloon drinks were seven times as expensive as in New York, but there were plenty of takers. The working girls, who soon flocked to the area from as far away as New Orleans or Paris, in the total absence of law enforcement took to carrying guns, a practice one British traveller called 'as disconcerting as hell. All the time she kept one hand on her blasted six-shooter.' Soon there were more than two hundred prostitutes working in Chagres alone.

The journey to Gorgona was along fifty miles of winding river, which took about three days. Most went in 'bungos', hollowed-out logs used for transporting bananas. If the water level was high, the boats could get as far as Las Cruces. Many of the travellers marvelled at their first sight of the lush, tropical jungle. 'The bright green at all times charms the beholder,' wrote Frank Marryat. 'The eye does not become wearied with the thick masses of luxuriant foliage, for they are ever blended in grace and harmony.' Alligators lazed in the shallows, and birds of every type and hue flitted through the treetops.

At Gorgona, originally a small town of a few hundred people, there was pandemonium. It quickly became a bottleneck in the flood across the Isthmus, as the eager gold hunters waited impatiently for scarce transport on the next leg of the journey. Such was the shortage and demand that to hire a riding mule could cost up to $20. Numerous tents and shanties sprang up on the hills above the town.

For the journey onwards to Panama City, the 'road' was diabolical. Over much of the twenty-one miles it was a trough rather than a track and if it was the wet season it would be swampy gunge, with mud up to 5 feet deep. Mules and horses, jolting over the rough, uneven cobbles, carried the travellers and their loads, but often it was local porters. Some of the men carried as much as 300 pounds over the journey of a day and a half. Many of the smarter travellers were even carried in chairs strapped to the bearers' shoulders.

The whole experience of crossing the Isthmus was for one American 'so like a nightmare that one took it as a bad dream – in helpless silence'. But at last the weary travellers could descend to the Pacific and Panama City, which they found had quickly realised the money to be made from the prospectors. 'The old ruined houses have been patched up with whitewash,' wrote Frank Marryat, adding that 'the main street is composed almost entirely of hotels, eating houses and "hells" [bars].' Most were American-owned and run. In the city much of the rushing stampede hit a brick wall, as there were at first insufficient ships to take the men on the next leg of their journey. Arriving steamers would be mobbed as a mass of people fought to get on board. The situation was made worse by the fact that most of the ships entering San Francisco lost their crews, who chose instead to join the gold rush. Soon abandoned ships lay seven deep in San Francisco harbour. The result was a huge throng stranded in Panama City, at one point more than four thousand. Those who could not find or afford a bunk set up filthy camps on the outskirts, where what they inevitably called 'Panama Fever' – malaria, yellow fever and dysentery – began to strike them down. Some were so desperate to leave that they even set off for California in canoes. The frustration was increased when men starting appearing on their way back to the East Coast, a lucky few carrying a fortune in gold dust and nuggets.

It was not just Americans who were setting up establishments to milk the transit trade. Mary Seacole, born in Jamaica in 1805 of a Scotsman and a freed black woman, arrived in Panama in early 1851, at the beginning of the rainy season. Seacole was a widow – she had briefly been married to a British trader, a godson of Admiral Nelson – and needed to support herself. To make ends meet, and unable to 'check' her 'disposition to roam', she had decided to join her brother who was running a 'hotel' and store in Las Cruces, at the mid-point of the trail.

Having negotiated an alarming boat ride up the Chagres, Seacole reached Las Cruces after three days. She was delighted to be met by her brother, but less impressed with his 'Independent Hotel': 'Rest! Warmth! Comfort! – miserable delusions!' The 'hotel' turned out to be a long, low hut built of rough logs containing a single room on each floor. The travellers from both ends of the trail had just arrived, and the place was packed. 'Those bound for the gold country were to a certain extent fresh from civilization,' wrote Seacole, 'and had scarcely thrown off its control, whereas the homeward bound revelled in disgusting excess of licence.' The next day, as some made their way to Panama and others on to boats heading downriver, Las Cruces was deserted for another week.

But before the passengers had been 'many days gone, it was found that they had left one of their number behind them . . . the cholera'. Mary Seacole went to see the body: 'the distressed face, sunken eyes, cramped limbs, and discoloured shrivelled skin were all symptoms which I had been familiar with very recently'. In 1850 there had been a catastrophic outbreak of cholera in Jamaica that had carried off up to fifty thousand people. Mary Seacole, who had been taught traditional medicine by her mother, had been kept busy caring for the sick. Now, with no doctor in as Las Cruces, she worked day and night as the disease inevitably spread. Her treatment consisted of 'mustard emetics, warm fomentations, mustard plasters on the stomach and the back, and calomel'. To quench her patients' thirst, she gave them water in which cinnamon had been boiled. Her successes in saving lives meant that soon she became known as 'the yellow woman from Jamaica with the cholera medicine'.

Seacole now set up a restaurant, grandly named 'the British Hotel', in a 'tumbledown hut'. It was highly lucrative as each week the town would be heaving with travellers. But soon she began to get worn down by the lawlessness of the Isthmus. She was frequently called on to treat knife and gunshot wounds, and despaired of the 'weak sway of the New Granada Republic . . . powerless to control the refuse of every nation which meet together upon its soil'. With the prospectors had arrived not only cholera, gambling and prostitution but also an epidemic of armed robbery. With increasing quantities of gold flowing back from California to the East Coast or Europe, there were rich pickings. There are many accounts of returning prospectors having their precious gold dust stolen from them. One reported of his disastrous trip in the spring of 1851: 'I knew nothing of the great risk in travelling alone, as the natives two years before appeared to me an exceptionally honest people. But . . . contact with American roughs had changed them to thieves and murderers, and the whole route . . . was infested with American, English and Spanish highwaymen.'

Occasionally, when squabbles at the gaming tables or in the bars got out of hand, the local 'soldier-police' were called in. These were, according to Seacole, 'a dirty, cowardly, indolent set' who 'bore old rusty muskets, and very often marched unshod'. The force of law was not helped by the chronic political instability on the Isthmus at the time. Between 1850 and 1855, there were no fewer than fourteen different governors of the province, and political violence added to the problems for the authorities.

Locals arrested were dealt with very severely, but Americans could usually bribe their way out of trouble. The protection of the law, was, Seacole reports, 'rather an expensive luxury'. On other occasions, a large group of heavily armed fellow Americans would turn up at the court to intimidate the judge into freeing their countryman.

More even than the violence, the chauvinism of the gold prospectors depressed Mary Seacole. Writing for a United Kingdom readership, she is undoubtedly as patronising to the Americans as she is ingratiating to the British. She even admits to a 'little prejudice' against all North Americans. For her the issue, more than anything, was slavery, which had been abolished in the British Empire in 1834.

'Knowing what slavery is; having seen with my eyes and heard with my ears proof positive of its horrors,' she writes, 'is it surprising that I should be somewhat impatient of the airs of superiority which many Americans have endeavoured to assume over me?' The lower ranks suffered the most: 'Terribly bullied by the Americans were the boatmen and muleteers,' she says, 'who were reviled, shot, and stabbed by these free and independent filibusters, who would fain whop all creation abroad as they do their slaves at home.'

It was not only gold prospectors but also the great and the good who used the Panama route, and Seacole is quick to acknowledge that she had 'met some delightful exceptions' to the American type, but on the whole the Americans travelling through the Isthmus were not good ambassadors for their country. Coming from a society in which slavery still flourished, they tailored their behaviour to non-white persons accordingly.

Occasionally, the Americans would get their comeuppance. Many of the public offices in Panama, in the priesthood, army, local government and judiciary, were held by blacks or mulattoes, and sometimes a judge would face down an American utterly confused and appalled to have a 'nigger' sitting in judgement on him. On other occasions locals would encourage and help the slaves of American travellers to escape. In response to the arrogance and bullying, as well as to the violence and lawlessness imported with the prospectors, anti-Americanism grew steadily. In a short time some Americans took to pretending to be French or British for their own safety.

Seacole moved to Gorgona, but it was not a success. Life was no better than in Las Cruces and the town was ravaged by floods and fires. Soon after she gave up and decided to leave the Isthmus. But there was to be one final indignity. She had bought a ticket on an American steamer but moments after she arrived in the saloon with her servant girl Mary, she was met by incredulous hostility. 'Don't be impertinent, yaller woman,' a female passenger, said to her. 'I never travelled with a nigger yet, and I expect I shan't begin now.' Mary was similarly abused and one lady passenger even spat in her face. Seacole appealed to the steward and then the captain, but to no avail. Her fare was returned to her and she was put back onshore. Two days later she boarded an English steamer captained by a friend of hers and was carried back to Kingston, Jamaica.

4

THE PANAMA RAILROAD

It was the acquisition by the United States of vast Pacific territories at the end of the Mexican War that had inspired William Aspinwall and his partners to establish steamer services from Panama City to San Francisco and from New York to Chagres. In this they were supported by the US Navy, and by promised payments from the federal government for carrying mail and officials to and from the West Coast. Crucially, they were also protected by the Bidlack Treaty.

With the interior of the United States largely unsettled and unpacified, the Isthmus was seen in Washington as a key strategic artery linking the two coasts of the country. Aspinwall also had a plan for a railway across the Isthmus, and with the gold rush in full swing there was new impetus to start the construction straightaway. Gold seekers were paying up to $100 to cross to Panama City, so the potential profits were there for all to see.

In spite of this, the initial stock offer was a disappointment with only about half of the $1 million worth being taken up by the public. Undeterred, the directors purchased the remainder themselves, and work started on surveying a route for what would be the first railway ever built in the tropics. The story of the construction of the Panama Railroad is part and parcel of the story of the canal: its triumphs and disasters, its effect on Panama, and the experiences of those working there look forward to the extraordinary events surrounding the canal construction.

The first and best choice for the Atlantic terminus of the railway was the old harbour of Portobelo, but a New York speculator had bought all the surrounding lands and held out for a huge sum. So the railway engineers were forced to choose Limón Bay, otherwise known as Navy Bay, where Columbus had anchored during his search for the mythical strait. It was an unfortunate beginning, which would have long-lasting repercussions, for the harbour provided little shelter from the occasionally fierce north winds – 'northers' – and the land bordering the bay was little more than a swamp. In fact, it was just about the wettest and unhealthiest place on the Isthmus. In the bay was an island, known as Manzanillo, and here it was elected to start work in May 1850. As a contemporary account has it, 'No imposing ceremony inaugurated the "breaking ground." Two American citizens, leaping, axe in hand, from a native canoe upon a wild and desolate island, their retinue consisting of half a dozen Indians, who clear the path with rude knives, strike their glittering axes into the nearest tree; the rapid blows reverberate from shore to shore, and the stately cocoa crashes upon the beach.'

Contracted as engineers were two men with experience of the tropics – they had recently completed a short canal in New Grenada – but even for them this was a forbidding prospect. 'It was a virgin swamp,' a contemporary wrote, 'covered with a dense growth of the tortuous, water-loving mangrove, and interlaced with huge vines and thorny shrubs. In the black, slimy mud of its surface, alligators and other reptiles abounded, while the air was laden with pestilential vapors, and swarming with sand-flies and mosquitoes.' As the island was cleared, a storehouse was erected, but it was found impossible to occupy on account of the insects, so the workers and American engineers was forced to live on an old wooden hulk anchored in the harbour. There, the myriad cockroaches and the constant movement of the ship in the unsheltered bay further weakened their spirits and constitutions. The huge demand for porters, muleteers and boatmen made it impossible to recruit sufficient local labourers (who were thought 'indolent and lazy' by the Americans anyway), so in June some fifty workers were brought from Cartagena in New Grenada, almost all descendants of the black slaves imported by the Spanish. The next month the same number of Irishmen arrived from New Orleans. There was surplus American labour in the United States at this time, due to the large number of men demobbed after the end of the Mexican War, but it was felt that they would demand too high a wage, and might be to prone to organising themselves.

Meanwhile the surveyors pushed on, locating the track, often wading up to their armpits in the deep marshes that lay beyond the island. One, it was reported, 'carried his noonday luncheon in his hat . . . and ate it standing, amid the envious alligators and water snakes'.

The surveyors did have one spectacular success, finding at a place called Culebra (Spanish for 'snake') a pass across the mountainous spine of the Isthmus at only 275 feet above sea level, when they had been expecting to have to site the line at 600 feet. But by now fever was carrying off the workers at an alarming rate, and in time the white members of the party 'wore the pale hue of ghosts'.

Nevertheless, in August the construction work was commenced. A decrepit steamer replaced the hulk, and Manzanillo Island was filled in enough to build a few more huts. As the jungle was cleared, so the clouds of insects lessened. But by the end of the first year, even though the workforce had increased to a thousand, only four or five miles of temporary track had been laid across the swamp on wooden trestles. Wooden docks were built on Manzanillo Island in April 1851, but by then the rainy season had restarted, the original capital was all expended, and the directors were compelled to keep the work going on their personal credit.

By October 1851, eight miles had been constructed and the railway reached to Gatún. But in New York the promoters began to doubt that the line would ever be completed. The value of the stock went into freefall, and the work came to a standstill.

It was a fierce south-westerly storm off the Caribbean coast of the Isthmus that saved the railroad project. In December 1851, two ships from Georgia and Philadelphia, filled with over a thousand hopeful gold hunters bound for the Chagres River, found themselves in serious trouble and were driven to take shelter in Limón Bay. The impatient passengers swarmed onshore and demanded to be taken inland by rail. The railraod engineers protested that there were no passenger cars, but the prospectors eagerly piled into wooden cattle trucks were transported the eight miles to Gatún. It cut a precious day off the transit and after that everyone demanded the same. The railway never looked back; and from then on, as fast as the beleaguered workers pushed on across the Isthmus, the California-bound passengers were right behind them. What's more, they were prepared to pay huge sums, sometimes more than $25, for their tickets. The initial rates set by the railway managers were, one admitted, 'intended to be, to a certain extent prohibitory, until we could get things in shape', but there were plenty of takers and the planned reduction in fares was never necessary. So from that day on, the 'goose began to lay golden eggs with astonishing extravagance'.

Early the following year, the Railroad Company formally inaugurated the growing town on Manzanillo Island, the Atlantic terminus of the route. They named it Aspinwall, after the company's American boss, but the locals were having none of this, passing a law calling the town Colón, after Christopher Columbus. For a while confusion reigned, as the local post service refused to deliver mail addressed to Aspinwall, and eventually the American name was dropped. Up the coast, the town of Chagres at the river mouth quickly lost its previous importance, declined and then disappeared altogether. Colón was now the key port and town on Panama's Atlantic Coast. For the hundreds of thousands of men and women who would come to Panama over the next sixty years for the railway or to build the canal, it would be their first sight of the Isthmus.

For almost all, it was a hugely dispiriting experience. Although streets and squares were laid out, there were no paved roads or sewers. The town was separated from the mainland by a small channel, known as Folks River, slightly above the tide level, where vultures hovered, attracted by the offensive and rotten stench that circulated with the breeze coming from the swamps. When travellers arrived from Panama or New York, the population of eight hundred would be doubled, and the town would come to life of a sort. An account published in 1855 describes how, 'The hotels – great, straggling, wooden houses – gape here with their wide open doors, and catch California travellers, who are sent away with fever as a memento of the place, and shops, groggeries, billiard rooms, and drinking saloons thrust out their flaring signs to entice the passer-by.' All the buildings, with the exception of the Railroad office and the 'British consul's precarious corrugated iron dwelling', were of wood, and many were on rickety stilts over the stinking morass that almost surrounded the island.

'I thought I had never seen a more luckless, dreary spot,' wrote Mary Seacole. 'Three sides of the place were a mere swamp . . . It seemed as capital a nursery for ague and fever as Death could hit upon anywhere.' Colón, the wettest and filthiest place on the Isthmus, was a death trap for the workers the company imported to push the railway project along. A huge recruitment drive saw arrivals from all over the world. Among the Europeans were Englishmen, Irishmen, French, Germans and Austrians. In 1852, one thousand Africans were brought in, and the following year some eight hundred Chinese workers were contracted. A number died on the way from Canton, on transport, according to a historian of the railway, 'as filthy and odorous as any slavers'. The terms of the contract were not far removed from slavery, either. The contractor was paid $25 per man per month, of which they saw about four dollars, with food and clothing thrown in. The men were expected to work up to eighty hours a week, even in the drenching rain. Right from the start the Chinese proved particularly susceptible to the local strains of malaria, and, as the mortality rate rose, many ran away from the works and started begging in Panama City. Progress on the work improved when the men were given opium, but soon afterwards Irish workers complained to a New York priest who was passing through the Isthmus, and he wrote an exposé in the New York *Herald* accusing the railway of trafficking in drugs. The supply was ended, and the combination of withdrawal from the drug and the depressive after-effects of malaria drove many of the unfortunate coolies to commit suicide by hanging, drowning or impaling themselves on sharpened bamboo poles. After less than a year, only some two hundred of the original shipment had survived and these shattered remnants were shipped off to Jamaica.

It was to this island that the Company now looked to solve their labour crisis. Jamaica, along with the other sugar islands, was in a chronically depressed state. The 1846 Sugar Bill had ended preferential treatment for West Indian sugar in the London market and by 1850 the industry was in crisis: the increasing absenteeism and uninterest of the planter landlords had led to short-termism, soil exhaustion and a lack of investment in new technology. Malnutrition was rife, worsening the effects of the devastating cholera epidemic of 1850. Thus there was a great response when the Railroad Company's recruiting agent, Hutchins and Company of Kingston, starting advertising for workers, promising food and doctors, and wages of 3s. 2d. per day minimum.

In July 1854, the Governor of Jamaica informed the Colonial Office that two to three thousand adult males had left Kingston for Panama, and by the end of the following year almost five thousand had made the journey. One newspaper editor claimed: 'We could name many persons who were walking the streets of the City for a long period of time, – literally starving because they could not get employment, – who are now doing well in Chagres.' The West Indians quickly acquired a reputation for being the best pick-and-shovel men and the hardiest of the imported workers.

A large proportion of the Jamaicans were more or less resistant to yellow fever, the disease most dreaded by Europeans on the Isthmus. Like Panamanians, many would have had a mild dose during childhood, thus becoming immune. But they readily succumbed to malaria and especially to environmental diseases such as pulmonary infections, tuberculosis and pneumonia, and many contracted this complaint on arrival, partly because they turned up in a malnourished state and thus had no resistance. It was thought that the West Indians would be used to work in the tropics, but for the Jamaicans it was nothing like home. The climate was totally different from that of the West Indian islands. Everything was strange. Even trees found in Jamaica were transformed in the water-soaked conditions of the Isthmus jungles into unrecognisable giants. They were also unfamiliar with the huge tarantulas, scorpions, snakes and land crabs that infested Panama.

Typhoid, dysentery, smallpox, hookworm and cutaneous infections were also endemic on the Isthmus. No record was kept of the number of workers who died, but estimates are from six to twelve thousand. The worst year was 1852, when a cholera epidemic killed unnumbered workers and all but two of the fifty American technicians then on site. One railroad historian reckoned that on average over the five years of construction one in five of the workers died every month. Another account written in 1912 describes how 'Workers who toppled over in the jungle [and] managed to drag themselves to the tracks . . . were picked up and taken to the hospital in Aspinwall . . . Others swallowed by the sinkholes or eaten alive by ants and land crabs, disappeared without a trace.' Certainly a problem emerged of disposing of the bodies. In fact, a grim trade sprung up in cadavers, pickled in barrels for medical schools in the States, the proceeds of which paid for the small and utterly inadequate Company hospital.

Food was scarce and expensive. Although strips of dried jerked beef were a staple, workers and whites alike ate monkey, iguana or snake stew to survive. Often it was best not to ask what was put in front of you in the so-called 'hotels'. Water was also hard to come by. In Panama it came from a spring about a mile outside the city, and was carried in earthen crocks holding three gallons, and sold for ten cents a crock, which was almost a fifth of the daily wage for a worker. Alcohol, in contrast, was cheap and plentiful. American technicians drank champagne cocktails for breakfast, laced with quinine as bitters. According to a local newspaper, it was this intemperance that led to many falling ill. 'In most cases,' the paper wrote, 'sickness and death have occurred from imprudence in drinking spirituous liquors, gluttony and a careless exposure to wet weather.'

The lawlessness of the gold trail was also a real problem for the railway construction. Finding the local police inadequate, the Company bosses made a secret deal with the provincial governor to set up their own force. In 1852 they imported a notorious Indian fighter and former Texas Ranger called Ran Runnels to lead a small but heavily armed company of about forty men, described by a contemporary as 'a bare-footed, coatless, harum-scarum looking set'. Brushing aside the local police force, they had the power of life and death on the Isthmus, liberally engaging in whipping, imprisonment and shooting. On several occasions dozens of men were hanged along the sea wall in Panama City. In 1853 the vigilante force broke up a strike by railway workers, in the process publicly flogging the Panamanian official who had been instrumental in organising the action. The force was also useful for providing surveillance against desertion of their contracted workers, to take up agricultural plots or to work in the better paid transit business. As well as withholding wages, the Company authorised lashings by overseers and the use of stocks to keep men on the job.

All the time the railway, tiny in length, but a massive undertaking considering the conditions, was steadily lengthened. By July 1852, the track had reached Barbaocoas, about halfway across the Isthmus, where it was to cross the Chagres River. At this point the work was returned to a contractor, whose first job was the construction of a bridge there, where the Chagres flowed through a rocky channel, some 300 feet wide.

It was the first taste of the power and unpredictability of the Chagres. A bridge was built, then promptly destroyed when a freshet swept away one of its spans. The Company was once again compelled to take the enterprise into its own hands.

Nevertheless, the bridge was rebuilt and in May 1854 the track reached Gorgona, and work started in Panama City heading up the Río Grande valley. Although incomplete, the railway was by now making serious money. In 1854, with thirty-one miles in operation, 32,000 were transported, and its gross income exceeded a million dollars. This was in spite of a falling-off of the stream of gold prospectors; by now the Isthmus was one of the major passenger routes of the world, and still the best way to get from the East to the West Coast of the United States, whoever you were. The Bishop of California crossed in December 1853 and left a vivid account of the 'pale and miserable' Irish workers, and the 'oaths and imprecations' of the 'ruffians' and 'women of the baser sort' he encountered at Las Cruces.

Setbacks continued even as the line neared completion. A 40-foot-deep cut was dug near Paraíso high in the mountains. When the first rain came, the surface became saturated and the greasy soil moved into the cut burying the railroad to a depth of some 20 feet. In December 1854 a hurricane from the north-east exposed the weakness of Limón Bay, destroying every ship anchored in the port. Nevertheless, at midnight on 27 January 1855, under a torrential tropical rain, two work crews met, bringing into existence the first railroad that crossed a continent. In many ways it was a heroic achievement, and that is certainly how it was viewed in the international press. The project was officially inaugurated on 15 February 1855, with elaborate celebrations. The *Aspinwall Daily Courier* described the whistles blowing from the heights to the Pacific lowlands, while for another English language paper, the *Star and Herald*, originally established in 1849, it was first and foremost a triumph of Yankee entrepreneurship ability.

Indeed, a precedent had been set of American engineering ingenuity and success in the tropics. More than that, Panama was now the site of an internationally important transport breakthrough. In an age where commerce and progress were the watchwords of the world, the excitement was irresistible. Even Mary Seacole, for all her reservations about the American builders, could not help but exclaim: 'Iron and steam, twin giants subdued to man's will, have put a girdle over rocks and rivers.' The official history of the railroad construction, published in 1862, eulogised that 'no one work . . . has accomplished so much, and . . . promises for the future so great benefit to the commercial interests of the world as the present railroad thoroughfare between the Atlantic and Pacific Oceans at the Isthmus of Panama . . . it forms a natural culminating point for the great commercial travel of the globe.'

Because of the railroad, the prospect of a canal had improved almost immeasurably. It was a giant step forward. Not only had the lowest pass in the entire Continental Divide outside Nicaragua been discovered at Culebra, but also the railway, with its 'slender feeler of progress', had opened up the interior. The railroad would serve as the right hand of the canal builders, and it offered a great advantage to Panama when the 'Battle of the Routes' would be fought with Nicaragua.

But perhaps the greatest lure that the railroad presented to those who dreamt of a trans-Isthmian canal was financial. It had cost a fortune to build – estimates are as high as $7–9 million, or $170,000 per mile – but for those who had put up the money the payback was huge. Even before it was finished, a third of the cost had been recouped. In a single month, March 1855, receipts topped $120,000. Fares remained incredibly high and steamship companies happily paid huge sums – one half of their entire freight costs – to unload on to the railway and then reload at the other end.

Panama was not the only trans-Isthmian transit route. The American entrepreneur Cornelius Vanderbilt, having tried and failed to raise the capital for a canal at Nicaragua, had established a transit for passengers there, which involved flat-bottomed craft sailing up the San Juan River and across the lake where they were transferred to carriages for the trip to the Pacific. Although many thousands chose that route, it didn't hurt the business at Panama. In the first six years after it was finished, the Railroad made profits in excess of $7 million. Dividends were 15 per cent on average and went as high as 44 per cent. Once, standing at $295 a share, Panama Railroad was the highest-priced stock listed on the New York Stock Exchange. Panama seemed indeed the golden Isthmus once again.

But amidst the euphoria there were plenty of lessons that later canal builders could have learnt – had they not been blinded by the financial and heroic success of the railroad. The appalling death rates, the political instability and labour troubles, the dangers of the Chagres River and landslides would be the bitter enemies of the next generations of West Indians, Americans and Europeans who went to 'the golden Isthmus' to build a canal. The experience of non-whites like Mary Seacole also looks forward to the confused and unfortunate racial attitudes of the American construction period. Indeed, the railroad, which had put Panama on the map as never before, was to turn out to be, for the people of the country, something of a mixed blessing.

In effect, the transit route had become a company town. Panama *was* the railway. The Company kept for itself the arrangement of accommodation and food for its employees, and everything was imported, just as all the expertise, labour, capital, materials and tools for the railway construction had been. In effect, the local businessmen lost their control of cross-Isthmian freightage and storage – Panama's primary resource – and the entire transit zone came under the control and ownership of British or American interests. In the same way the New Grenadan authorities found themselves outmanned and outgunned by the railway's men, and deferred to the foreigners. The dollar replaced the peso as the common currency and English became almost as common as Spanish. Attempts to impose a toll for the local economy on tonnage on the railroad was defeated. The railroad and foreign interests, backed by the almost continual presence of the US Navy, under the terms of the 1846 Bidlack Treaty to keep the transit open, were just too strong.

In the early 1850s Panamanian landowners had made small fortunes renting or selling properties for hugely inflated sums. But with the opening of the railway and the falling away of the gold rush, demand for accommodation and other services collapsed and the Panamanian economy went into a deep slump, even as the railway prospered.

It was boom and bust for the poor of the country as well. The demands of the gold rush had led to a bonanza for the country's muleteers, oarsmen and porters, but the opening of the railway had ended those trades forever. In the meantime, thousands of men had flooded to the two terminal cities looking for work causing chronic overcrowding and attendant filth and disease. Between 1842 and 1864, the population of Panama City had swollen from five to thirteen thousand and a massive slum had grown up outside the city walls. With the completion of the railway, there was widespread unemployment. In fact, the lot of the majority of Panamanian citizens had actually worsened.

The huge influx of black workers also caused a great challenge to the region's delicate social structure and narrow, class-based political system, and hopeful labourers were still arriving from Jamaica even after the main construction work was finished, leading for calls for migrants without work to be turned away at Colón. The loose social stratification based on colour had been magnified and solidified by the advent of the Americans. One observer described the situation in Colón in the mid-1850s: all the railroad officials, hotel and bar keepers were white, mainly American; the 'better class of shop-keepers are Mulattoes from Jamaica', while 'dispensers of cheap grog, and hucksters of fruit and small wares are chiefly negroes'.

Influenced by the spread of European liberalism, and inflamed by the exclusion of local people from the prosperity of the railway, there was growing popular opposition to the foreign presence in Panama. In many ways Washington replaced Bogotá not just as the real power in Panama, but also as the focus of growing nationalist efforts to rid the Isthmus of empire-building foreigners.

Panama was now essentially an American protectorate. This did not go unnoticed around the world, and in 1857 Britain suggested that this new state of affairs was in violation of the Clayton–Bulwer Treaty, offering instead to be one of a triumvirate of 'protecting powers' alongside the US and France. Showing early signs of the abandonment of the idea of a 'neutral canal', the Americans rejected the solution, citing the Bidlack Treaty as taking precedence.

New Grenada's rule on the Isthmus weakened further during the six-year civil war from 1857, essentially a struggle between federalist, secular Liberals and centrist, clerical Conservatives, complicated by the regional rivalries of the scattered population. Together with economic depression, the 1860s in Panama saw constant turmoil or revolts, *coup d'états* and revolutionary conspiracies that so depleted the treasury that at one point mid-decade many public schools were forced to close through lack of funds. In response to separatist convulsions, as well as the 'mob' threatening the railroad (or, more precisely, foreign or white elite Panamanian interests), US troops were landed five times during the decade, sometimes at the request of Bogotá, sometimes not.

Between 1861 and 1865 the United States was, of course, fighting its own civil war, and the Panama route was used several times for moving troops, materials and bullion from coast to coast. In the build-up to the war, and during the armed conflict itself, America's European rivals were quick to take advantage. The French Emperor, Napoléon III, had long been obsessed with Central America, dreaming of control of a canal, and of a buffer against the alarming expansion of the United States. In 1861 and the following year, there were expeditions to Mexico from Spain, Britain and France to demand payment of debt. But for France it was more than that. Troops were landed, and control established over much of the country. An Austrian Habsburg, Maximilian, was established as Emperor. Never recognised by the United States, he was overthrown in 1867 after the withdrawal of French troops decimated by disease. But a legacy of deep distrust of French activities in the region had been firmly established in the United States. This would have profound effects on later canal efforts.

The United States leadership emerged from the Civil War determined to reverse creeping European intervention in their backyard and to point the US in an outward-looking and expansionist direction. The vague aspirations of the Monroe Doctrine now became national dogma, and from being a defensive strategy it became a licence for US intervention throughout the hemisphere.

The directions of the expansion of American influence and interest – southwards to Peru and Chile, producers of valuable raw materials such as nitrate of soda, copper and tin; and westwards to the newly opened up markets of China and Japan – focused attention on the need for a trans-Isthmian canal. It was now becoming a cornerstone of American strategic ambitions, and part and parcel of the country's Manifest Destiny.

In 1866, the Senate requested that the Navy report on possible canal sites. Rear Admiral Charles H. Davis, having consulted the mishmash of Spanish, French and other sources, came back with nineteen possibilities, including six in Panama and three in Darién, in modern-day Colombia, based on the Atrato River. However, he suggested that investigations on the ground should start in Caledonia Bay on the Atlantic side and in the Gulf of San Miguel on the Pacific.

Late the same year, the US minister in Bogotá negotiated a concession with Colombia for the exclusive right to build a canal, but the treaty, and further efforts over the next three years, never satisfied both sides. The Americans demanded a large measure of control and freedom of action in the transit zone, while Bogotá was fearful of further loss of sovereignty on the Isthmus, and tried to limit the number of US employees or troops in Panama at any one time. On the Isthmus itself, the failure of these efforts was dismaying and further fuelled secessionist temperament.

As the civil war raged in Colombia, sometimes sweeping into its northernmost province, then sweeping out again, Panama faced the prospect that even its key business (albeit foreign-owned and controlled), the railroad, had seen its best days. The opening of the transcontinental railroad in the US on 10 May 1869 ended the Panama Railroad (PRR)'s monopoly almost overnight. The previous year had seen record revenue for the business, but after that the line on the graph heads ever southwards. Arrogant mismanagement of the railroad did not help, either. Panama sank further into poverty and sporadic anarchy, with US troops landing twice more in the 1870s and Colombian soldiers in 1875.

However, the opening of the transcontinental railway did nothing to quieten calls in the US for a trans-Isthmian canal. Never before had it been so plain that the United States was now truly a continental power, with two oceans separated by an enormous but, on paper, highly avoidable distance. The inauguration in 1869 of Ulysses S. Grant as the eighteenth American President brought further new momentum to US canal policy. For one thing, Grant had actually been on the Isthmus. In his first address to Congress, the new President laid out his canal ambitions and soon established a new Inter-Oceanic Canal Commission – headed by Admiral Daniel Ammen – to investigate possibilities, and conduct and weigh new surveys of all the possible routes.

So for the first time, every possibility, however remote, was to be meticulously and systematically explored. What's more, the methodology would be the same everywhere, so real comparisons would now be possible. At last myth would be separated from reality about the true prospects of an interoceanic canal.

5

THE COMPETING ROUTES

The first Grant expedition to be equipped and leave the United States for the Isthmus headed out from Brooklyn Navy Yard on 22 January 1870, commanded by thirty-three-year-old Lieutenant Thomas O. Selfridge. 'The Department has entrusted to you a duty connected with the greatest enterprise of the present age,' read his orders from Admiral Ammen. The weather was clear and bright. The destination was Darién, seen by Admiral Davis's report as the most favourable site. On the map it looked simple. The distance from Paterson's Caledonia Bay to the Gulf of San Miguel, the probable route taken by Balboa, was only forty miles and large rivers promised to reduce further the necessary length of a canal there.

Because of this, Darién, a hundred miles south-east of the present-day canal, had been the focus of the majority of the canal schemes and explorations of earlier in the century. In 1850, an Irish doctor, Edward Cullen, had announced in England that he had found there a short and convenient canal route from Caledonia Bay to the Gulf of San Miguel, with a break in the Continental Divide at only 150 feet above sea level. He also claimed that the previous year he had painlessly walked the route from coast to coast in a few hours. But subsequent investigations by a British-funded but internationally manned surveying team had found no pass across the Continental Divide at this point at less than 1300 feet above sea level. During the course of the exploration, several teams got lost and were either ambushed by hostile Cuna Indians or died of starvation or disease.

'It is proved beyond all doubt,' the leader of the surveyors had reported, 'that Dr Cullen never was in the interior, and that his statements are a plausible network of fabrications.' The Irish doctor wisely disappeared, though he emerged later as an army medic in the Crimea. It seems that his entire story was a gigantic lie, a dream built on wishful thinking, which had cost the lives of more than a dozen men.

For Selfridge, though, it was still worth checking. For two months his team, heavily armed to deter Indian attacks, searched the high mountain range near the Atlantic coast, but found no sign of Cullen's pass. From there he moved on to San Blas a hundred miles west in the direction of Colón, the slimmest part of the Isthmus at only thirty miles at its narrowest, but with a 1000–1500-foot-high mountainous spine. British explorers had been here in the 1840s, but had been turned back from the interior by the Cuna. Unable to gauge the height of the Divide correctly, there had been plenty of optimism about a low 'very remarkable' depression. In the 1860s an expedition here had been funded by Frederick M. Kelley, a Wall Street millionaire.

Kelley, described in his later years as 'mystical and imaginative', was clearly a man in the grip of an idea. As well as corresponding with and meeting the by now ancient Humboldt, he toured the capitals of Europe looking for support for a canal, and over the next twenty years spent an estimated $150,000 of his own money on surveys in Darién that, with hindsight, would never have led anywhere.

Although also turned away before completing the journey coast to coast at San Blas, Kelley's explorers had come up with a plan involving a tunnel seven miles long. Selfridge's team almost crossed the Isthmus, quickly discarding either a sea-level or lock canal. Even if the long string of locks needed to lift ships to over 1000 feet could be built, every lock canal needs a supply of water at its summit elevation. Each time a lock is used, thousands of gallons of water pass 'downstream'. At the top of the mountains of San Blas there were only trickling streams. The only option was for a tunnel, but this time the estimate was at ten miles long. Nevertheless, because of the overall shortness of the route, Selfridge refused to rule it out as he led his exhausted team back to New York.

He was back in Darién at the end of the year, still chasing shadows. On 29 December 1870 he sailed to the deep, beautiful Gulf of Urabá into which flows the Atrato River. This was where Balboa had founded the now-abandoned town of Santa María more than three hundred years before. The area had been explored with a view to using the navigable section of the Atrato for part of a cross-Isthmian waterway, which offered an excavation of as little as twenty-eight miles. The area had been crisscrossed by French and British naval officers in the 1840s, and in 1852 one of the hard-bitten Panama Railroad engineers, John C. Trautwine, had been employed by Kelley to explore the tributaries of the Atrato, where there were rumours of a 'lost canal' linking, deep inland, two rivers flowing to opposite oceans. There was no sign of the 'lost canal'. Trautwine's gloomy final report was shrugged off by Kelley, who sent three more expeditions into the area, but none reported any more favourably.

Selfridge himself was now succumbing to the wishful thinking, or 'excessive optimism', that is a recurring part of the canal story. He was much taken by the

Atrato basin and allowed himself to imagine that 'it [would] one day be covered with sails from every clime'. Selfridge had been joined by fellow naval officer Edward P. Lull, who carried out a hydrographic survey of the river's gulf, while Selfridge himself headed upstream. He eventually came up with a plan that involved more than twenty locks and a tunnel five miles long. Tunnels were particularly problematic, even without the geological considerations. They had to be high enough for the tallest sail-bearing masts, but there was also the problem presented by the accumulation of steam from engine-powered vessels. Nevertheless, Selfridge was keen, and further expeditions were sent to take another look.

Meanwhile, separate expeditions had set off to investigate the other possibilities suggested by Humboldt, Davis or both. On 11 November 1870, a party of surveyors landed at Tehuantepec, the narrowest point in Mexico, where the Isthmus is 130 miles wide, with a low range of mountains. This had been investigated as a possible canal site by the Spanish as early as 1555, and post-independence had been surveyed by Mexican engineers, one of whom suggested a canal with no fewer than 161 locks. As elsewhere, concessions were granted, then sold on, but no work was ever started, except that a primitive overland transit was established by an American company. Although it was the possible canal site nearest to the United States, the surveyors found little else to recommend it. The lowest pass was over 750 feet above sea level, and the only possible canal would be 144 miles long and need an impossible 140 locks.

Compared with Nicaragua, the Mexican mountainous divide was enormous. In 1851 a Nicaragua survey had been commissioned by Cornelius Vanderbilt while he was busy setting up his transit route to compete with Panama for the Gold Rush traffic. Led by Orville Childs, who was assisted by a young Cuban-American, Ancieto G. Menocal, the survey had been meticulous and successful, finding a pass only 153 feet above sea level in the narrow neck of land between Lake Nicaragua and the Pacific. This was more than 100 feet lower than any pass in Panama. Childs had suggested a canal with fourteen locks, but Vanderbilt had been unable to raise the capital.

There were two Grant expeditions to Nicaragua. The first got off to a calamitous start, when the commander of the expedition was drowned while attempting to land. But a second expedition headed by Selfridge's previous companion Edward Lull and the Cuban-born Menocal conducted a thorough survey, running a line of levels from coast to coast, measuring river flows and depths, and preparing tables, maps and charts. They suggested a canal with ten locks on either side, costed at nearly $65 million.

It was not until the beginning of January 1875, after a further expedition had explored other routes around the Atrato River, that the Panama survey team left

New York for Colón. From the surveys conducted for the railway, the Panama route had been the best known at the outset, so it was left to last. The survey was led by Lull and Menocal, and for two and a half months some one hundred men explored and measured the land between Colón and Panama and up into the Chagres watershed. Nevertheless, both Lull and Menocal had pretty much decided on their preference for the Nicaragua route, and they were unable to imagine a realistic design for a canal at Panama. The line of their proposed waterway largely followed that of the present-day canal. It was to be a lock canal, with a summit level of 124 feet, reached by twelve locks at either side. The problem with Panama, they correctly judged, was the volatility of the Chagres River. They suggested that what was needed was a huge, 2000-foot-long viaduct sufficiently high to allow the floodwaters to pass under it. But the river was required as well, to supply water to the summit level, so the engineers decided that ten miles of aqueducts would have to be built. In all, it was just too impractical and expensive.

When the Commission reported in February 1876, there was a firm majority in favour of a route through Nicaragua. Tehuantepec was dismissed out of hand. Darién was too remote and wet; and the various Atrato schemes were too ambitious. All myths about 'lost canals' and low passes in those areas had been exposed. Panama was too expensive, and what's more, 'the deep cut would probably be subject to land-slides, from which the Panama Railroad has suffered seriously, and the canal would be exposed to serious injury from flood'.

From then on, for the next twenty-five years, US canal policy was firmly focused on Nicaragua, which became lodged in the minds of the public as well, who were beginning to demand an American canal under American control, rather than the neutral and open-to-all arrangement favoured by the Commission. In 1877, treaty negotiations were started with the Nicaraguan government. In contrast, the US did not even have a diplomatic representative in Bogotá. Slowly, draft treaties were drawn up, while private companies jostled for the concession, and Ancieto Menocal prepared another, even more thorough, Nicaragua-bound surveying party.

Then, in May 1879, the United States public read in their papers news from Paris that upset all previous plans. Le Congrès International d'Etudes du Canal Interocéanique had approved the building of a canal at Panama and events were moving fast. It seemed that there would be a canal, but it was going to be built by the French.

PART TWO

The French Tragedy

'We are, gentlemen, soldiers under fire; let us salute the comrade who falls in the battle, but let us think only of the fight of tomorrow and of victory'

– French engineer and lobbyist Philippe Bunau-Varilla

6

'LE GRAND FRANÇAIS'

Paris in the spring of 1879 was awash with frenetic financial speculation. Money was more plentiful and mobile than ever before. The expansion of the railway and telegraph networks had brought new branch banks to provincial France, and the reach of those seeking to raise capital had widened considerably. Penalties for debt and bankruptcy had been eased and restrictions on issuing shares had been lifted. In the Paris Bourse, young stockbrokers noisily wheeled and dealed, betting on pretty much anything and often making huge sums for themselves. One such was the artist Paul Gauguin, who would later work on the Panama Canal. In 1879, he made 30,000 francs, a fortune at the time.

Much of the activity had been triggered by the creation of a new investment bank, the Union Générale, whose founders set out to raise money from the country's Catholic majority. They were not averse to playing on the anti-Semitic and anti-Protestant sentiments of their potential investors, and the Union's original prospectus was even blessed by the Pope. The approach was a great success, and the bank launched in 1878 with an astonishing capital of 25 million francs, an indication of the amount of money sloshing around the country.

It had been an extraordinary recovery from the disaster of August/September 1870, when the troops of Napoléon III had been routed by the Prussians. The Second Empire had collapsed immediately. There followed a grisly siege of Paris, and with the surrender of the city in January there across the bloody terror of the Commune. At the end of the war, France lost the ore-rich provinces of Alsace and Lorraine and was given a heavy indemnity, designed to cripple its power for a generation.

But France paid off the debt ahead of time, and soon overtook pre-war industrial production. In spite of endemic political instability, everywhere there was optimism and energy, a spirit of *revanche*. France would be great again, and would recover her prestige not through fighting, but through, as Victor Hugo put it, 'astonish[ing] the world by the great deeds that can be won without a war'. In the spring of 1878 Paris held a great exposition, which covered sixty-six acres of the city and attacted thirteen million visitors. It cost a fortune, but it demonstrated to the world France's recovery and new ambition.

The embodiment of this spirit of *revanche* was the builder of the Suez Canal, Ferdinand de Lesseps, called by Gambetta '*Le Grand Français*'. Untainted by the events of 1870, he represented a new patriotism based around, not war, but achievement for all mankind. De Lesseps had followed his father into the diplomatic service, and in 1832, aged twenty-seven, had been appointed vice-consul at Alexandria and served in various roles in Egypt for the next five years. During this time he be friended the son of Egypt's Viceroy, and also became interested in the idea of a canal linking the Mediterranean to the Red Sea.

After several positions in Europe, de Lesseps left the diplomatic service in 1849, but returned to Egypt six years later as the guest of the new Viceroy, his friend Said Pasha, who had just succeeded his father. Although de Lesseps had neither expertise nor money, after just over two weeks' stay he had persuaded Said Pasha to sign the concession that gave the Frenchman the right to build the Suez Canal.

A plan was drawn up by two French engineers and approved by an international commission of civil engineers, but there were plenty of critics of the scheme. In Britain Robert Stephenson said he believed it to be a folly on a huge scale, and Palmerston, now Prime Minister, called it 'an undertaking which, I believe, in point of commercial character, may be deemed to rank among the many bubble schemes that from time to time have been palmed upon gullible capitalists'. In fact, the British feared the alterations in the status quo that the canal would bring. Other doubters at home and abroad predicted that the trench would fill with sand from the desert as quickly as it was dug.

But nothing could dishearten Ferdinand de Lesseps. Supported by the Emperor Napoléon III and the Empress Eugénie, who was a cousin of his, de Lesseps succeeded in rousing the patriotism of the French and obtaining by their subscriptions more than half of the required capital of 200 million francs. The other half of the shares was taken up by the Egyptian government, who also forced thousands of local labourers to work on the project in conditions of semi-slavery. The excavation operations through the desert took nearly eleven years, during the course of which numerous technical, political, and financial problems had to be overcome. The final cost was more than double the original estimate, but the canal opened to traffic on 17 November, 1869. De Lesseps, then sixty-four, was world famous.

Celebrated as the greatest living Frenchman, he was honoured around the world. In 1870 he visited England, and *The Times* eulogised him as a 'man eminent for originality enterprise courage and persistence . . . moral qualities of the highest order'. At a banquet in his honour at Stafford House he was toasted by both Gladstone and Disraeli, and six days later more than twenty thousand people filled London's Crystal Palace for a reception in his honour.

He was praised for having the unshakeable confidence to believe in his project even when virtually no one else had done, but in France in particular he was also seen as a man of the people for having built the canal with the capital of 25,000 small investors, rather than the big Jewish banks. Following Suez he was never out of the Paris papers, and the fact that he had a pretty, young wife and a brood of adorable children made him perfect fodder for the new illustrated magazines. When in the early 1880s he ascended to the pantheon of the French Academy, he was praised thus: 'You exercise a charm. You have that supreme gift which works miracles . . . the true reason for your ascendancy is that people detect in you a heart full of sympathy for all that is human . . . people love you and like to see you and before you have opened your mouth you are cheered. Your adversaries call this your cleverness; we call it your magic.'

There was a special magic, too, about the nature of his achievement in Egypt. In a stroke, India had been brought nearly six thousand miles closer to Europe. Furthermore, in a Victorian Age determined to take on and conquer the world with innovation, engineering and entrepreneurship, the Suez Canal was the perfect marriage of commerce, transport and industry, the forces that were changing the world. It was also seen as a great civilising achievement, which 'will help to wed the whole universe into one great unit, politically, industrially, and religiously', as was written at the time. 'The two sides of the world approach to greet one another . . .' wrote a British commentator in a similar vein.

But it was not just about high ideals. By the late 1870s, the benefits of the canal to European manufacturers were more than apparent, as the raw materials of the East, as well as their markets, had been brought so much nearer. More than anything, it gave Europe a critical advantage over the frighteningly fast-growing economy of the United States' East Coast. For the canal's original backers, too, it was a bonanza. Initially the share price of de Lesseps' company had slumped. It soon turned out that his promises made during the construction for the traffic through the canal had been wild overestimates. But by 1879 a 500-franc ($100) share was trading for over 2000 francs, and the company was paying dividends of 14 per cent. Suez had made a lot of people rich.

Bathed in public adulation, de Lesseps, although by any reckoning now an old man, was not going to content himself with just the Suez Canal. 'Is it not a glorious thing for us to be able to carry out . . . vast projects,' he wrote, 'thus affirming the progress made by our race and age, in which all obstacles seem to have disappeared.' In 1873 he became interested in a project for uniting Europe and Asia by a railway to Bombay, with a branch to Peking. He subsequently got involved in a hare-brained scheme to break though a low-lying ridge in Tunisia to create a huge inland sea in the Sahara.

Other French writers, artists and thinkers shared this outward-looking energy and ambition. Jules Verne's *Around the World in Eighty Days* was the greatest popular literary success of the decade. On the left, thinkers radicalised by the events of the Paris Commune dreamt of a future without nation-states or empires, in which the whole world, driven by the engine of Progress, would live 'like brothers', as Elisée Reclus wrote. Reclus was one of the founding fathers and the 'soothsayer' of the French anarchist movement, which would erupt into violence in the 1890s, but his training and the foundation of his reputation was as a geographer. In the 1870s, much of which he spent in enforced exile, he produced a mammoth new geography of the world, which moved from the formation of the earth to the future of man, synthesising his geographical and social views.

In fact, geography was hugely in the ascendant at this time, with societies for its study springing up throughout France and international conferences attracting delegates from all over the world. In 1871, Reclus inspired the holding of a conference in Antwerp. It was just two years after the opening of Suez, so it was natural to have an American canal on the agenda. What's more, Reclus had actually been to Panama several times in the 1850s when he was in his twenties, during another period of exile. At the conference it was, ironically, an American army officer, Wilhelm Heine, who brought the delegates up to date on the efforts of French explorers, who had focused, like others, on Darién.

Back in 1864 a young French naval lieutenant, Henri Bionne, had published a pamphlet speculating about the existence of a realisable passage through the Isthmus between the southern part of the Gulf of San Miguel and the Atrato River. Much of the waterway, he suggested, could consist of existing rivers, the Tuyra and the Atrato. This led to a French explorer, Lucien de Puydt, reporting the following year that he had found a pass only 150 feet above sea level on a similar, but even better, route. Bionne never claimed to have visited Darién; de Puydt admitted that he had seen the low pass, but not actually reached it. In turn, de Puydt was followed up by Anthoine de Gogorza, who had been born in America of French parents and lived in Cartagena. Having visited Spain to research the ancient maps, he tried to interest US companies in sponsoring further explorations. Failing at that, he secured backing from a group of French capitalists and sent an expedition, which after fourteen days recommended a sea-level canal close to Bionne's suggested route, linking the Tuyra and Atrato rivers in Darién. The highest point, the group's leader claimed, was only 290 feet above sea level.

Wilhelm Heine, who was attached to the US embassy in Paris, was then sent by de Gogorza's backers to Darién to verify the claims. On his way south, he stopped off in Caledonia Bay in March 1870 while Thomas Selfridge was there. He was given two of the Selfridge party and taken by boat down to the Atrato delta. He was unable to proceed up the river, or land on firm ground because of the absence of local boatmen, but declared himself satisfied by what he had seen from his position offshore. 'At last the truth about the Isthmus of Darién is being revealed by M. Gogorza's discovery,' he announced to the Antwerp conference the following autumn. 'Doubt is no longer possible.'

The conference was impressed, tabling to recommend de Gogorza's 'achievement' 'to the attention of the great maritime powers and of all scientific societies'. The ball had started rolling. Four years later there was a further international conference, this time in France and under the auspices of the Société de Géographie de Paris, timed to coincide with a huge geographical exhibition in the Louvre, which was attracting more than ten thousand people a day. At the conference, the American canal question was at the top of the agenda. De Puydt, Bionne and de Gogorza presented their respective southern Darién schemes and an American delegate described the lock canal plans of Menocal and Lull in Nicaragua and Panama.

The star turn, however, was Ferdinand de Lesseps, who had become a leading light in the Paris Geographical Society. He spoke about his experience with the Suez Canal, but then went on to lay out for the first time his ambitions for an American waterway. Even at this early stage, he said he hoped that it would be possible to construct in Central America a canal *à niveau*, at sea level without locks. Swayed by the arguments of the French explorers, he announced that his personal preference was for a route in southern Darién. There was another crucial advantage to this area that had nothing to do with surveys or geography: it was outside – just – the limit of the Panama Railroad's transit monopoly as granted by their concession from Bogotá.

Following the conference, the Society established a special committee, with de Lesseps as president, to study the canal question. The careful surveys of the Americans were insufficient, it was decided. There were two Colombian delegates included, and it was hoped to involve scientists from all over the world, with the idea of further exploration being jointly undertaken and financed. It was an idealistic beginning.

But at this point a syndicate emerged and offered to fund the work itself. Led by General Istvan Türr, a Hungarian who had fought with Garibaldi during the unification of Italy, the group, which called itself the Société Civile Internationale du Canal Interocéanique de Darién, included eminent figures from the worlds of letters, industry and finance. Henri Bionne was a member, as was Lucien

Napoléon Bonaparte Wyse, a twenty-nine-year-old naval officer, brother-in-law to Türr and married to a wealthy Englishwoman. Wyse had experience of exploration on the Isthmus and was distantly related to his famous namesake. There were some less salubrious characters, too, as would emerge later. In total their capital was 300,000 francs, represented by sixty shares of 5000 francs each. De Lesseps was not a shareholder, but he was a friend of Türr and a number of the others.

Türr's first move was to send Anthoine de Gogorza back to Colombia to obtain permission to conduct a survey in southern Darién and a provisional concession to build a canal there. This was obtained in May 1876, and it shows the influence of de Lesseps' thinking that it stipulated that the canal should have neither locks nor tunnels. Six months later, a seventeen-man survey team had been assembled. In charge of the party was Lieutenant Wyse, with Armand Reclus, younger brother of Elisée, the geographer-anarchist, second in charge. A naval officer, thirty-three-year-old Reclus was an old and close friend of Wyse. Their actions over the next twelve months, on two long and arduous forays to the Isthmus, would determine the entire history of the Panama Canal. In contrast to everything that had gone before, the findings of these expeditions would at last result in actual construction.

7

THE FRENCH EXPLORERS

Reclus published an account of the two Wyse surveys in 1881. It was based on a diary he kept at the time, and was finished before the French canal construction was underway. Reclus writes in an extremely detailed way about the expeditions. He comes across as very perceptive in his observations, even-handed and friendly with those he meets, and something of a romantic. Reclus would return to Panama as the local head of de Lesseps' canal company. In many ways his approach sets the tone for the French attitudes to Panama and its people. There is very little about the technical aspects of the trip, suggesting that, at least initially, he was not over qualified for the task. He does, however, report with pride at the end of the book that the expedition's findings were adopted as the blueprint for the de Lesseps canal.

On the 12 November 1876 they left St-Nazaire on the *Lafayette*, a two-masted steam sailer. They were to voyage to Colón, cross the Isthmus, and then head from Panama City to the Pacific coast of southern Darién to look at de Gogorza's route. The secretary of the first expedition was Oliviero Bixio, the thirty-five-year-old nephew of one of Türr's comrades-in-arms from Italy, described by Reclus as a hugely well-liked and intelligent companion. He brought with him from Italy a young engineer called Guido Musso, like Bixio friendly and good-looking. And like Reclus, they had both volunteered their services out of friendship for Wyse and a desire to be part of the '*oeuvre grandiose*'.

Also on board was a doctor, a young Hungarian engineer, a very senior French civil engineer, an English geologist called William Brooks, as well as other specialists from France. The *Lafayette* arrived in Limón Bay on 21 November, and after two days in the 'disgustingly dirty' town of Colón the party took the railway to Panama City, where they were met by two Colombian government engineers, Pedro Sosa and a Scotsman, A. Balfour, who were to help with the surveying, and a French explorer from the de Gogorza expedition of 1866. 'Compared to Colón, Panama is a real paradise,' Reclus noted as he enjoyed the city's 'air of grandeur with its eight or ten churches and convents in ruins, its palaces, its prisons, its arsenals of another age, its giant fortifications'. In the original walled city, a 'few of the old Spanish residences in Moorish style remain, where thick walls protect from the heat and where fountains constantly maintained freshness in the vast patios'. Everywhere, Reclus noted, was evidence of the Panamanians 'dislike of uniformity'. But the city had sprawled inland from the old city, and in the suburbs there was squalor to match anything in Colón.

Conditions in the slums had been worsened by a severe fire early that year, but Reclus still found the Panamanians optimistic that a future canal would bring them great riches. In the meantime, Wyse had established a base at the Grand Hotel, run by a French-Panamanian, Georges Loew. According to Reclus, 'travellers who lack curiosity (and there are lots of them) spend their whole time in Panama within its bounds'. All the 'gentlemen of the town' met in its huge downstairs bar, 'veritably the Stock Exchange of Panama, the place where most important business is transacted'. In one corner Felix Erdman, an American-born banker and the richest man in Panama, held court, betting on dice for cigars. In another, the English geologist Brooks quickly established himself on the roulette wheel, while also proving his prowess and impoverishing his companions playing poker and billiards.

The 'laziness' or 'inertia' of the Panamanian people is a recurring complaint of visitors from Europe or North America from throughout the construction period. In the absence of any large agricultural or manufacturing industry on the Isthmus, there was indeed little of the tradition of disciplined wage labour as had been established elsewhere. Unemployment was widespread as well. But the climate also dictated a different pace of life to that in cooler climes, as Reclus quickly gathered. For him, it was all summed up in a joky hymn to the ever-present hammock: 'Perfidious friends, more dangerous than the climate, debauchery and drunkenness! You find them everywhere . . . they beckon to you; they wait for you. They rock you deliciously in that steamy atmosphere as the torpor following a meal takes hold of you . . . as with half-closed eyes one follows the spiralling blue smoke from a cigarette . . . Soon you can't escape, you spend your entire days there. The most active of men soon turns into an indolent dreamer overtaken by inertia.'

Panama's other speciality, the French party were discovering, was pomp and ceremony, and when the surveyors left Panama City on 11 December it was to the cacophonous strains of a military band and fireworks, and the applause and speeches of a host of local worthies. The State President, General Rafael Aizpuru, a populist leader deeply distrusted by commercial and foreign interests in the city, was to accompany them as far as the Gulf of San Miguel, and they also had on board a party of local machete men and two translators. Their doctor, however, had succumbed to fever and had to be left behind.

For three days they made their way up the Tuyra River and then established a base camp in a small Indian village. While Bixio organised the stores, Reclus, accompanied by the Scotsman Balfour, began a hydrological survey of the Tuyra, measuring depths, currents and curves in the river. Reclus was enjoying himself. On the coast the weather was dry and there was a cooling breeze. As the work got underway, Reclus allowed himself to imagine what the outcome might be: 'Our hope is to fill these waters with all the ships which will bring world commerce to the canal: clippers with their huge hulls, three-masters letting their white sails fill to the breeze, busy steamships ploughing through the water with all the force of their steel lungs, brigs, all sails set, vessels of every nation.'

The excitement continued as Reclus worked upriver. 'The varied landscape unfolding before our eyes, mighty rivers, streams of fresh water, dangerous rapids, dark and mysterious forests, hidden gorges, peaks from which the view stretches out over unknown territory . . . All that, enriched with the thousand adventures of a journey, keeps the spirits up, and you can endure, almost enjoy, the hardships and the misfortunes.'

Worst of the hardships were the mosquitoes that attacked the men all night long, leaving them tired and itching in the morning. In fact, other problems and setbacks were steadily mounting: as they went inland the dry weather on the coast gave way to intermittent rainfall; everyone seemed to be ill with something or other; their helpers from Panama City suddenly demanded more money to continue. The worst blow so far came when Bixio, who had spent much time wading chest deep in rivers and streams, contracted pneumonia and died. Reclus was devastated, and from now on the jungle lost its air of wonder and became instead monotonous and eerie, a 'green devil'. The Englishman, Brooks, was the next to die. Weakened by the blood loss resulting from vampire bat attacks, he caught dysentery and passed away at the end of January.

The exploration continued but it soon became clear that the surveys by de Puydt and de Gogorza had been yet more wishful thinking, 'swarming with errors', as Reclus put it, in contrast to Selfridge's far more reliable data. Reclus was sent with a small party to try to find one of the 'low passes' promised, but found surveying in the thick jungle almost impossible. On some days they covered less than a thousand yards. Deep in the jungle lurked a menace even worse than the mosquitoes – blood-sucking ticks that left suppurating wounds and imparted a variety of debilitating diseases. At one point Reclus's legs were so badly affected that he could no longer walk. Then, in late March, the rainy season restarted early, bringing the unfortunate expedition to an end.

The journey home saw the death of another of the party, the young Italian Musso, whose enthusiasm for the great idea of the canal had lifted everyone on the expedition. Wyse and Reclus arrived back in Paris thoroughly disheartened. When they met de Lesseps and the rest of the Société's Committee of Initiative, they made it clear that a sea-level canal in Darién was out of the question without a huge tunnel, and even a lock canal would require a tunnel as well. This was not to de Lesseps' liking. He had made up his mind that only a canal *à niveau* would do, and he urged Wyse to return to the Isthmus and widen his investigations. De Lesseps later told American reporters that he ordered Wyse specifically to investigate a sea-level canal along the line of the Panama Railroad, suggesting that he had given up on Darién and decided to bite the bullet about the PRR's local transit monopoly.

Reclus's wife, who had been alarmed at seeing her husband return from the jungle exhausted, emaciated and covered with scars from insect bites, urged him not to go on the second expedition. But Reclus had been reanimated by de Lesseps' infectious enthusiasm, and his friend Wyse, offering a thousand francs a month and the possibility of future employment with the forthcoming canal company, persuaded him to be his second-in-command once more. The party, with many of the survivors from the first trip on board again, left for Panama at the end of 1877.

In spite of the supposed instructions from de Lesseps, the second Wyse expedition headed first for San Blas, then back to the Tuyra River in southern Darién. But ten days into the latter expedition, which was hampered as before by hardships and disease, Wyse returned to Panama. There, he travelled along the route of the railroad, before ordering Reclus to cut short his explorations in Darién and return to Panama. Wyse, for his part, was now planning a journey to Bogotá, where, according to rumours, opinion was moving towards cancelling the Türr concession.

While urging Reclus to secrecy, in particular ordering him to play down the disadvantages of the San Blas and Darién routes (presumably to give the PRR the impression that there were other options), Wyse outlined his instructions for his colleague's work between Colón and Panama. From what he had seen, Wyse had decided a preference for linking the Chagres and Río Grande valleys, effectively following the route of the railroad. The valleys were the lowest and longest in the region, had been comprehensively surveyed, and were served by the railway. 'You should start preparations immediately to study the route of an inter-ocean canal at sea level, but with a tunnel,' Wyse wrote to Reclus. Waiting for Reclus in Panama were a 'succinct version' of the American engineer Lull's report (though not the unpublished accompanying maps and charts), the maps made by the Frenchman Garella in the 1840s and the American railroad survey from the

1850s, in all pretty comprehensive information. The operations, Wyse suggested, should be finished by the end of April.

The key component of the rough plan envisaged by Wyse was a tunnel through the Continental Divide to link the two valleys. So sure was he that this was necessary that he instructed Reclus to skip a detailed survey of the part of the route that would be underground. In Reclus's own account, admittedly published after the adoption of the Colón–Panama route by the subsequent International Congress in Paris, the Panama choice suddenly becomes the favourite, and the San Blas and the Darién routes fallbacks if difficulties 'political or otherwise' prevented the adoption of the railroad line as first choice. According to him, Wyse's trip to Bogotá was primarily to remove from the concession restrictions that it had to be a sea-level canal without locks or tunnels.

Wyse and Louis Verbrugghe left for Bogotá on 25 February, the same day as Reclus arrived back in Panama. Speed was vital as President Aquileo Parra, known to be sympathetic to the Türr–Wyse syndicate, was about to retire from office. As well as removing the restriction on tunnels, it was also necessary for Wyse to negotiate a new concession that would allow a canal to be built right on top of the American-controlled railway route.

It took the two men only twelve days to complete the four-hundred-mile journey, which, because of the isolation of Bogotá and the intervening impassable Darién jungle, usually lasted up to four weeks. After going as far as possible by steamer, they then rode for up to seventeen hours a day to the high Andean plateau. The day after their arrival, Wyse was granted an audience with the Foreign Minister Eustorgio Salgar.

Reclus, it seems, was not such a good horseman. On 4 March, just as all the preparations for the new survey reached completion, he fell off his mount while riding just outside Panama City. He was confined to bed, while his deputy for this task, the Colombian Pedro Sosa, started without him, looking at a river some ten miles down the coast from the Río Grande at Panama City. The bad luck continued when another severe fire broke out and spread throughout Panama City's old quarter. Although Reclus was able to save a good deal, many of the expedition's maps, notes and instruments were lost.

Nonetheless, on 11 March Reclus joined Sosa on his exploration of the rivers along the coast from Panama City, to cover all the possible routes and to find out as much as possible about the hydrography of the Panama area. At last, at the beginning of April, they were back in civilisation, and ready to look at the railway line route from Colón to Panama City. Here, in contrast to their previous explorations, there was transport and provisions at hand, although as they worked along the line they were hampered by the loss in the fire of Sosa's tachometer. In other places, where the canal, for reasons of curvature, would need to follow and adjacent valley, the surveyors were again coping with thick vegetation, which entailed arduous clearance with machete or fire.

After a week, Sosa, ill with dysentery, returned to Panama for five days, but the work continued until the early rains at the end of April made further measuring impossible. To his good fortune, Reclus encountered a friendly PRR engineer calculating river flows and rainfall who was happy to share his data. For ten days Reclus collected together his measurements and papers ready for Wyse to look at when he returned from Bogotá. On 1 May he was finished and sailed for France.

Reclus's work in the river valleys of the Chagres and the Río Grande, which largely determined the route of the French canal, remains highly controversial, and would be viciously attacked by the enemies of the canal, in particular the Americans. There is no doubt that Reclus's cheerful and positive account is somewhat partial. He underplays Sosa's leaving the expedition after a week, and fails to mention the agonising earache he himself suffered, which as much as the rains brought the surveying to an end. Certainly, it was a mistake to leave so little time for what turned out to be the most important part of both the expeditions. Astonishingly, Reclus's description of surveying what would be the actual route of the canal takes up only the last eight of a four-hundred-page account of the two expeditions. Perhaps this illustrates Reclus's exhaustion and dissatisfaction after the travails of Darién and the earlier loss of his friends Brooks, Bixio and Musso.

Meanwhile, Wyse had more clear-cut success. Meeting the Colombian President on 14 March, he was given all encouragement and presented a draft contract the very next day. Five days later, after only minor amendments, the deal was signed. Under the terms of the 'Wyse Concession' the Türr syndicate was granted exclusive right to build an interoceanic canal through Panama. As a provision of the deal, the waterway would revert to the Colombian government after ninety-nine years without compensation. It was also stipulated that the exact route of the canal should be determined by an international commission of competent engineers, just as had been the case with Suez. The syndicate could sell the concession to a private company but not to a government. Colombia was to receive a percentage of the gross revenue rising from 5 per cent over time, with a minimum amount of $250,000 per annum. It was left to the concession holders to negotiate 'some amicable agreement' with the Panama Railroad concerning its rights and privileges.

With this potentially valuable piece of paper in his pocket, Wyse returned to Panama, picked up the details of Reclus's work there, and took a steamer for Nicaragua. There he travelled the route so extensively surveyed by the Americans, then proceeded to the United States where he met Lull, Menocal and

Ammen. He got to see their unpublished Panama survey plans and maps, and was promised copies.

Back in France, Wyse and Reclus worked on collating their information. The date of the 'International Commission', as demanded by the Wyse concession, was set for May the following year, and in the meantime Wyse, Türr and Reclus, in concert with Ferdinand de Lesseps, prepared a series of plans to present.

Reclus, in particular, was concerned about the dimensions of the massive tunnel and the problem of the Chagres River, suggesting further hydrographical studies. Wyse had to chase him to finish his final report, writing of the syndicate's head Istvan Türr: 'You wouldn't believe how annoyed he is at the delay you are causing.'

Just before the conference Wyse put together a table of seven options, to be decided by the delegates. Four were in Darién or San Blas, all of which required extensive locks, tunnels or both. The fifth and sixth options were the Panama Railroad Colón–Panama route. One was Lull's twenty-five lock canal from the Grant surveys, the other was at sea level with ('or even without') a 5- to 8-kilometre tunnel. Wyse estimated this could be built in six years at a cost of 500 million francs. The last option was in Nicaragua, with twenty-one locks. The disadvantages of this route were stressed: the complete absence of suitable ports, political instability and the fact that the land over which the canal would pass was disputed between Nicaragua and her southerly neighbour, Costa Rica. In his conclusion, Wyse was unambiguous about his preferred option: the sea-level Panama route, and in his part of the published report Reclus expanded this to include five variations, one of which, seen as uneconomical by its author, dispensed with the tunnel altogether. Although Wyse acknowledged the difficulties presented by the Chagres River, he stressed that this was outweighed by the advantages this route offered, which included, somewhat surprisingly, the political stability of the province of Panama.

In all, some one hundred pages of maps, notes and data were collected together. To the great disappointment of Reclus and Wyse, the detailed tables and maps from the United States' Grant surveys had, in the end, not been forthcoming from the Americans. All versions of the canal were designed to accommodate what was then the largest ship in the French Navy, which had a draught of 7.9 metres. The tunnel would be 24 metres wide and rise 34 metres above the water level to allow ships through without lowering their masts. For Wyse and Reclus, the construction of such a massive tunnel was not the impossibility it might seem now. Seven years earlier had seen the completion of the Mont Cenis Tunnel through the Alps. Thirteen kilometres long, it was built with the help of key new technologies – dynamite (electrically ignited) and pneumatic drills. An even longer Alpine tunnel, the St Gotthard, was currently under construction, and in England work had started on a seven-kilometre tunnel under the River

Severn. In 1875, bills had been passed in the French and British governments for the construction of a tunnel under the Channel, an astonishing distance of about thirty miles. Geological surveys for the project had started in France the following year.

As the Le Congrès International d'Etudes du Canal Interocéanique of international engineers assembled in Paris on 15 May 1879, many of the leaders of these projects were present. From Britain came Sir John Hawkshaw, in charge of the Severn Tunnel; Ribourt, one of the engineers of the St Gotthard Tunnel was there, although as an observer rather than delegate. In all there were 136 delegates from twenty-two nations, including financiers and businessmen as well as civil and military engineers. The American contingent included Ammen, Menocal and Selfridge. All had been personally invited by de Lesseps. The delegates included a large proportion of Frenchmen – over seventy – and among this group were a number of *Le Grand Français*' Suez cronies. One was Abel Couvreux of Couvreux, Hersent, the giant contracting firm that had helped build the Egyptian canal.

By this stage de Lesseps seems to have insinuated himself into every aspect of the canal planning process. Armand Reclus ends his account of his two expeditions with praise for de Lesseps. There was, he says, 'no more glorious name'. In his private letters, though, he was more extact about the appeal of *Le Grand Français* – as a raiser of capital. For Wyse, who would later fall out with the great man and be banished from the project, visiting de Lesseps at his mansion in Chesnaye, was, he wrote to Reclus, 'an inconvenience'. It was Wyse who did the deal with the Colombian government and had led the surveying expeditions, but all the talk was of the involvement of Ferdinand de Lesseps. Nevertheless, Wyse knew he needed the support of a man of de Lesseps' stature if his plan was not to join the host of schemes that had come to nothing.

De Lesseps had a proven track record from Suez. He never professed to be a trained engineer. But he was someone who could make things happen. More than anything, he was a communicator, one who could conjure up both the necessary capital and the dedication among the workforce that a project like the Panama Canal would demand. He had the reputation, energy and charisma to turn fantastical schemes into reality. In many ways, after so many disappointments and false starts, he was just what was needed to deliver to Panama its destiny as a crossroads of the world. Inevitably, though, his strengths were also his weaknesses. His confidence was not only in himself, but also in his age. His lack of concern about the extraordinary challenges of the task ahead had its foundations in his faith in the serendipitous nature of emerging technology. He was sure that the right people with the right ideas and the right machines would somehow miraculously appear and take care of all the seemingly impossible challenges, just as had happened at Suez.

De Lesseps was not alone in riding a wave of confidence in engineering and technological progress. As well as the amazing Alpine tunnels, the previous decade had seen the creation of London's sewer system and the first electric streetlights; 1869 had witnessed the opening of the Union Pacific Railroad across the United States, as well as the Suez Canal. The Brooklyn Bridge was under construction in New York. Increased world trade had driven the development of new iron-screw ocean steamers. Alongside spectacular recent advances in chemical and electrical science, cheap steel, made by the Bessemer or Siemens–Martin Hearth methods, together with new techniques in precision manufacturing, had made possible a huge range of new goods – bicycles, clocks, sewing machines – at affordable prices. Improvements gave birth to further progress and commerce, in a seemingly endless virtual circle. All this was celebrated in international 'expositions', such as the American Centennial in Philadelphia in 1876, which in its machinery hall showcased cutting-edge new technologies. Most impressive was the gigantic 1500 horsepower Corliss Steam Engine, taller than a house, which alone powered thirteen acres of machinery in the great hall.

Dreams of ever more powerful machines, of cheaper and quicker transport and communication, of opening up the world to increased trade and 'civilisation', gripped the public imagination as an endless stream of amazing new inventions and mind-boggling ambitious engineering projects were introduced to the world. De Lesseps' Panama Canal, which embodied so much of this forward-looking confidence, would surely, it was believed, take its rightful place among these great achievements.

8

THE FATAL DECISION

The news from Paris, carried on the newly laid submarine cable from Jamaica, was keenly followed in Panama. At home, however, they had their own concerns. At the end of the previous year there had been an armed uprising against the State President, 'who for some time past,' reported the British consul Hugh Mallet back to London, 'has become obnoxious to the coloured people from whom he has hitherto derived his principal support'. A district judge and two others were killed during violence in Panama City's sprawling and volatile suburbs. The president resigned and was replaced, but on 17 April, a month before the conference opened in Paris, there was another attempted revolution led by soldiers of the Colombian garrison. 'Tragic scenes,' the New York *Tribune*'s local correspondent reported, 'fighting in the streets and many persons killed.' Although the conspirators surrendered, martial law came into force. 'Business has been paralysed,' wrote Hugh Mallet, 'and trade diverted from the Isthmus by constant apprehensions of danger to life and property.' Disorder continued even as the delegates in Paris were debating the future of Panama. On 24 May there were violent scenes, including gunfire, in the provincial assembly building. Two weeks later, General Rafael Aizpuru, who had been State President on Reclus's first visit to the Isthmus, led a revolt of his supporters among the racially mixed lower classes in Colón, while Benjamin Ruiz, another radical, tried unsuccessfully to seize power in Panama City.

None of this political turmoil seems to have registered with the delegates at the Paris conference who met in the afternoon of 15 May in the sumptuous surroundings of the headquarters of the Société de Géographie on Rue Saint Germain, in Paris's Latin Quarter. It was the best time of year to be in Paris: the sun was shining and the young chestnut trees that lined the street were in new leaf. Crowds had gathered to see the multinational delegates, clad in top hats and morning coats, alight from their carriages and make their entrance into the beautiful, lofty auditorium. Here, there was seating for nearly four hundred: de Lesseps, his officers and the titular head of the congress, a grand French admiral, de La Roncière-Le Noury, were on the small stage. The front rows were reserved for delegates, while all other seats were taken by spectators.

The opening session was purely ceremonial: the admiral made some remarks, then de Lesseps stood to great applause, welcoming the delegates 'with all his heart'. Particular welcome was given to Rear Admiral Daniel Ammen, the head of the American delegation, who was made a vice-president, and seated on the right of de Lesseps.

The real work of the Congress would be performed by five committees, investigating different aspects of the challenge ahead, from potential tolls to issues of navigation. As work got underway, an early conclusion of the financial or operational aspect of the undertaking was the overwhelming desirability of the canal being at sea level, a feeling shared by the majority of the delegates. A sea-level canal could be open twenty-four hours a day, without any restriction on the number of ships passing. Only one ship can pass through a lock at one time, with serious implications on operational revenue. For de Lesseps there was another financial imperative: a lock canal would just not capture the imagination, and therefore the savings of the public in the way that an 'Ocean Bosporus' would.

The most important of the five groups was the Technical Committee, who would decide the location, cost and the type of canal that should be built. The day after the opening ceremony, Ammen was asked to start the debate for this committee, but as the trunk containing the American maps and charts had been delayed, Selfridge was the first to take the floor, extolling the virtues of a sea-level route in Darién that he himself had visited. Immediately, the American delegation began to squabble among themselves. To Ammen, Selfridge was off-message: there was no good route in Darién, as the Grant Commission had decided. So why did Selfridge push for this? It is a recurring theme of the canal story that time and again those in the grip of the 'great idea' were also passionately committed to a certain type and location for the canal. For de Lesseps it had to be a sea-level canal, whatever the cost. For Menocal, Lull and Ammen, it had to be in Nicaragua using the lake, which presupposed a lock canal. Selfridge had been the leader of the American surveying effort in Darién; if it was built there it would be, in a way, his canal, and he was just as blind to the huge disadvantages of the area as de Lesseps would prove to the monumental folly of attempting a sea-level canal in Panama.

The next day Ammen and Menocal laid out their arguments for a lock canal in Nicaragua. This was followed by Wyse, outlining his various plans for a sea-level canal from Panama City to Colón. By now, all other options, from Mexico

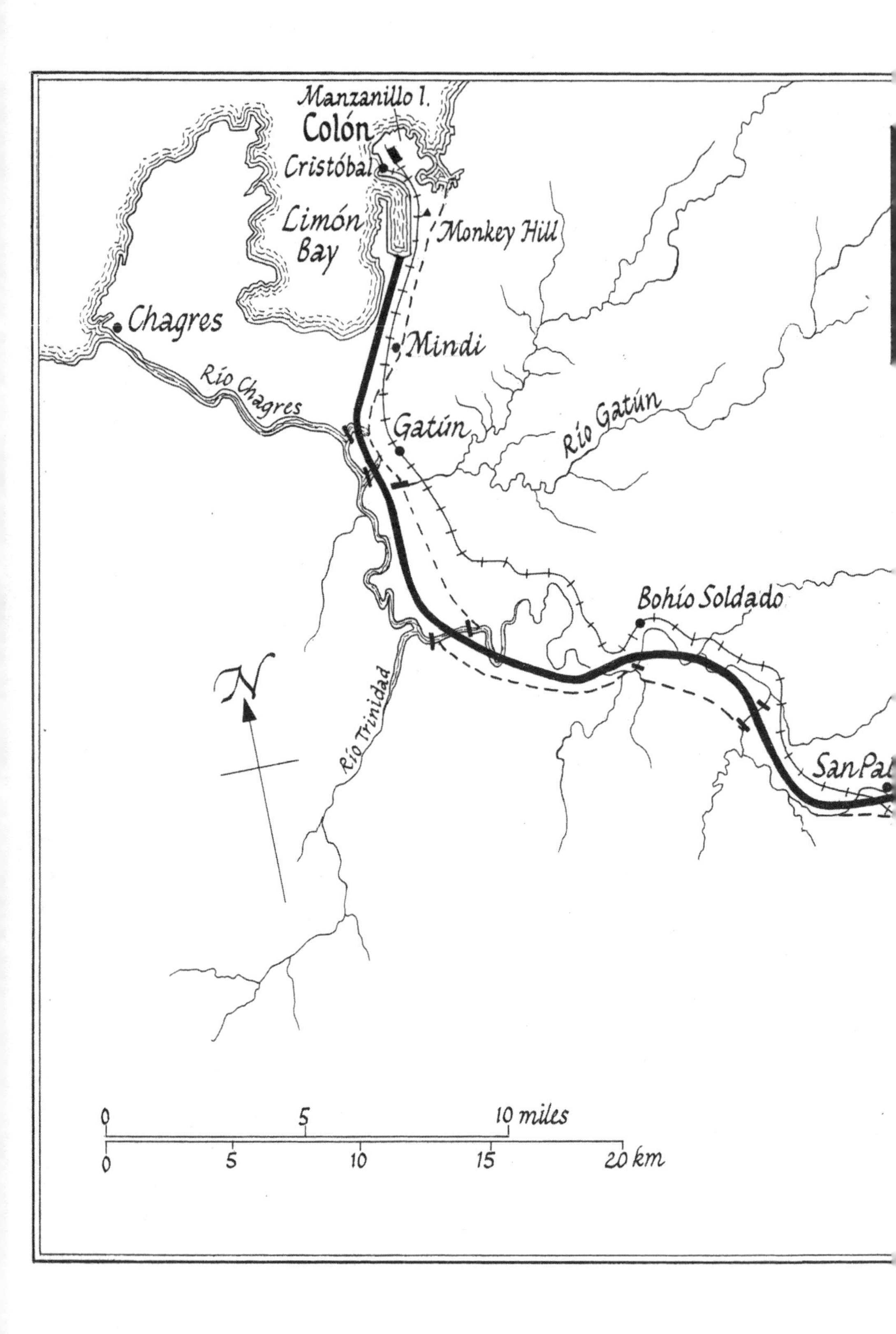
Manzanillo I.
Colón
Cristóbal
Limón Bay
Monkey Hill
Chagres
Mindi
Río Chagres
Gatún
Río Gatún
Bohío Soldado
Río Trinidad
N
San Pa
0
5
10 miles
0
5
10
15
20 km

The French Sea-level Plan

ompare with the modern canal route shown on the frontispiece)

Original railway Colón—Panama

River barrages

Planned river diversions

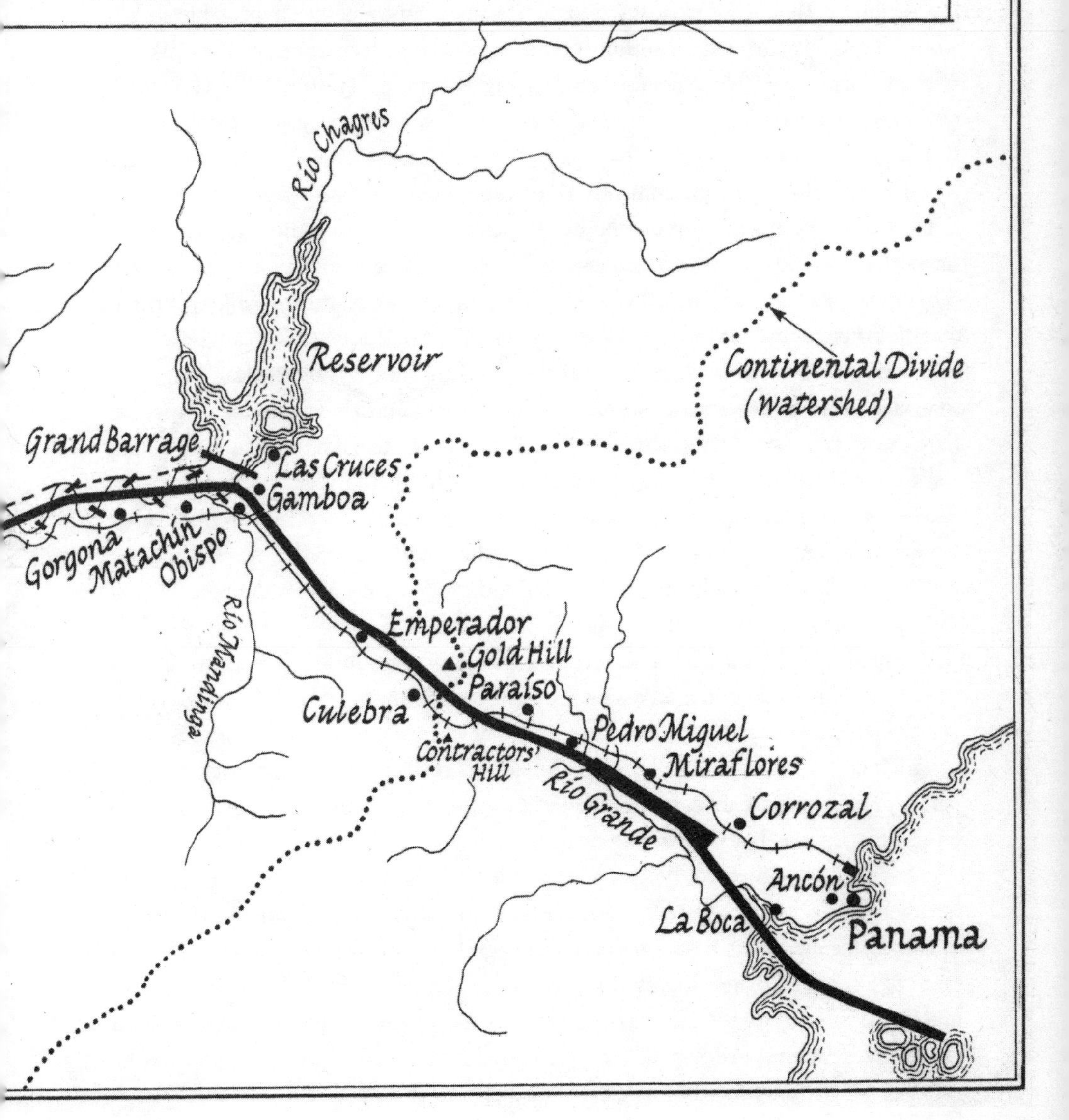

through San Blas to Darién, had been pretty much discounted; the battle was between Panama and Nicaragua, which corresponded, to a large extent, to a battle between the French and the American points of view. On 20 May Menocal went through his arguments against Wyse's scheme for Panama and outlined his own plan involving the huge aqueduct over the Chagres and some twenty-four locks. He made it clear he thought this too expensive and impractical. But a sea-level canal was impossible, he argued, because of the volatile Chagres River, which would have to drop over 40 feet into the canal, causing a dangerous cataract. Another objection was then raised by Ribourt, actually a Frenchman. He reckoned that the tunnel envisaged by Wyse and Reclus, if possible at all, would alone cost over 900 million francs and take nine years to build. Even if it was shorter than what he was attempting through the Alps at St Gotthard, the diameter of the ship tunnel was many times what was required for a railway tunnel.

Following this, two sub-committees of the Technical group were formed: one to deal with the tunnel versus open cut question at Panama; the other to investigate the cost and engineering issues behind the Nicaragua lock plan. Very quickly the former, staffed by some of the most eminent engineers of the Congress, ruled out the construction of a sea-level tunnel and also concluded that a canal *à niveau* 'involved so much uncertainty . . . that it would be impossible for any engineer to arrive at an approximate estimate of its probable cost'. The much-respected British Severn Tunnel engineer, Sir John Hawkshaw, also added his opinion that a sea-level canal at Panama would have to 'provide for the whole drainage of the district it traversed [which] . . . would require the canal to be 160 metres wide'. In addition, any tunnel would be entirely filled with water during the regular freshets of the Chagres. Ironically, it was flooding that very nearly doomed Hawkshaw's own great project, although this was still a year away.

According to Menocal, the report and the comments of Hawkshaw were 'met by the friends of Lieutenant Wyse with the affirmation that M. de Lesseps would positively refuse to accept the presidency of any other canal company except that of Panama – a statement which had the desired effect of bringing back to their party some who had deserted to the sides of the best engineers, who seemed to be in favor of the Nicaragua Canal.'

For the de Lesseps party, things were clearly beginning to get out of hand. So on Friday 23 May, *Le Grand Français* 'threw off the mantle of indifference', as one delegate wrote, and convened a general session. Striding confidently in front of a large map, he addressed the Congress. He spoke spontaneously, without notes, in simple, direct language, and with great conviction, if not abundant knowledge, making everything sound right and reasonable. The map, which he referred to with easy familiarity, clearly showed that the one best route was through

Panama, as it had been, he argued, for centuries. It was the route that had already been selected to develop the cross-Isthmus railroad. There was no question that a sea-level canal was the correct type of canal to build and no question at all that Panama was the best and only place to build it. Any problems would resolve themselves, as they had at Suez. His audience was enthralled.

Wyse and Reclus, however, had been stung by the criticisms of their Panama project, and, to their credit, refused to participate in the wishing away of the problems of the Chagres River and the massive excavation required by an open-cut, sea-level waterway. Later the same day, they presented a new plan that had been devised by Baron Godin de Lépinay, a senior French civil engineer and a delegate at the Congress. Although de Lépinay had never been to Panama, he had worked in South America on a railway project, where most of his workforce had been killed by yellow fever. De Lépinay's idea was to create an 'artificial Lake Nicaragua' in Panama by damming the Chagres as near to the Atlantic as possible as well as the Río Grande near the Pacific. In between would be created a vast artificial lake some 80 feet above sea level, accessed by a small number of locks. In a stroke the amount of excavation would be limited to a short and relatively shallow cut through the Continental Divide, and the Chagres River would be tamed. Amazingly, the de Lépinay design contained all of the basic elements that ultimately, after years of mistakes and heartbreak, were designed into the current Panama Canal.

De Lépinay himself, although favouring Panama over Nicaragua, predicted that an attempt at a sea-level canal would be a catastrophe. For a start, the cost, he argued, would approach a thousand million francs, a debt which could never be serviced by a private company. Secondly, working in a tropical climate such as Panama's was totally different from anything experienced by the vast majority of the delegates: iron rusted in no time, labour was scarce and unreliable, coal and provisions were expensive. And then there was disease. De Lépinay shared the current, though mistaken, belief that tropical fevers were caused by a poisonous emanation – 'mal air' (hence 'malaria') – from rotted vegetation in the ground. The huge excavations would release this, he argued, to devastating effect. To persuade European or American technicians to work there would require tripling their normal salaries, and the death toll, which he estimated at fifty thousand, would require constant replenishment of the workforce. In perhaps the most telling attack on de Lesseps and the confidence that underwrote the whole Congress, he turned the comparison with Suez on its head: 'They want this canal to be made after the model of the Suez Canal, that is to say, without locks – and yet its natural conditions are so very different. In Suez there is no water, the soil is soft, the country is almost on the level of the sea; in spite of the heat, the climate is perfectly healthy. In tropical America there is too much water, the rocks are exceedingly hard, the soil is very hilly, and the climate is deadly. The country is literally poisoned. Now to act thus after the same fashion under such different circumstances is to try to do violence to nature instead of aiding it, which is the principal purpose of the art of engineering.'

Wyse later admitted that this plan was put forward in large part to spike the guns of the official American Panama plan as outlined unenthusiastically by Menocal, as it required far fewer locks. If there were to be a lock plan for Panama, at least it would be a French one. But by the time he came to write his account of the conference in late 1885, the plan has far higher visibility and he claims that he investigated its potential on all his visits to the Isthmus in the early 1880s, and was convinced of its merits.

Many of the delegates at the time were impressed by de Lépinay's vision, although this included neither de Lesseps, for whom locks were totally unacceptable, nor the American contingent. With some justification, Ammen complained that the plan seemed to have been conjured out of thin air – it hadn't even been mentioned ten days earlier – and Menocal pointed out that none of the necessary surveys, particularly in the placing of the all-important dams, had been carried out. He also argued that the new lake would be large enough neither to provide water for the locks in the dry season, nor to deal with the Chagres when the rains arrived. Although costed by the Technical Committee as cheaper than both the Menocal Nicaragua plan and the Panama sea-level scheme, de Lépinay's idea disappeared from the reckoning.

Meanwhile, de Lesseps worked feverishly behind the scenes, twisting arms and calling in favours to secure the votes he needed for his dream of an 'Ocean Bosporus' at Panama, as the arguments between the two routes, and the two types of canal continued in the Technical Committee, becoming more heated and personal by the hour. 'A great deal of interest and feeling [was] manifested,' Ammen wrote, 'amounting, at times, to disorder.' On 27 May Selfridge spoke for the last time, again infuriating Ammen and Menocal by attacking the Nicaragua option. It would be impossibly expensive to improve Greytown harbour, he argued, and, what was more, Nicaragua was prone to earthquakes and volcanic activity, the first mention of what would later be a crucial factor in deciding the American Panama Canal. On Wednesday 28 May, the committee finally made its recommendation after a stormy session during which twenty of the delegates, nearly half the committee, had staged a walkout. The next day, as heavy rain fell, the full Congress met to hear the committee's report and to vote on its recommendation. The hall was packed as it was announced that a decision had been made: 'the committee, standing on a technical point of view, is of the opinion that the canal, such as would satisfy the requirements of commerce, is possible across the Isthmus of Panama, and recommend especially a canal at the level of the sea.' The cost of the work, including interest on the capital, was estimated at 1200 million francs, or about $240 million, with completion in twelve years.

Soon after it was put to the vote. It was highly public and highly charged. In alphabetical order each delegate stood up and voted on what was essentially the de Lesseps plan. Henri Bionne, whose 1864 paper had started the whole process of French exploration and who was the secretary of the Congress, read out the names. Daniel Ammen abstained, saying that only professional engineers should be entitled to vote. By the time it was de Lépinay's turn, seventy-six of the delegates had voted, forty-three for, with most of the others abstaining. De Lépinay, the French canal's own Cassandra, rose to his feet, declaring 'In order not to burden my conscience with unnecessary deaths and useless expenditure I say "no!".' He sat down to widespread boos and jeering. Next up was de Lesseps himself. 'I vote "Yes!",' he declared in a loud voice. 'And I have accepted command of the enterprise!' The room erupted in cheering and applause.

When it came to his turn, Menocal abstained, on the grounds that 'the resolution is indefinite as to what system of canal should finally be adopted'. But in the end the resolution passed with 74 in favour and 8 opposed. Afterwards, de Lesseps told the conference, 'It has been suggested by my friends that after Suez I ought to take a rest. But I ask you: when a general has just won one battle and is invited to win another, why should he refuse?'

His adoring public in the onlookers' seats were ecstatic. At the end of the Congress, the chairman, the elderly Admiral de La Roncière, underlined that it was *Le Grand Français* himself who had carried the day: 'May that illustrious man, who has been the heart and soul of our deliberations,' he declared, 'who has captivated us by charm, and who is the personification of these great enterprises, may he live long enough to see the end of this work, which will bear his name forever. He has not been able to refuse to assume its command, and in so doing he continues to carry out the mission which has made him a citizen of the whole world.'

With the benefit of hindsight there were warning signs everywhere. Thirty-eight delegates had conveniently absented themselves from the vote and sixteen abstained. The predominantly French pro votes did not include any of the five delegates from the French Society of Engineers. Of the seventy-four voting in favour, only nineteen were engineers and of those, only one, the Colombian Pedro Sosa, who had a vested interest in the Wyse concession, had ever been in Central America.

To the Americans the whole affair had been an appalling travesty. At the heart of their displeasure was outrage that anyone else, particularly the French, should dare to presume to build a canal in their backyard, in what was effectively a

United States protectorate. On his return to the US, Menocal wrote an article for the *North American Review* entitled 'Intrigues at the Paris Canal Conference'. It was introduced by a short piece by the journal's editor, which gives some indication of the vitriol of the Americans. As far as the editor was concerned, it was all a conspiracy among hard-up prominent Bonapartists, who had lost their means of subsistence with the fall of Napoléon III. 'A careful examination of the names of the French delegates to the Canal Congress shows how entirely it was packed with subservient friends of the fallen dynasty; nor is it well to overlook the fact that the shares of the Türr company were largely held by them.' 'These people once went to Mexico to seek their fortunes in a Franco-Mexican empire,' he goes on. 'It seems passing strange that the conspicuous defeat of those plans, which embraced the destruction of the American Union, should have failed to teach them some degree of caution before affecting to . . . tamper with American interests in America.'

This suspicion predated the conference. In February 1879 Wyse had travelled to Washington to ensure American attendance at the Congress. He had met President Hayes and Secretary of State William Evarts. The latter was openly hostile to any further French 'adventures' in Central America – it can't have helped that Wyse had 'Napoléon' in his name – and might well have been instrumental in blocking Wyse and Reclus's access prior to the Congress to the detailed American maps and surveys.

When Menocal and Ammen reported back to Evarts at the end of the conference, they confirmed his misgivings, saying the decision for Panama had been determined not by 'relative consideration of natural advantages', but from 'personal interests arising from the concession'. If Nicaragua had been chosen, then the Wyse concession would have been worthless. It had been a foregone conclusion, a stitch-up, what another American delegate, William E. Johnston, called 'a comedy of most deplorable kind'.

Menocal, for his part, was sure that the whole Panama project was doomed to early failure. But this was just as much a worry the scheme's completion. Johnston, among others, feared that if de Lesseps could get the stock into the hands of a mass of ordinary Frenchmen, as he had done with Suez, the canal's inevitable failure would lead to the intervention of the French government. In terms of the sacred Monroe Doctrine, this was worse even than a French-controlled private company meddling in American affairs.

Ammen and Menocal's reports were widely covered in the US press, the New York *Tribune* being particularly concerned by the 'shenanigans' in Paris. In government, too, there was disquiet about the outcome of the conference, a New England senator moving a resolution to the effect that the US 'viewed with serious disquietude any attempt by the powers of Europe to establish under their protection and domination a ship-canal across the Isthmus'. The New York *World*, which would be another implacable opponent of the French canal, went further, declaring that the project 'prefigures for us an era of complications and difficulties in regard to the foreign policy and the commercial relations of this country more serious than any we have had to deal with in the last twenty years of our history'.

In Panama itself, where civil and political disturbances continued through June, the news from Paris was warmly welcomed, though not without a certain amused detachment. On 29 May, the *Star and Herald* reported: 'The Wyse Panama route is the one likely to be adopted. Indeed, many said before the Congress met that the acceptance of this route was a foregone conclusion.' The following day, reporting the final vote of the Congress as soldiers enforced martial law in the streets outside their offices, the same paper added, 'It may amuse some of our readers in Nicaragua to know that a recent publication cites another objection to their route, and that is the 'political instability' of the country. We presume the author was satirizing Colombia at the time.' Nonetheless, to those in Panama it seemed that at last prosperity was around the corner.

In France, de Lesseps swung into action. Raising 2 million francs from selling 'founders' shares' to a syndicate of 270 rich and influential friends, he started negotiations to buy out the Türr syndicate, including their concession and all their maps and surveys. The deal, for 10 million francs, was concluded on 5 July 1879 and for the Türr syndicate it was almost all profit. Next, de Lesseps embarked on a whirlwind tour of France. As with Suez, he aimed to raise the starting capital, set at 400 million francs, directly from the public.

However, times had changed in France, and the power of the financial institutions and the press had risen considerably since the 1860s. This time, the banks showed their displeasure at being cut out of the lucrative issue by organising a campaign against the canal venture. Costs had been underestimated, the venture would never pay, suggested Marc Lévy-Crémieux, vice-president of the powerful Franco-Egyptian Bank. Emile de Girardin, proprietor of the mass-circulation *Petit Journal*, was another opponent. Panama's climate was a death trap, it was argued; plus, the Americans would never allow work even to start. It was rumoured, too, that the vibrant de Lesseps was now in his dotage and had lost his winning ways. 'The financial organs were hostile,' de Lesseps later told American reporters, 'because they had not been paid.'

Consequently, the issue, on 6 and 7 August, was a severe disappointment. Only 30 million francs of the desired 400 was raised. Almost anyone other than de Lesseps would have given up right then.

9

THE RICHES OF FRANCE

Public subscriptions live or die on confidence. It did not take a promoter of the experience and skill of de Lesseps to realise that some urgent public relations were now needed to allay fears about the Panama project. To calm concerns about the technical and practical issues of the canal, de Lesseps announced that he would himself go to Panama together with a Technical Commission of international experts. He would even take his young wife and three of his children with him, showing that fears about disease were unfounded. To counter claims that the United States opposed the project, de Lesseps would tour that country reassuring officials and drumming up support.

The first move, however, was to set up a bi-monthly journal, the *Bulletin du Canal Interocéanique*, which aimed to counter the 'lies' of the press with stories alleging, among other things, that Panama was 'an exceedingly healthy country'. The magazine, which included extracts from favourable press coverage in France and abroad, was first published on 1 September 1879.

The following month a young engineer, Gaston Blanchet, from Couvreux, Hersent went to Panama to investigate the route for his bosses. Already, the Franco-Belgian company was being lined up to be the contractor for Panama. Blanchet himself had made his name building for the firm a metal bridge over the Danube near Vienna. When he returned from Panama with a favourable report, additional engineers were sent out in November to take test borings. In December, de Lesseps himself boarded the *Lafayette* at St-Nazaire bound for the Isthmus. The expedition was funded by Couvreux, Hersent, and on board was Abel Couvreux's son as well as Blanchet, who was particularly keen to return to Panama as he had met and fallen in love with a local girl, Maria Georgette Loew, the daughter of the proprietor of the Grand Hotel. Also on the trip was Louis Verbrugghe, Jacob Dirks, a Dutch engineer who had built the Amsterdam Canal, Henri Bionne and a number of other technicians.

Lieutenant Wyse had been promised by de Lesseps that he would head up the operation in Panama. But no such appointment had actually materialised, and he had not been invited along. In the end he paid for his own passage, and engaged during the voyage in several heated rows with de Lesseps. The latter was not one to share leadership responsibilities or glory; Wyse was effectively dropped.

The arrival of de Lesseps in Panama was one of the biggest news stories the local press had ever covered. 'Mr Lesseps' enterprise,' the *Star and Herald* proclaimed, 'will . . . take rank with Columbus' discovery of America.'

The *Lafayette* entered the harbour of Colón a little after 3.00 p.m. on 30 December. The party could not have arrived at a more clement, or misleading, moment: the dismal rainy season had just ended, the skies were cloudless and the north-west trades blew pleasantly in from the Caribbean. Coming alongside the wharf, it was met, the paper reports, by 'the committee of reception appointed by the government, the delegation from the State Assembly, and a large number of invited citizens'. At a little after 4.00 p.m. the landing stage was put on board, and all repaired to the spacious saloon of the *Lafayette* for formal addresses of welcome. One of the reception committee was an American diplomat, businessman and journalist Tracy Robinson. He remembered de Lesseps, in perfect Spanish, responding 'very pleasantly' to the welcome, 'wearing the diplomatic smile for which he was noted. He was then over seventy years of age, but was still active and vigorous: a small man, French in detail, with winning manners, and what is called a magnetic presence. When he spoke the hearer would not fail to be convinced that whatever he said was true.' All the time, the 'enthusiasm from every class knew no bounds,' the *Star and Herald* wrote. 'The flags of all nations were displayed, with the notable exception of that of the United States, and the reception can be said to have been a decided success.'

The next morning, together with his entourage of 'distinguished engineers', de Lesseps 'made an examination of the harbour', all the time holding forth about his enthusiasm for the project. Tracy Robinson remembers him invariably concluding every phrase with the assertion 'The Canal will be made.' '"There are only two great difficulties to be overcome",' the *Star and Herald* reported de Lesseps as saying, '"the Chagres River, and the deep cutting at the summit. The first can be surmounted by turning the headwaters of the river into another channel, and the second will disappear before the wells which will be sunk and charged with explosives of sufficient force to remove vast quantities at every discharge."'

At ten o'clock that same morning, a party arrived from the United States. This included Trenor Park, the diminutive boss of the Panama Railroad. On his visit to the United States prior to the Paris Congress, Wyse had met Park to discuss the 'amicable agreement' stipulated by the Wyse concession from Colombia. Park, a successful Wall Street speculator, knew he had Wyse over a barrel – if an arrangement was not forthcoming the Wyse concession was worthless. Wyse had offered to buy the railroad, and Park, who personally owned 15,000 shares, was happy to sell – at $200 a share, twice the market value. Several other Railroad board members were on the trip, along with a journalist from the New York *World*, José Rodrigues, and two American engineers who had agreed to be part of the Technical Commission: an army engineer called W.W. Wright, and Colonel George Totten, who had been in charge of the construction of the railroad back in the 1850s.

An hour and a half later, the entire party left Colón for Panama on the railroad. Halfway across, the train was met by the President-elect of the State, and 'a fine lunch was provided on the train, with wines &c., &c., and everything gave great satisfaction'. Yet another reception committee of local dignitaries and generals was waiting at Panama station in a specially erected tent. Also in attendance was, it seemed according to Rodrigues, 'every one of the [city's] 14,000 inhabitants . . . shouting, struggling to get a glimpse of the distinguished guest'. Following more speeches, the party was conveyed in carriages to the Grand Hotel. Houses along the route were decked in French and Colombian flags, and no expense had been spared in cleaning up the city in honour of the 'Great Engineer'. According to an occasional correspondent for the New York *Tribune*, 'Such an air of neatness and real cleanliness has not pervaded this city of pigs and smells within the memory of its oldest inhabitants.'

That night the hotel hosted a state banquet. The only woman present was Madame Louise-Hélène de Lesseps. She had been only twenty-one when she had married de Lesseps, then sixty-four, in 1869. She was as beautiful as the first Madame de Lesseps had been witty and brilliant. According to Tracy Robinson, she 'gave éclat to the occasion . . . Her form was voluptuous, and her raven hair, without luster, contrasted well with the rich pallor of her Eastern features.' After the dinner, dancing and singing, spilling out into the plaza, went on for the most of the night.

The next morning, the first of the New Year, de Lesseps was up early, in full regalia, for the inauguration of the new President, Damaso Cervara. After that, making good on his promise to dig the first spade of earth for the Panama Canal, de Lesseps organised a special ceremony at which his seven-year-old daughter, Ferdinande, would do the honours of turning the first sod. The symbolic act was to take place at the mouth of the Río Grande, scheduled to become the Pacific entrance to the future canal.

The steam tender *Taboguilla* took de Lesseps and a party of distinguished guests – which included the British consul Hugh Mallet and his son Claude – three miles to the site on the Río Grande where the ceremony would take place, following appropriate feasting and festivities on board. However, since late arrivals had delayed the *Taboguilla*, the Pacific tide had receded such that the vessel could not land at the designated site. Undaunted, de Lesseps was ready with a solution. He had brought a shovel and pickaxe with him from France especially for the occasion. Now, declaring that the act was only symbolic anyway, he arranged for his daughter to strike the ceremonial pickaxe blow in an earth-filled champagne box. Each member of the Technical Commission then in turn took a swing at the box, and the whole work was blessed by the Bishop of Panama. The passengers landed back at Panama City at 8.00 p.m., according to a local newspaper, 'unanimous in their expressions of gratification with their delightful trip'.

The following day, it was down to business. The Technical Commission was run by Dirks and Totten, while the detailed work was parcelled out among the more junior members, which by now included Pedro Sosa. All the surveyors reported in to Dirks and Totten every three days. De Lesseps took a hands-off role, except for impressing on the men one vital factor. As the *Tribune*'s correspondent reported, 'His mind is unalterably made up on one point – he will have nothing to do with a canal with locks.'

As the surveys continued, de Lesseps threw himself into the coincident celebrations of the new President, the New Year and the new canal venture. On 3 January there was horse racing: 'Mr Lesseps is an accomplished horseman, and joined in the pleasures of the day with all the zest and activity of youth,' reported the *Star and Herald*. On other days there were excursions with his children, two boys aged nine and ten, who, with their sister, had also caught the eye of the locals. Robinson called them 'as dark as Arabs and as wild'. On 10 January, there was another ceremonial commencement of the canal work when a huge dynamite charge was exploded near the summit of the Continental Divide. Champagne flowed once more.

A great highlight of the trip was the wedding of Gaston Blanchet and Miss Maria Georgette Loew, the beautiful daughter of the owner of the Grand Hotel. For observers, it was the perfect embodiment of the new friendship and shared future between Panama and France. Held in the cathedral, 'bright with myriad lights, and gay with the presence of a multitude of ladies and gentlemen of the best of our Panama society', it was blessed with a performance at the piano by the French vice-consul and singing by Madame de Lesseps herself, 'with great sweetness and expression'. The party afterwards, inevitably held at the Grand Hotel, continued well beyond midnight.

There were balls and dinners most nights. All were resplendent with toasts and speeches. 'You have seen the soldiers of Count de Lesseps,' Verbrugghe announced to a reception given by Colombian officials. 'We have the faith to move mountains, or at least, in the present age, rend them asunder.' No one could believe the energy and enthusiasm of Ferdinand de Lesseps, now in his seventy-fifth year. He would dance 'all night like a boy', noted Tracy Robinson, and still be on the train at seven on some expedition 'fresh as a daisy'.

In this atmosphere, optimism about the work of the Technical Commission abounded. De Lesseps proclaimed that the job at Panama would be easier than Suez, writing to his thirty-nine-year-old son Charles, in charge of affairs at home, that he couldn't believe that it hadn't been done already. The local press, too, reported that the new geological survey had shown the estimates of the Paris Congress to be pessimistic: 'The engineering difficulties . . . are being solved one by one,' wrote one paper, 'and when the present surveys are concluded the project will assume so positive and practical a shape that financial men in Paris and elsewhere will not hesitate to give it the attention and support it deserves.'

'Elsewhere' to all intents and purposes meant the United States, who remained the ghost at the banquet. Everyone knew that the US was de Lesseps' next destination, and the *Star and Herald* anticipated a great welcome there for the builder of the Suez Canal. The publication of the commission's report on Panama and de Lesseps' proven charm would, the paper concluded, gain 'the moral, if not the material support of the people and Government of the United States.'

The same issue of the newspaper reported the provisional findings of the Technical Commission, which, it emphasised, were based on maximum costs. Further details were to be worked out during the trip to New York. With only a few minor modifications and improvements, the open-cut plan and route of Wyse and Reclus were approved. There would be a 40-metre-high dam at Gamboa to retain the Chagres, along with a channel running alongside the canal for the regulated flow of the river to the sea. Another smaller channel would keep river water from the canal on the opposite side. At Colón, a two-kilometre breakwater would be constructed to protect the port, while at the other end of the canal a tidal lock would be required.

A study of the stability of rocks had led to the adoption of slope for the canal of 1 to 1, or 45 degrees, except on the summit division at Culebra, where a slope of a quarter to one was considered sufficient because of the hard rock. In the Culebra Cut, the canal would be 24 metres wide at the bottom, 28 metres at the waterline, with a water depth of 9 metres. Elsewhere, the canal would have a water line width of 50 metres, but otherwise similar dimensions.

The cost of diverting the Chagres was increased substantially from the Congress estimate, as was the amount of spoil to be removed, now judged to be some 75 million cubic metres. Nonetheless, the overall increase in cost was slight, helped by the reduction of the contingency fund from 25 to 10 per cent. Not including the expense of purchasing the railroad, interest on the capital, administrative outlay and the sums due to the Türr syndicate and the Colombian government, the cost was now estimated at 843 million francs ($168,600,000). The comparable figure from the Congress had been 765,375,000 francs ($153,075,000). The time of construction was reduced from twelve years to eight. In Paris, the *Bulletin du Canal Interocéanique* crowed that, in all, the work would be easier and quicker than had been previously believed.

This overall sum seemed eminently affordable. The Congress had estimated tonnage through the canal at six million a year. At 15 francs per ton, this would bring in 90 million annually, producing 10 per cent a year on a capital of 900 million francs. If 15 francs a ton seemed high compared to Suez's 8 francs, the current cost per ton, it had been pointed out, of moving freight on the railway, including loading and unloading, was nearly 80 francs.

However, there is no doubt that, overall, this was an optimistic estimate. With hindsight, the slopes deemed sufficient were wildly wrong; the plan for the Chagres would prove a minefield; furthermore, the reduction of the contingency fund to hold down the cost seems to be taking 'putting a favourable gloss' to the point of dishonesty. In contrast, US costings under the Grant surveys had always included a contingency fund of 100 per cent, a sensible move considering that the Suez Canal, for example, had gone over its original budget by some 128 per cent.

The de Lesseps party left Panama on 15 February. They depart, wrote the *Star and Herald*, 'vastly to the regret of the people of the Isthmus of all nationalities . . . during their stay on the Isthmus they have taken possession of the public heart to an extent never before witnessed among us. The importance of Mr de Lesseps' mission, a preparatory step to one of the greatest undertakings the world has ever witnessed, did not prevent that courteous and affable gentleman from taking a vivid interest in our rather dull isthmus life, into which he and his amiable lady have infused new vigor and animation.' He headed for New York, 'with heartiest best wishes'.

There were dissenting voices. Wyse, cast out from the project, wrote from Panama to Reclus on 24 January: 'Blanchet does nothing . . . the tachometers are broken. There is total disorder, waste, and disorganisation.' De Lesseps, Wyse alleged, was making a nuisance of himself by trying to grope '*les tétons des femmes*'. The American journalist Rodrigues, who was there for only twelve days, is another cynic. In 1885, he published a highly critical account of the French project, by which time the de Lesseps dream was starting to unravel. Blessed with hindsight, he describes the visit in early 1880 as dominated by 'sham and charlatanism'. Even Tracy Robinson, who was so impressed with de Lesseps, and believed that 'from first to last he was perfectly conscientious and honest', pronounced him 'too old, too eager, too vain of the glory it would add to his already great reputation'.

De Lesseps' expedition to Panama and his imminent arrival in the United States had created a great deal of comment in the newspapers of that country. In general, they were cynical about the realistic prospects for the sea-level canal, concerned about the repercussions for the favoured Nicaragua option, and hostile to the perceived international weakness of the United States that the European-led project revealed. It was the United States, said the *World*, who had 'proper predominance over the seas to be united by any Isthmian or Central American ship-canal'. What was needed was massive and accelerated naval expansion, along with naval stations near the future canal. The *New York Times* thought that the sea-level canal was practicable and desirable, but its cost would make it uneconomic. Nevertheless, it praised the energy and determination of de Lesseps compared to his rivals in the States: 'While they protest and discuss and reflect and hesitate, he goes ahead.' The New York *Tribune* agreed: 'All this looks like business.' The paper interviewed the Panama Railroad boss, Trenor Park, on his return from Panama in late January. Park, who must have been rubbing his hands together at the killing he knew he could make by selling the railroad to the French, announced that he had gone to the Isthmus with 'grave doubts of the success of the project', but returned reassured by de Lesseps: 'He is certainly a great man, and his enthusiasm is contagious.'

As the day of de Lesseps' arrival neared, the *Tribune* urged decisive action: 'Now is the time for the Government to make up its mind what to do . . . If it merely continues to make no sign, foreign capital will get committed, and then foreign Governments will be drawn in to protect the capital of their citizens, and the Monroe Doctrine will have disappeared like a morning fog . . . We are unwilling that the successors of Napoleon III, should attempt another foothold on the soil of the continent, even in the seductive guise of a peaceful triumph of engineering skill.' The answer, according to the *Tribune*, was to get the Nicaragua project up and running as soon as possible, and thereby see off the interlopers' Panama scheme.

De Lesseps arrived in New York on 24 February. On the journey from Colón he had completed the report of the Technical Commission, cutting the estimate to 658,600,000 francs ($131,720,000). Half this amount, he announced, was reserved for American subscribers.

De Lesseps never expected to raise this amount in the United States. Before his trip he wrote to an American friend, who was busy setting up meetings, that he expected the Americans, and, to a lesser extent, the British, to fall for 'inaccurate objections'. The money for the project, de Lesseps believed, would come from France, 'where one is used to working for the civilisation of the world'.

He had never visited the United States before, and, as in Panama, de Lesseps enjoyed himself enormously. He set up court in the Windsor Hotel, which flew a Tricolour from its mast in his honour. There, he was waited on by a committee from the American Society of Civil Engineers, who took him to see Hell's Gate and the East River Bridge. Later he visited the uncompleted Brooklyn Bridge, the Erie Railroad Station and the docks and grain elevator in Jersey City. In the evening, there were receptions given by the Geographical Society, the French expatriate community, and on 2 March there was a great dinner in his honour at Delmonico, which had been decorated with elaborate confectionary symbolising the achievement of the Suez Canal – sphinxes, dredges, elephants and bales of goods. One of the welcoming speakers was John Bigelow, the publisher and diplomat. Everywhere he was honoured as the builder of Suez, and the press were fascinated by his energy and charm.

As de Lesseps had predicted, there was little interest, however, in investing in the Panama project, and even at dinners with handpicked guests there were dark mutterings about the Monroe Doctrine, and predictions that a foreign government might take over the canal. In fact, while de Lesseps had been in Panama, the French government had sent a clear message to Secretary of State Evarts, formally notifying him that 'the enterprise of M. Lesseps is of an entirely private nature and has no political color or protection at all', as the Panama *Star and Herald* reported of the news. 'This dumfounds the speculators, adventurers, contractors and others,' said the paper, 'who have spread . . . rumors of war within six months, and of a tremendous European alliance against the United States.'

At Delmonico's, de Lesseps tried to answer the fears of his audiences: 'He spoke rapidly in a distinct but quiet tone,' reported a journalist onlooker. 'Occasionally when criticising or indirectly ridiculing theories opposed to his own, he dropped into a sort of chuckle.' It had to be at Panama, rather than Nicaragua, because it had to be a sea-level canal; his project had nothing to do with any government, so the Monroe Doctrine did not apply, de Lesseps argued. The outfit would be based in Paris only because it was there that company laws gave the best protection to shareholders. 'I have offered America 300,000 of the shares. If she takes that much she will have a controlling voice in the enterprise,' he told a *Tribune* reporter the following day. 'But even though no shares are sold here, I shall still build my canal . . .'

From New York, de Lesseps travelled to Washington. There he met President Hayes and Secretary Evarts and appeared before the House Interoceanic Canal Committee. He was received politely enough, but it was plain that his scheme infuriated everyone from the President down. On 8 March, Hayes declared in a special message to Congress: 'The policy of this country is a canal under American control. The United States cannot consent to the surrender of this control to any European power or to any combination of European powers.' So what if the Clayton–Bulwer Treaty committed the US to shared control with Great Britain? This should 'by just and liberal negotiations' be altered. 'The capital invested by corporations or citizens of other countries in such an enterprise must, in a great degree, look for protection to one or more of the great powers of the world,' Hayes continued. 'No European power can intervene for such protection without adopting measures on this continent which the United States would deem wholly inadmissible. If the protection of the United States is relied upon, the United States must exercise such control as will enable this country to protect its national interests.' The canal, he proclaimed, will be 'virtually a part of the coast line of the United States.'

'When M. de Lesseps gets ready to leave Washington to-morrow,' said the *Tribune* the following day, 'he will have no reason to complain that he has been left in the dark as to the attitude of the Government of the United States toward the interoceanic canal project.'

But, as ever, de Lesseps was undaunted. Ignoring the threat of takeover implicit in Hayes' statement, he welcomed the interest shown by the American leadership in his project. Of course, as he had said time and again, no European governments would be invited to become involved in the canal. But if the United States, clearly the major power in the region, wanted to offer protection to the capital he was planning to raise, then that was to be celebrated. The next day he expressed 'his delight with the President's message', 'because it would certainly be advantageous to have the protection of the United States during the work, and after the opening of the canal'. He had just sent a message to his son, which would be printed in the *Bulletin*, saying that the 'President's message assured the political stability of the canal'. It was just what his home audience wanted to hear.

From Washington, de Lesseps travelled to Boston, Chicago and nearly twenty other US cities. However, everywhere he went, although there were flattering attentions, there was still a marked paucity of financial backing.

Back in Europe, de Lesseps threw himself into a new round of speeches and lecture tours around France, Britain, Holland and Belgium. Everywhere there were reassurances that the United States was on board. 'In these provincial tours,' reported one infuriated American critic, 'he everywhere gave the impression that the governments of France and the United States were equally favourable to the enterprise; the flags of the two nations were everywhere united over his head when he spoke.'

De Lesseps was in Liverpool on 29 May, specially invited to be present at a banquet in honour of the Queen's birthday. The 150 guests included foreign consuls and numerous naval officers in full uniform, as well as all the principal merchants of the city. At the town hall, 'loyal toasts [were] enthusiastically drunk [and] a glee choir was in attendance'. American opposition to the scheme, de Lesseps announced, was 'a phantom and a bugbear'. As he had for the United States, de Lesseps offered to keep back some shares especially for British investors.

In Britain, however, he found a cautious response. There was certainly support for the idea of a canal. In a leading article written in response to Hayes' statement to Congress, *The Times* drew a picture of the 'few miles of oozy quagmire and jungle' separating the Atlantic and Pacific at Panama as 'a heavier tax on the industry of mankind than a war or a famine'. The paper also strongly criticised the US President's statement: 'an inter-oceanic canal,' it wrote, 'would form as much or as little a part of the European coastline as of that of the US . . . All that Europe wants is that a block of earth which it is growing to regard as it might a sunken ship in the Medway . . . should be cleared away, whether in the manner proposed by M. de Lesseps or in some other manner.' Interestingly, the President's claim of local hegemony was both resented and acknowledged. 'That the US, by furnishing the money, should obtain a special right to watch over the safety and peaceable use of the new channel is what Europe, and particularly Great Britain, would most of all desire.'

For British investors, if de Lesseps' claims that the US was now on side were widely believed, he was less successful in persuading them that the climate of Panama was 'salubrious'. 'It is a region,' wrote the London *Standard*, 'endemic with tropical diseases and notoriously more pestilential than any part of the desert of Suez.' Also 'In Panama [de Lesseps] has no gangs of Fellaheen forced to work for scant wages, no enthusiastic Khedive willing to command the resources of the state for the benefit of the undertaking.'

In France, though, it was a different story. It was incorrectly rumoured that Couvreux, Hersent had signed a contract to build the canal for even less than de Lesseps' most optimistic estimate. No one rushed to deny this. In August, a new subscription was announced, to be held at the beginning of December. On offer would be 600,000 shares of 500 francs each, a total of 300 million francs, which was all the French legislature would authorise. The shares were expensive when 500 francs was nearly a year's wages for about half the working population of France. But the terms were attractive – 25 per cent down with six years to pay off the difference. During the time of construction, shareholders were to receive 5 per cent on their paid instalments. Once the canal was completed, they were to get 80 per cent of the net profits. A huge publicity campaign was launched: much was made of the fact that 500-franc shares in Suez were now worth more than 2000 francs and paying a dividend of 17 per cent; there were special picnics; de Lesseps was everywhere, staging conferences and banquets, urging the purchase of shares as a patriotic duty; there were advertisements trailed from hot-air balloons; handbills of various eye-catching colours were pasted on every highway; purchases from shops were sent home with advertisements attached; a silver medal was offered to every individual to whom five shares of stock were assigned.

Even more importantly, de Lesseps had now decided to play on the terms of those who had wrecked his first share offer in August the previous year. This time, the banks would handle the subscription. A syndicate of commercial and investment houses was formed by Marc Lévy-Crémieux, a vicious opponent of the first issue. For the price of a 4 per cent commission, or 20 francs a share, the opposition of the financial community melted away. The press, too, was brought on board. One of the chief opponents the first time round, Emile de Girardin of *Le Petit Journal*, was offered and accepted a place on the Company's board of administrators. Elsewhere simple payments were made to editors and journalists, to a total, it emerged much later, of some one and a half million francs. Papers previously scathing about the project were now falling over themselves to find rhetoric sufficient to describe its attractions. 'Capital and science have never had such an opportunity to make a happy marriage,' the *Journal de Débats* announced. 'Success . . . is certain,' said *Le Gaulois*. 'One can see it, one can feel it.' *La Liberté* proclaimed: 'The Panama canal has no more opponents . . . Oh, ye of little faith! Hear the words of Monsieur de Lesseps, and believe!'

The opening of the sale of the stock on 7 December marks one of the most extraordinary moments in the history of finance capitalism. Within three days more than 100,000 people had subscribed for 1,206,609 shares, more than twice the number available. Eighty thousand were small investors, buying one to five shares each. Sixteen thousand were women ordering shares in their own names. Mothers sent de Lesseps their children's savings wrapped in handkerchiefs. Whether or not they had actually met de Lesseps and been a victim of his hairdryer charisma, the fever of the Great Idea had clearly taken grip of the entire country as no great financial enterprise had ever done before. Together with the influence of the press, de Lesseps' trips to Panama and the United States, or more exactly, his version of them, had won over all the doubters. The riches of the ordinary French family were now committed to the great project.

A French commentator later wrote: 'At that time they realized the poetry of capitalism . . . This is private enterprise, this is the shareholders' democracy which is gradually changing the face of the world and setting humanity free.' As it would turn out, as the share issue stoked the fires of speculation on the Bourse and kept the whirlwind of dealing spinning along nicely, a massive time bomb had been laid under the financial structure of the Third Republic.

*

Two weeks later, Armand Reclus was offered the job of *Agent Supérieur* of the canal company to head up the operation in Panama. He left for the Isthmus on 6 January 1881, together with his deputy, the lawyer Louis Verbrugghe, who had accompanied Wyse on his trip to Bogotá. In Paris, a sumptuous headquarters was purchased for a million francs. The first shareholders' meeting was held on 31 January with more than five thousand people in attendance, and on 3 March the new company, the Compagnie Universelle du Canal Interocéanique was officially incorporated. De Lesseps was president, Henri Bionne was secretary general. Charles de Lesseps, who had urged his father not to take on such a mammoth task but who had promised his support when he saw the old man was unmovable, was appointed a director.

On the same day was held the second general meeting of the Company. Wasting no time, De Lesseps outlined the programme for the year ahead: to clear the line of the canal of vegetation, to study the hydrography of Colón and Panama bays, their tides, currents and winds; to build houses for the accommodation of the employees and hospitals for the sick; and to build workyards. All preliminary work was to be completed by October, when Culebra would be attacked, and in November and December dredges would start work on the soft soil on the lower parts of the line. De Lesseps announced that the canal would be open to traffic by 1888, predicting that the job would be far easier than what he had achieved at Suez. His comments were greeted with cheers and applause.

News of the incorporation of the Compagnie Universelle reached Panama about a week later. 'The company now has a legal existence and a name,' wrote the *Star and Herald*. 'It is no longer in the realm of inchoate projects like its rival institution the Nicaragua Canal project, but is a solid and substantial entity, commanding unequalled resources and unrivalled influence. The news is welcome, and we hasten to offer it to the public.'

In the United States, however, the success of the subscription fanned rising fears about an open ship lane through the Isthmus. 'The worry is that it will weaken the United States strategically,' wrote the New York *Tribune*. The only option, the paper continued, would be 'to exercise such control over it as will prevent the passage of the fleets of any Nation with which we may be at war.' What this would mean would be an abandonment of the long-held policy to avoid 'such vast and costly naval armaments as are kept afloat by England, France, Germany and Spain . . . Otherwise we would find our Western coasts, from San Diego up to Sitka, exposed to attack within a few weeks after the breaking out of hostilities with any country that can keep a formidable squadron in the West Indies . . . Suppose the canal was open and a sudden quarrel were to arise between the United States and Spain,' the paper wondered prophetically. 'What is to prevent Spain from sending a dozen ironclads through the Isthmus to bombard our California towns?'

Such was the success of de Lesseps' moneyraising that the spectre of the canal had changed from being distant and uncertain to an almost done deed, and the dire strategic repercussions of a open waterway under the control of a hostile power were also debated in the House of Representatives, where it was suggested that the United States 'insist on acquiring from Colombia the territory through which it runs, in order to be able to fortify its mouths and control its operations in time of war.' But, as the *Tribune* reported in mid-February, such was the new wealth of the Compagnie Universelle, that it was surely only a matter of time before they controlled the bankrupt Colombian government. To its fury and frustration, the United States seemed powerless.

10

'*TRAVAIL COMMENCÉ*'

On 1 February, de Lesseps read to an enthusiastic French press a telegram from Reclus on the Isthmus, which the journal *La France* described as 'eloquence in a few words': '*Travail Commencé*'. It was thrilling news.

Reclus and Verbrugghe had arrived in Panama at the end of January, together with thirty-five engineers, five of whom brought their wives with them. As always, there was a warm welcome from the people of Panama. Also of the party was Gaston Blanchet, the representative of Couvreux, Hersent, who were to be contracted to do the actual work, although under the overall control of Reclus. The giant engineering firm had an excellent reputation based on their business on three continents. But none of their directors had ever worked in Central America. Alarmingly, Abel Couvreux had made a speech in Ghent at the end of 1880 saying that the so-called deadly climate of Panama, 'was nothing but an invention of the canal's adversaries.' In March 1881 a contract was signed between Couvreux and the Company that allowed for two years of preparation, assembly of plant, and preliminary studies, which would determine the costs for the contract for the actual excavation. The directors of Couvreux were old friends of de Lesseps from Suez days, and the preliminary contract was loosely drawn and generous.

Following the arrival of the first party from France, almost every boat from Europe brought its share of engineers, office workers or adventurers hoping to profit from the vast new project. At the beginning of 1881, Henri Cermoise was twenty-two years old and recently qualified as an engineer. French technical schools, run by and for the state, were the finest in the world, with rigorous entrance examinations, strict rules and a rigid, theoretical approach. After his gruelling training as a civil engineer, Cermoise was keen to see something of the world outside 'lecture theatres and blackboards'. Having had his application to work in Panama accepted, he sailed from St-Nazaire on the hard-working

Lafayette on 6 March 1881, on the third shipment of personnel to the Isthmus. To gain employment with the Company had been difficult, he writes, and he was aware that the journey was long and hard, that he ran the risk of meeting yellow fever 'nose to nose', and that the tropical sun could kill you 'like a cannonball'. '*Mais, bah!*' he writes, he was going all the same.

Those words – '*Mais, bah!*' – typify the bravery and foolhardiness of the French years. Whatever de Lesseps, Abel Couvreux or the *Bulletin* said, it was common knowledge that Panama was an extremely dangerous posting. But like a young man volunteering for service in a war, there existed for Cermoise and many of his contemporaries a belief, firstly, that the worst would always happen to someone else; that their country and the general progress of humanity demanded that they take the risk; that at the end of the day, amazingly, they were prepared to die for the great idea of the canal.

Cermoise's account of his journey to Panama is full of excitement and exuberance. After a storm in the Atlantic, on the sixteenth day of the journey they entered calm seas. It became hot and sunny, and the air was now 'charged with the perfumes of tropical earth'. There were more stops, at Martinique and on the South American mainland, before at last the *Lafayette* neared Colón. All that morning the passengers remained on the deck, excitedly scanning the horizon with their binoculars. Soon a long blue blurred line appeared in the distance. Then, little by little, the wooded summits of the Isthmus came into view. Gradually, everyone fell silent. 'A great thoughtfulness took hold of us all,' writes Cermoise, 'even those least given to contemplation. Silently, we thought of these lands where we were going to engage in the great scientific battle, and where, like in all battles, there would be the wounded and the dead.'

The party went ashore, and at the railway station were met by Gaston Blanchet. 'He was tall,' writes Cermoise, 'with an energetic look about him.' He seemed straightforward and kind. Blanchet had a list of his new arrivals, but when he went through it he found that most of the men were not what he had been expecting. Either through bureaucratic mix-ups, or because those applying to work in Panama had lied about their qualifications, Cermoise and a friend, Montcenaux, he had made on the voyage, turned out to be the only two qualified engineers in the entire group. Blanchet was furious.

Nonetheless, squeezed in amidst sacks of rice and other provisions, the group travelled on the train to Panama City. Cermoise describes the low-lying marshy land behind Colón, continuing through Gatún, Bohío Soldado and Buena-Vista, native villages along the line comprised of huts 'built of bamboos thatched with palms or oleanders'. After Gorgona there was a gradual rise up to Matachín, where the short, limp vegetation gave way to a taller and more solid mass of green. The railroad was single-track, and at Matachín they had to wait on a siding for the train coming in the opposite direction to pass. They had time to disembark, and were offered an array of provisions including water that was 'twenty-five degrees in the shade', and hard-boiled lizard and iguana eggs. Even more daunting was their first look at the jungle close up. The forest here reached right to the edge of the line. It seemed utterly impenetrable, a 'horrifying tangle of trees, a wall of creepers'. At last on the move again, they climbed further, to the top of the Continental Divide.

At Emperador they saw a gang drilling for rock samples, the first signs of canal work in progress. Then they descended, the railtrack taking alarmingly tight curves down the slope to the Río Grande valley and Panama City.

The party was conveyed to the Grand Hotel, where Cermoise was in for another shock. Hobbling down the main stairs on two walking sticks was a man with a bright yellow face. After a few moments Cermoise recognised him. It was a friend of his from his student days called Tansonnat. They greeted one another, surprised to have met in this faraway place. Tansonnat explained that he had been one of the first arrivals with Reclus and Blanchet, and had been in charge of a group of workers in the jungle. There, he had caught a bad fever. Cermoise thought he looked extraordinarily sickly, but Tansonnat insisted that he was on the mend, undaunted, and looking forward to returning to work. That evening Cermoise was introduced to Armand Reclus. To Cermoise he was already a legend, having been involved in the canal project from the earliest surveying days. The *Agent Supérieur* welcomed him warmly.

Reclus's first priority in these early months was to secure sufficient roofs to protect workers and machinery from the rains, due at the beginning of May. Time was of the essence. Within a week of the arrival of the first French engineers, Colón had been transformed into a busy port. A new wharf was speedily constructed, and ships started docking every day carrying prefabricated wooden buildings from New Orleans and countless railway sleepers and rails. At the same time, all sorts of machinery started arriving from the United States and Europe: drills, locomotives, wagons of all types, dredges, barges, steam shovels and cranes. All were transported in bits and had to be reassembled.

Louis Verbrugghe was now in charge of recruitment, a job in which he had local experience running his family's plantations in Colombia. The initial workers were from nearby – from Darién, Cartagena, or from the Jamaican community left over from the construction of the railroad. They were set to work expanding Colón's port facilities and assembling buildings – machine shops, a sawmill, wooden cottages for the white technicians and larger barracks for the workers. The Grand Hotel in Panama City was purchased by the Company and refurbished as its headquarters. Along the line of the railway the scattering of native villages were to be transformed. At Emperador a huge clearing was made for a work camp. Outside Gatún on the Chagres a new settlement, grandly called Lesseps City, was to be established.

After a week or so to acclimatise, Henri Cermoise reported to Blanchet to start work. Blanchet outlined the progress so far. The initial party of engineers, he said, had found much of the proposed canal line covered by impenetrable virgin forest. While the general route was known – which valleys the canal would follow – details such as where the axis of the canal passed from one valley to another were still to be determined. In addition there was much drilling to be done to discover the type of rock or soil that would have to be removed. Therefore all of the engineers were to be employed either in surveying and taking soundings along the axis of the canal, or in the central office collating the reports as they came in from the field. Blanchet offered Cermoise the choice of work: 'I hadn't come to Panama to be stuck in a study again,' he writes. 'The idea of the virgin forest, with tigers, crocodiles swirled round in my head; the life of a pioneer, penetrating into the unexplored depths of this isthmus was an irresistible temptation.' His friend Tansonnat, who had actually already worked in the interior, opted for a head office job, and Cermoise took over his old work gang. The next day, still with his friend from the boat over, Montcenaux, he set off for Gamboa.

This is where the Chagres River, what Cermoise calls the 'implacable enemy of our great enterprise', met the line of the canal at right angles. From Gamboa through to Barbacoas, the river followed the same route as the planned waterway. While other work parties set about clearing a 50-metre-wide strip along the line of the canal and began to take detailed measurements, it fell to Cermoise and Montcenaux to survey the site of the huge dam planned to regulate the flow of the river at Gamboa. Leaving the train at Matachín, the two men were punted upriver in a hollowed-out canoe. Twenty minutes later they arrived at Gamboa and took possession of the two huts, one for '*chefs*' and one for the thirty or so workers, which constituted the camp established by Tansonnat. There they met a Belgian, Blasert, a veteran of the North American West, who was acting as quartermaster and administrator of the work camp. Cermoise found him an impressive figure. 'He laughed at the climate, the snakes and at Yellow Fever . . . he considered himself invulnerable.' Blasert had even brought his wife and ten-year-old daughter with him to the camp. The latter 'passed her days *à vagabonder* barefoot and bare-headed' around the camp.

It was immediately apparent to Cermoise that this work would have 'nothing in common with what one does in Europe . . . When we saw the thick forest which covered the mountains we were thoroughly daunted.' The first task was to begin clearing pathways through the jungle towards those summits that seemed the highest. The plan was for the crucial dam to be constructed between two hills, each about 250 metres high, and some 500 metres apart. In Europe, comments Cermoise, both could have been surveyed in a morning. But here in the tropical jungle it was another matter altogether. The jungle was so thickly matted that one could see only a few yards in any direction, hopeless for taking measurements. The heat and humidity, like 'a steam bath', sapped the strength, making legs and arms as heavy as stone. The narrow pathways gradually cleared in the jungle were 'suffocating [as] the high branches met each other overhead, forming a vault which kept out the light and the air'.

Most of the workers were local mulattoes, who were instinctively hostile towards Europeans, and to taking commands, often replying, Cermoise notes, with the declaration, 'I am a free man.' They had to be handled with great tact, and at one point the Europeans, fearing a mutiny, took to carrying arms and guarding their hut by night. On the whole, however, Cermoise is generous in his praise for his workers. 'In spite of the faults, the Colombians . . . did us great service with the clearing.' They knew the forest, and were experts with their machetes. 'Bamboos, creepers, even trees fell before them like hail.' Sometimes, however, the party would meet a colossal tree that, because of the hardness of its wood, would take a whole day to cut down by axe. Progress was slow.

Few animals, apart from the odd parrot, were encountered in the jungle. However, there were plenty of snakes. Tansonnat had reported killing more than a hundred during the clearing of the space for the camp alone. The locals seemed to be adept at despatching them with a single machete blow, but it still required constant alertness as huge specimens could fall on the men from branches above their heads. Most feared were the coral snake and the *mapana*, otherwise known as a bushmaster. The coral snake is quite small, some 24 inches long, and brightly coloured with black, yellow and red bands. It was most usually encountered in the early morning or at dusk. Its bite contains a neurotoxic venom which attacks the nervous system and is frequently fatal. The bushmaster is much larger, up to 10 feet long, with enormous fangs that cling to the victim the better to inject its venom, which is hemotoxic, killing by destroying red blood cells causing internal bleeding and rapid tissue and organ degeneration.

These snakes soon became attracted to the camp, or more exactly to the vermin that attacked the party's stores and feasted on their waste. Great care had to be taken by the quartermaster, Blasert, when fetching provisions from their storehouse. The Blasert family shared the single officers' hut with Cermoise, Montcenaux and a Colombian translator, Jeronimo. All slept in hammocks suspended in the corners. Madame Blasert was heavily pregnant, and the Frenchmen urged her to return to Panama City, but she and her husband insisted there was plenty of time. Soon after Cermoise's arrival, however, she went into labour. With the assistance of an 'ancient black crone', she successfully gave birth in the tiny hut to a second daughter. All the while the two young French engineers tried to sleep in their hammocks, 'smoking cigarettes when the din became too loud'.

The team of Europeans became very close, and Cermoise enjoyed the challenges of the work, despite the privations. Everyone got stomach illnesses from drinking the river water, and fevers came and went regularly. For the Europeans, this could prostrate them for days, although, Cermoise notes, the locals, although just as susceptible, seemed to be able to recover after only a few hours. Vampire bats and tarantulas, which took to climbing down the ropes of their hammocks, tormented them at night. The jungle also teemed with ticks and *niguas*, small insects that would burrow under the skin to lay their eggs, threatening to cause gangrene. Evenings would be spent digging them out of each other.

At night, the jungle came to life. 'Above all, there was an invasion of insects,' Cermoise writes. 'With each step, one's foot crushed hundreds of them; with each movement of the hand, one picked up a fistful, and with each nod, one's face brushed swirls of them flying in the darkness. One breathed them in, as one went along! Moreover, the flame of a lamp was extinguished within minutes under the heaps of their small corpses . . . a monstrous buzzing filled the forest and rose all the way to the sky, while in this clear tropical night the huge trees . . . Flamed with millions of fireflies.'

At the end of April, just before the rains were due, Reclus carried out a tour of inspection. At Colón, he was pleased to see a great deal of activity. Jamaican workers were busy creating new port space by filling in the marshes to the southwest of the town. Nevertheless, such was the volume of material being unloaded daily that there was considerable confusion. Just outside Colón at Monkey Hill could be seen the first actual excavation in progress, where rocks and earth was being dug out to provide the filler for the marshes below. This had generated great excitement, and was the destination for American visitors curious to see the French actually at work on their great project. At Gatún, Lesseps City was well established, with wharves for unloading from the river and adequate shelters for the workers. He inspected other work camps at Emperador and Culebra, and at La Boca, the Pacific terminus of the canal, new wharfs had been constructed and a railway line laid to nearby Panama City.

But, as he explained in his report to the Company in Paris, there were great difficulties as well, many of which, indeed, would also be encountered in the early years of the American project. The first criticism that Reclus makes is levelled at the choice of men. The executives and technicians chosen and sent from Europe by Couvreux were, he said, both insufficient in number and, above all, mediocre in quality. On 30 April, Reclus wrote from the Grand Hotel to Charles de Lesseps in Paris that the contractors seemed to be sending over men who 'would be accepted nowhere – people who have dabbled all over and have never done anything well – nutters, drunks, incompetents etc. They send us travelling salesmen for mechanics, for blacksmiths men who've never been behind a forge.' Cermoise's group, it seems, had been pretty typical. In the terminal cities and the camps, Reclus wrote in the same report, he had seen far too much liquor, both purchased and home-made, as well as endemic gambling with dice and cards. All combined to produce frequent violent disputes.

Addressing the question of the railroad, Reclus reported that his 'personal relations with [the manager] M. Woods continue to be most cordial but that doesn't prevent frequent unpleasant incidents from happening'. It was essential, Reclus wrote, that the Company purchase the railroad as soon as possible. He even suggested that the railroad bosses were being deliberately obstructive in order to hold the Company to ransom and secure the best price for their business.

In fact, negotiations had been going on for some months. Trenor Park, the boss and majority shareholder in the Panama Railroad, was now holding out for $250 a share in cash. The value of the railroad had been falling steadily since the opening of the transcontinental US railroad in 1869, and the real share value was nearer $50. It was a 'hold-up', and there was nothing the French company could do about it. In the end, to cover all expenses and the railway's own sinking fund, de Lesseps parted with just short of $20 million, nearly half his start-up capital. It was a severe early blow to the finances of the new Company. One-fifth of the balance was to be paid every year with 6 per cent interest due on the rest. The Panama Railroad Company retained possession and management of the property until the whole amount was paid, and a majority of the seats on the board of directors, who still sat in New York. In spite of the buyout then, the railroad would remain American for now, which would have important political effects.

Park, who retired from the board but put his son-in-law in charge instead, cleared $7 million on the deal. However, interestingly, he retained a strong interest in the venture, and according to his daughter never doubted that the de Lesseps project would succeed. He visited the Isthmus several times before his untimely death in December the following year, on board a steamer from New York to Colón.

As well as tourists and grateful railroad shareholders, the United States government also maintained a keen interest in the canal, and kept up serious efforts to take further control of the Isthmus. In turn, the various European diplomats in Washington, Paris, Panama and Bogotá kept an eye on what the Americans – and each other – were up to in the region. The previous year, President Hayes, to give body to his aim of 'a canal under American control', had sent US naval vessels to investigate sites for coaling and naval stations on either side of the

Isthmus, near the future canal termini. The move angered the British, who saw it as incompatible with the Clayton–Bulwer Treaty, and was blocked by Bogotá after a furious popular anti-American reaction in Colombia. Nevertheless, the controversy carried into the next year, with Congress, in March 1881, voting $200,000 specifically for the establishment of permanent military bases near the canal.

In the meantime, the US envoy to Colombia, Ernst Dichman, did everything in his power to persuade the Bogotá government that their country was 'menaced by a grave danger'. 'Turning away from the United States which had been her firm friend, ally, and protector,' he said, 'Colombia recklessly and ungratefully concludes with an adventurous Frenchman, a contract for the opening of a Canal.' Dichman pressed for an updating of the 1846 Bidlack Treaty that would allow the US to establish a permanent garrison in Panama, while Secretary of State Evarts demanded that the United States be given the right to veto any canal concessions, future, present or past.

Seeing 'all alliances with the United States as an exemplification of the fable of the wolf and the lamb', Colombia started sounding out European capitals over a multilateral guarantee of the waterway and considered denouncing the 1846 treaty. For the aggressive and anti-British new US Secretary of State James Blaine (later to be nicknamed 'Jingo Jim' by the press) this was totally unacceptable. In June 1881 he wrote to the British Foreign Minister, Earl Granville, citing the Monroe Doctrine and condemning the plans, which he said came close to 'an alliance against the United States'.

Blaine's domestic audience was delighted, and in Congress there was a growing determination to act unilaterally on the canal question. The Clayton–Bulwer agreement that forbade such a move was debated in the House and referred to as 'a singular and ill-omened treaty', which should be abrogated.

None of this went unremarked upon in London when it came to reply to Blaine's June letter. For a long time, there was a haughty silence. Blaine was an upstart troublemaker, it was felt – he had already waded into disputes with Britain over other matters. In the meantime, Granville confidentially sounded out the European capitals about this possible international guarantee that had so annoyed Blaine.

One by one, the replies came in from Britain's European ambassadors. Every power was in favour, in principle, but no one wanted to make the first move. France was keen to provide support for beleaguered Colombia, but could not be at the forefront as it was a French company at work on the Isthmus. In September, Britain's Madrid representative reported that Spain 'would be glad to see England and France take joint measures to check the pretensions of the United States' Government with regard to the interoceanic canal, but that Spain hesitated to place herself "*en premiere ligne*" in opposition to the United States, in view of the consequences which might ensue in the island of Cuba, where a fresh insurrection could easily be fomented by American influence.' In Germany, Bismarck, who still held German ambitions in check, declared himself neutral in the matter, and said that it was a question for the Clayton–Bulwer Treaty between the US and Britain.

In November a reply eventually made its way to Washington from the British Foreign Office, taking issue with details of Blaine's pronouncements and slapping down the pretension of the Monroe Doctrine. By now, Blaine, encouraged by favourable press support at home, had written to the British again on the subject, this time asking that the clause in the Clayton–Bulwer Treaty that forbade the fortification of the canal be revoked. Because the United States had no navy to speak of, he argued, the only way that their vital strategic interests could be protected was through the establishment of permanent military power on the Isthmus itself. Otherwise, the primacy of the Royal Navy would make British control of the waterway a done deed. Privately, Blaine made a contingency plan to build a railway through Central America to Panama to 'enable the United States to keep military possession of the canal in the event of a war with Great Britain'.

But for now, Britain was not prepared to be bullied by a power with huge potential, granted, but no real international strength. Granville was not to be moved, replying that since Great Britain had complied with her side of the Clayton–Bulwer Treaty – something many Americans disputed – it would be 'manifestly unjust' for the Americans to request abrogation.

Soon after, as the new President, Chester Arthur took control, Blaine was out of office and out of favour. This coincided with a popular feeling that unilaterally breaking a treaty with Britain would have been unwise. 'Did Blaine want war?' asked the newspapers, mirroring the sudden popular recoil from pushing the mighty European powers too far. 'Mr Blaine had overshot the mark and misjudged public sentiment,' decided the *Herald*, only days after backing his aggressive approach.

If the Americans remained stalemated by the Clayton–Bulwer Treaty, they were still responsible under the original 1846 Bidlack Treaty to keep the transit passage on the railroad open, and, more often than not, there would be United States Navy gunboats standing off Panama City and Colón. By now, Colón was unrecognisable from six months earlier. There had been an explosion of wooden huts and shanties out into the swamp around Manzanillo as the town's population doubled. In the harbour there was a constant coming and going of steamers, and on the reclaimed land to the south of the island was a grid of warehouses, offices and residences, many in a grand style. For their wealthy European or

American residents, there were now three new French restaurants and a new French hotel with a Parisian chef imported from New York.

In Panama City there was a never-ending round of banquets, dances and ceremonial occasions. In fact, the canal company found itself paying for all sorts of public events. In 1881, Bastille Day was added to the calendar of *fêtes* – a three-day party during which the Company paid for the distribution of liquor to the people, bullfights and other entertainment. '*Viva el canal*', '*Viva los franceses*' responded the Panamanians. On top of all this, there was a constant stream of journalists and '*gros bonnets*' (big hats) as Reclus called them, arriving during the dry season on every steamer. In order to fill the pages of the bi-monthly *Bulletin*, so important for maintaining confidence with the investors in France, the Company appointed, at great expense, several renowned journalists who came to the Isthmus to write up uncritical articles. The '*gros bonnets*' were also often influential figures from the world of politics, finance, commerce or industry, who would return filled with admiration for the colossal project and its promoters. However, they, too, all shared a common interest: each man collected a handsome cheque in exchange for his sycophancy. Moreover, some of the less scrupulous among them would have no qualms about leaving substantial unpaid bills for hotels or excursions, debts that had to be picked up by the *Agent Supérieur*, much to his annoyance.

Reclus was also becoming increasingly exasperated by his colleagues in Couvreux, the contractors, in particular with Blanchet. Because it was unclear exactly who was responsible for what, there was growing tension between the two leaders of the project. In June 1881, convinced that he was being denigrated by his rival, Reclus wrote an ultra-confidential report highlighting Blanchet's weaknesses. 'He works a lot,' Reclus reported, 'but plunges ahead without having studied the question. He asks for material from France without knowing where to put it.'

Reclus demanded that he be given far greater control over accounts and payments, but by now it seems that his own relations with the office in Paris were highly strained, and nothing was done.

In the meantime, Blanchet and his men pressed on. By the middle of 1881, there were some two hundred technicians and about eight hundred labourers, and the number was increasing. The engineers were not only from France but also Germany, Britain, Switzerland, Russia, Poland and Italy. Many of the mechanics were American, who came along with equipment purchased in the United States. The workers were not only Colombians, but also Cubans, Venezuelans and West Indians.

In Gamboa, Cermoise's camp was transformed by the influx of a considerable number of new personnel and a batch of prefabricated buildings from the

United States. 'Everyone had his own room!' he exclaims. Now in charge was a distinguished French engineer called Carré, who brought with him an expert Belgian cook and much improved provisions. There were even regular deliveries of ice. Because of the importance of dealing with the problem of the Chagres, the Gamboa camp was now a key site on the Isthmus.

By the middle of September, Cermoise's surveying work was completed. While Carré remained behind to start mapping the giant area to be flooded by the dam, Cermoise was given a new job in Panama City, while his friend Montcenaux was sent to Gatún, the most notoriously sickly area of the Isthmus. After nearly seven months in the bush, Cermoise writes, 'We said goodbye with a certain sadness to this corner of the world where we had more than once shivered with fever, but also where we had passed many good days, busy with cheerful work in the company of devoted friends.' His new task was to work on the detailed maps and charts being prepared from the reports now flooding in from all along the line. As well as measuring levels, the engineers were sinking five great wells to test the ground along the summit of the Continental Divide. The greatest was some 150 feet, three times deeper than any well drilled during de Lesseps' visit early in the year, but still less than half the planned depth of excavation. The findings encouraged the canal planners – there seemed to be a lot less hard granite-like rock than had been factored into their costs. Before, drills had hit solid rock in several places, sometimes at 20 feet or less, and stopped. Now, with improved drill bits, the rock was passed through and in some cases found to be only 2 foot thick, an 'angular boulder of dolerite'. Below, there was 'brown clay and pulverized rocks, seamed with diverse colours.'

In fact, the ground was nothing like anything the majority of the geologists had ever seen before, bearing no resemblance whatever to the terrain in Europe mined and dug by many of those present. The unique geological history of the Isthmus, with the land bridge sinking below sea level and then rising again in a series of cycles and a long record of ancient volcanic activity, had created bewilderingly complex strata, including layers, at various angles, of breccia, limestone, coral, carb, sand, gravel, volcanic lava and clay. In the forty-odd miles from Colón to Panama City there are six major faults, five substantial volcanic cores and seventeen fundamentally different types of rock. Every well told a different story. But the engineers focused on the positive: at least it was not all solid rock. In fact, the surveys had said it all: dig here, and you do not know what you are going to find. As it turned out, it would have been easier had it been solid rock.

Blanchet himself turned his attention to the rivers, establishing observation posts on the Chagres, Trinidad, Obispo and the Río Grande; these were equipped with fluviographs, which confirmed the challenge that the rainy season would bring to the successful construction and running of the canal, with rivers rising 20 feet in as many hours and their rate of discharge increasing overnight from 3000 to over 60,000 cubic feet per second.

By October 1881, much of what de Lesseps had outlined back in March had been achieved. That month the *Bulletin* published details of the 'second campaign' for the next twelve months, which included settling the question of the dam for the Chagres, the digging of a waterway between Colón and Gatún and the removal of 5 million cubic metres of spoil from Culebra, the point of maximum elevation. In addition, all the necessary machinery was to have been ordered.

But not all was going to plan. Engineers had been experimenting with different excavating machinery and had found that the plant that had built the Suez Canal was proving too light for the heavy clay of the Chagres valley. Even more worrying was the question of labour. Only one in ten newly arrived labourers had remained on the job for more than six months. There were two thousand men at work at the end of the year, but de Lesseps had promised there would be ten thousand. Worst and most ominous of all, though, it was becoming obvious that the Isthmus was nothing like the healthy place that de Lesseps and Abel Couvreux had promised.

As early as March 1881, only two months after the arrival of the first French engineers on the Isthmus, the Panama *Star and Herald* reported that 'Mr. de Lesseps contemplates making up what is short in the labor supply on the Isthmus and the neighbouring coast states of Colombia, with laborers from the West Indies. Barbados and Jamaica are spoken of as the principal source of supply.'

After Emancipation in 1834 many former slaves, keen to establish their independence and a different life from that of plantation labour, opted to set up small plots, producing food for themselves and for market. Sometimes their cash income would be boosted by seasonal work on the big plantations. But the British islands had not prospered in the years since the construction of the railroad. Since the then all-time-low sugar price in 1850, the value of the colonies' principal export had fallen a further 30 per cent as refiners in the United Kingdom started to source their sugar from heavily subsidised European beet. It was impossible for West Indian growers to obtain credit, and the collapse of the industry depressed the rest of the islands' economies, lowering wages and decreasing the money in circulation. Furthermore, the new peasant farmers suffered a heavy and unequal burden of taxation through duties on imported staples such as rice and codfish. They had no representation – you had to be earning at least £20 a year to be allowed to vote in Jamaica – and the 1860s onwards saw a campaign to evict squatters from their smallholdings. In addition, most of the freeholds on which the ex-slaves had settled were under five acres – large enough for one family perhaps, but not enough to afford a livelihood for their sons and daughters, much less their grandchildren.

In response to all this, a tradition of emigration grew up throughout the islands. Blacks remained condemned by the West Indian social structure to a permanent lower-class status, denied real recognition of their freedom, autonomy or even humaneness. Agricultural labour, with all the stigma of slavery attached to it, remained the mainstay of the islands, and was chronically poorly paid. Only by going abroad could a Jamaican or Barbadian find levels of reward for labour sufficient for their needs and lower levels of abuse.

Nevertheless, it is wrong to see the emigrants solely as passive objects, or victims of these conditions. In spite of everything, black West Indians had developed a strong sense of independence and personal dignity. The trend towards emigration also points to the ambition of the ex-slaves' descendants to 'better themselves' – to see the world outside their small island, and to earn enough to improve their conditions at home, to be masters of their own destiny. The Caribbean basin offered many good opportunities for work, not only on the Panama Railroad in the 1850s, but on other railway projects as well, in goldfields and metal and rubber industries, in logging or in plantations being set up throughout the area. The islands also exported teachers, missionaries and ministers as well as colonists to previously unsettled areas.

In many ways the emigration to Panama during the French period is part of this pattern. There were particular 'push' factors, such as the severe drought in Jamaica in 1879, but the project, in the eyes of the emigrants, also offered great opportunities. They made the journey to Panama not just to escape poverty at home, but just as much to 'see the world', 'learn a foreign language' or 'seek adventure'. The chance of work on the canal was seen as a means of truly freeing themselves through their own efforts.

Nevertheless, the 1880s represent a quantum leap in West Indian emigration. Although records are incomplete, there seem to have been some five hundred British West Indians working on the canal by October 1881, 40 per cent of the total workforce. The following year and for the rest of the project this percentage rose to sixty, as the overall workforce ballooned to over twenty thousand by 1884.

Early in 1881, the Company took on the services of Charles Gadpaille, a Jamaica-based Frenchman, to handle recruitment in the islands. Straightaway, he started posting advertisements in the local press. One such read:

A trip to Colón?
Wanted immediately!
10,000 labourers
for the
Panama Canal Company.
No indenture. Passengers returning when they like.
Both passage and food given.
$1.50 to $3.00 a day.*
Medical care given when sick.
Apply to Charles Gadpaille
Hincks Street,
Agent, Panama Canal Company.

Daily wage rates in Jamaica for a field labourer in the 1880s were between sixpence and two shillings, less for women and children, so this was a great offer. If it seemed too good to be true, that is because it was. Gadpaille had no right to make many of these promises, as shall be seen. As well as newspaper advertisements, the agent posted flyers and sent runners into rural areas to drum up recruits.

Gadpaille concentrated his efforts in Jamaica, Barbados, St Lucia and Martinique. Although the Company might have favoured the French-speaking Martiniquans, in fact they were less than impressed with workers from the French Caribbean islands, considering them 'pretentious, and always complaining, for they had been ruined by the political customs in vogue in the old French colonies'. So the vast majority of the workers on the French canal came from Jamaica, with St Lucians the next largest group. The first arrivals, in 1881 and 1882, tended to be skilled labourers and artisans who were not involved in cultivation on small freeholdings, but had drifted into the towns and cities as they were displaced from estate labour. After this first period, the typical migrant was male, an agricultural worker, twenty-five to thirty-five years old.

Henri Cermoise was highly impressed with the new arrivals from Jamaica. 'They were excellent workers,' he writes, 'much more active and energetic than the Isthmians and easier to manage.' Back at home, some of the planters were less than pleased at seeing their pliant labour force leave the country, but for now the governments of the islands saw the money that would be sent home outweighing the disadvantages to the planters. The Governor of Jamaica's annual report of 1882 commended the returning emigrants, 'bringing with them money with which they arrange their affairs and aid their families'.

* Wages here, as throughout the French period, are in Colombian silver dollars, one being equivalent of about US$0.65–0.70

But for every labourer returning with this pockets rattling with coins, there was another who came back in a different state, or didn't make it back at all. Some Jamaican workers, Governor Musgrave reported in 1882, 'are left when ill to die in the streets of Colón'. His main complaint was that his government was 'put to large expenses for the relief, burial, or return to the colony of any natives whose case comes before the British consular authorities. And, moreover, many of those who return are so broken down by Chagres Fever, and other disorders, that they become a burden upon the community, and the poor rates.' It was clear that, already, disease was taking a toll on all sections of the workforce.

11

FEVER

The beginning of the project's first wet season had seen the initial serious outbreaks of disease. The first high-profile death among the thousand-odd employees occurred in the second week of June 1881, soon after beginning of the rains. A distinguished and experienced engineer named Etienne died on 25 July, supposedly of '*ataxie cérébrale*' – 'a fit of the brain'. On the Isthmus at the time, on a two-week tour of inspection, was thirty-nine-year-old Henri Bionne, the Company's secretary and the man who had started the ball rolling for a French canal. On 9 July he had dined in Henri Cermoise's mess hut at Gamboa for which the excellent Belgian cook had pulled out all the stops. 'He drank to our success on the Isthmus,' remembered Cermoise, 'we drank to his good luck.' Bionne boarded a boat home for France on the evening of Sunday 24 July and at the captain's table that night he reported himself feeling poorly and without appetite. The following afternoon he was visited in his cabin by Georges Hopkins, the ship's doctor, who found him in the throes of a violent shaking fit, which was followed by a high fever. According to the doctor's report, reprinted in the *Bulletin*, Bionne was given quinine, and, the next day, a mustard bath. After this he felt better, and even had a little to eat. The doctor was much encouraged that his patient seemed to show no signs of the symptoms of yellow fever. But on the Thursday morning, it became obvious that he was far from well. The doctor reports: 'He got up, dressed, and went to the captain's cabin, and sat down to have a meeting with him, being evidently in a state of delirium.' After that, he was given more quinine and mustard treatment, but 'his state worsened quickly; after a fit, he fell into a coma during which he died . . . the last symptoms indicate kidney failure'. His body was hastily disposed off, shunted overboard in the Gulf of Mexico.

The doctor's diagnosis was that Bionne had died of a breakdown of his nervous system. For the benefit of readers of the *Bulletin*, he was keen to stress that it had not been yellow fever. This illness held a particular fascination and horror for Europeans and North Americans. Emerging in the Caribbean in the 1640s, supposedly from the Mayans on the mainland, or, as recent theories suggest, from mosquito stowaways on the slave ships from Africa, the disease spread to Barbados and Cuba, where it killed one-third of the island's inhabitants. Outsiders in the region seemed particularly vulnerable. In 1665, it claimed the lives of all but eighty-nine of an English squadron of 1500 stationed in St Lucia. Spaniards on imperial duty carried it back to their home cities, where it wrought havoc. For the next two hundred years, the disease also came and went along the United States' southern and eastern coasts over a hundred times, on one occasion killing more than five thousand in the Mississippi valley. In 1793, the city of Philadelphia was decimated by yellow fever, or 'yellow jack' as it became known. Famously, Napoléon's Polish legion of 25,000 men, sent to the area to recapture Haiti from Toussaint L'Ouverture and to re-establish control of France's North American empire of Louisiana and New Orleans, was wiped out by yellow fever and retreated, vanquished, home.

It is an almost uniquely distressing, disgusting and terrifying disease. There is still no cure, apart from treating the results of the disease, such as kidney failure, and in the 1880s a strong adult would have only about an even chance of surviving an attack. At that time it was treated with whisky, mustard seed, brandy and cigars. If you do survive, you are then subsequently immune. Thus the disease, to flourish into an epidemic, needs an influx of non-immune subjects. Caused by an *arbovirus*, a small virus transmitted by the bites of certain mosquitoes, the early symptoms include headaches, loss of appetite, and muscle pain. A high temperature follows, accompanied by severe back pain, which many described as like being on the rack. After that comes a burning, agonising thirst, the telltale jaundice as the face and eyes yellow, and the dreaded '*vomito negro*' – vomiting up choking mouthfuls of dark blood, as the virus causes liver and kidney failure and multiorgan dysfunction and haemorrhage. The brain is often affected as well, producing delirium, seizures and coma. The medical shock, caused by extreme fluid loss, can in itself be fatal.

Part of the terror of the disease was in its mystery, how it arrived from nowhere, created havoc and then just as inexplicably disappeared. As well as there being no effective treatment, there was little idea of how the disease was transmitted. Some doctors suggested that it was due to a certain wind off the sea, others were sure it was some sort of fungus. One insisted that it came from eating apples. Most agreed, though, that it was airborne, and as it was so often found around ports, that it had to have something to do with mud, filth or dead animals. Worst of all for causing infection, it was believed, was the patient himself, or anything that he had touched. Victims were shunned, given a hasty burial and their clothes destroyed.

Fevers came under many names on the Isthmus. As well as yellow fever there is mention of *calentura*, miasma, the shakes, blackwater fever, the chills, *paludisme*, ague, pernicious fever, putrid fever, and, particularly nasty, 'Chagres' or 'Panama Fever'. Yellow fever was the most feared by whites, but these versions of malaria were actually the biggest killers.

Caused by a parasite that is transmitted to humans through the bites of an infected *Anopheles* mosquito, malaria remains a huge scourge at the beginning of the twenty-first century, killing around a million people a year in Africa alone. The parasites migrate to the liver and then infect red blood cells, where they multiply, rupturing the cells, then spread further. The weakening and disruption of the body's blood results in many symptoms that are similar to those of yellow fever: uncontrolled shivering, chattering teeth, high fever, sweating and a burning thirst, headaches, nausea and vomiting, muscle pain and anaemia. This can lead to jaundice, convulsion, coma, rupture of the spleen and subsequent massive haemorrhage. If you lived, you would be severely weakened mentally, as well as physically. Unlike yellow fever, malaria attacks confer no instant immunity and can recur to those who survive, often killing on the third or fourth attack. The only effective treatment was the administration of quinine, a palliative made from the bark of the cinchona tree, a trick learnt from the Incas of Peru. This stops the disease's progress by interfering with the growth and reproduction of the parasites in red blood cells. But when malaria patients stop taking quinine they relapse. This was more common than might be thought: quinine not only tasted disgustingly bitter, it also had side-effects including nausea, painful earache, deafness and, most dangerous of all, hypoglycemia.

For centuries, people had believed that malaria was caused by 'miasma' – toxic emanations from the rich corruption of tropical soil. That is why de Lépinay had argued during the Paris Congress that to dig over so much earth in Panama would be particularly dangerous. In the early 1880s, recent studies near Rome had led to the theory that it was a bacterium in the soil, made airborne when released, that caused the malaria symptoms. Although isolated individuals had suggested that neither yellow fever nor malaria were airborne but were transmitted by mosquitoes, and numerous experiments with this idea were being carried out in Cuba as early as 1881, the development and acceptance of the theory were some years away. Until that time, the 'miasma' theory held sway, and the Isthmus was consequently a death trap.

Two days after Bionne had boarded his boat to France, Blasert, Cermoise's 'indestructible' Belgian friend from the Gamboa camp, put his wife and children on the steamer back to France. The very next day he took sick, with yellow fever according to Cermoise, and died soon after. He, too, had been at the Bionne dinner.

As others of his companions starting sickening and dying, including the much-praised Belgian cook, even Cermoise's optimism and humour faltered. 'There was a dismal period for the administration,' he writes. 'It seemed as if a wind of death was blowing on its employees.' Even at Emperador, high in the mountains and seemingly the healthiest spot on the line, there was a bad outbreak of yellow fever. 'The situation looked bad,' says Cermoise. 'These successive deaths . . . had shaken our courage, striking the imaginations of even the bravest men; everyone anxiously began thinking of steamers home; in a word, we were stuck by one of the those moral weaknesses from which a panic is born.'

Armand Reclus was away in Paris, having left Louis Verbrugghe in charge. On 5 October 1881, the lawyer wrote to France: 'At the moment, the state of health conditions in Panama is distressing: an upsurge of disease is occurring . . . the morale of our personnel is a bit shaken by the sudden deaths . . . Natanson and Marinovitch are leaving Panama. All the pretty promises they made to Abel Couvreux have been broken.' This panic, the fear of disease, was almost as bad for the project as the actual fatalities.

Inevitably, rumours reached Europe, but de Lesseps, addressing a Geographical Congress at Vienna that month, insisted: 'No epidemic of maladie had manifested itself at Panama. Only a few cases of yellow fever had appeared, and these had been imported from abroad.' But even worse was to come. In November, Gaston Blanchet, whose marriage the previous year had made him a popular figure in Panama, became shivery and feverish while on an expedition mapping the headwaters of the Chagres. He made it back to Panama City, but died two days later. 'Mr Blanchet's death is an irreparable loss to the Company,' reported the British consul back to London. 'Operations will be almost entirely at a standstill until his successor arrives.'

The 'casualty figures' from the French construction period were argued over at the time, and have been ever since. Contemporary American newspapers hostile to the project doubtlessly exaggerated their reports, claiming that among the couple of hundred white technicians alone nearly seventy had perished in the first twelve months. The Company retaliated by ridiculing the figures. The best estimate is that about fifty men died in the first year, from an average workforce of about a thousand for this period. Many more, though, were incapacitated by illness.

As part of his pitch to investors in the canal, de Lesseps had promised up-to-date hospitals would be built to serve the workers on the project. And he was as good as his word. A hundred-bed hospital was constructed in Colón, and work started on a huge five-hundred-bed establishment on Ancón Hill, a salubrious and breezy spot high above Panama City. Huge sums were spent – a million dollars on the Colón hospital and more than five and a half million dollars at Ancón, on a seventeen-building complex which included its own fresh water supply, a vast laundry, an abattoir and a farm that provided the patients with an abundance of milk, eggs and fresh vegetables. Outside the ward windows, patients could enjoy a well-laid-out garden, irrigated on a terrace system, bright with herbs and flowers. Inside, as a contemporary noted, 'the hospital rooms are so vast and well-ventilated that, even in the ones occupied by Negroes stricken with marsh fever, visitors with the most acute sense of smell could not detect the slightest odour'. It was by some distance the best hospital anywhere in the tropics.

A further half a million dollars was spent on building an extensive sanatorium on the island of Taboga, about an hour and a half by steamer out into the Bay of Panama. In its lavish and beautiful surroundings, employees could convalesce after a time in one of the hospitals, or just take a break from the feverish climate of the mainland.

In charge of the medical operation was the former head of the sanitary division at Suez. He ran a team of six doctors, several of them English, and thirty nurses from the order of St Vincent de Paul, led by Sister Marie Roulon. Their care, though medically primitive by later standards, was much praised. 'She is one of those rare women whose personal zeal is contagious,' wrote the New York *Herald*'s occasional Panama correspondent of Sister Roulon in October 1881. 'Every one of her Sisters has caught the trick of her cheery kindness . . . When all healing is unavailing, they make even the scorched death of yellow fever easy if such a thing can be.'

But in November the rains ended, and as the pools of stagnant water on which the mosquitoes depended to hatch their young dried up, so rates of infection fell away. In addition, according to Henri Cermoise, the skilful and courageous leadership of Louis Verbrugghe did much to calm exaggerated fears. But now that the small window of dry weather had returned, it was imperative that work be pushed ahead with as much urgency as possible.

By November 1881, Henri Cermoise's task of mapping the precise axis of the canal was completed and it was time to set it out on the ground, with a line of stakes either side of the area to be excavated. Together with Montcenaux, he was given a 10-kilometre section to mark out in the area of Gorgona. Cermoise was happy to be back out in the field, and to be reunited with his friend, who, as Cermoise had predicted, had caught a fever while working near Gatún, and had very nearly died. At first, writes Cermoise, Montcenaux had presented the symptoms of the dreaded *vomito negro* – yellow fever – but as he had survived, Cermoise deduces, it must have been something else. 'There's only one certain way to diagnose fever,' he writes. 'Did he die? Then it's Yellow Fever. Did he recover? Then it is only an attack of bilious fever.'

At Gorgona they camped out in an open shack in the middle of the village until, fifteen days later, a prefabricated house was sent up to them by the Company. In the meantime, they started the job of clearing vegetation from their ten kilometres. They were all too familiar with this arduous task from their surveying work at Gamboa, but this was now on a different level. Previously, their '*tranches*' had been fairly haphazard – they had been able to bypass obstacles or particularly enormous trees – but now they had to stick exactly to the route on their maps as well as create a much wider and more complete clearance.

Missing the Belgian cook, Cermoise now found the available provisions scarce and expensive. As elsewhere on the Isthmus, the local retailing (the single shop) was run by a Chinese gentleman, in this case married to a local. As more and more foreign workers arrived on the Isthmus, so prices climbed with demand.

Many of the people turning up, Cermoise complains, were still of dubious qualifications or even competence. At one point a carpenter working on the accommodation requested thirty nails from the workshop in Panama, carefully carving a piece of wood in the exact dimensions and size of what he required. A fortnight later the order duly arrived, to exactly the right dimensions, but each one made of wood and so utterly useless. In all, there was a feeling of muddling through.

Among the imported workers, reported the *Star and Herald*, there was also dissatisfaction. At the beginning of January 1882 a general strike broke out in Colón, based on demands for $1.50 rather than $1.20 a day. The following day, 6 January, the strike was 'in full blast', and had taken in workers on the railway, the steamers and the canal itself. Fearing wider trouble, the American consul summoned the US Navy. The stand-off continued for a week, with the works of the Isthmus at a standstill, while the Company tried in vain to import enough new men to cover the gaps. On 13 January, the bosses offered $1.35, but the workers stuck to their guns, citing the huge rise in prices of provisions over the previous twelve months. The *Star and Herald* sympathised with this, but urged the strikers to do the right thing and take the offer, as soon the Isthmus would be flooded with labour and their negotiating power would be gone. Finally, on 14 January, the paper could report the end of the strike: 'The railway wharf is again a scene of life and animation . . . The price paid for labor by the Railway Company is $1.50 per day.' When the US warships arrived two weeks later, all was peaceful.

On 20 January 1882, the first spade-load came out of the actual line of the canal at Emperador. This was fitting, as here was the highest position on the canal route, the point at which the depth of the trench to be dug to reach sea level was over 350 feet. The French called the area '*la section de la grande tranchée*'.

A spectacular explosion in the presence of a number of invited guests launched the work, after which the great and the good retreated to Panama for a banquet followed by a gala dance.

It was a great boost for the project that actual excavation was underway, but for the next four months of the dry season the company was unable to get mechanical diggers into place and operational. The work fell to some seven hundred men, mostly Jamaicans, toiling away with pick and shovel, so progress was slow. Much of the overall workforce was still employed with erecting buildings, laying track and improving the access to the site.

From Matachín to Gatún work was still needed pegging out the line, which was not entirely cleared of vegetation until May. In February, Henri Cermoise and Montcenaux had been sent to San Pablo, an isolated spot offering few diversions apart from watching the trains go past and hunting for iguanas and local wild turkeys. Two months later, just as their time there was coming to an end, Cermoise suddenly felt 'invaded by a persistent tiredness'. He had a headache, could hardly eat, and was left indifferent by even the most succulent iguana eggs.

For him, the attack seemed inevitable. Montcenaux had done his time while at Gatún, now it was his turn. Soon, he felt dizzy, then overcome with aches and was unable to stand up. After that he suffered a high fever for two days, during which, 'unfortunately, my reason left me on several occasions'. He was convinced that he had been taken with yellow fever. Montcenaux, however, did not panic, dosing his friend with quinine and calling for help from a passing train, which took Cermoise to Panama City.

To his relief, at the hospital Cermoise was diagnosed with *calentura*, rather than yellow fever. But for fifteen days he suffered a high fever and delirium, never sure if he was awake or asleep. When he regained his senses, he felt lucky compared to the man he saw in the bed next to him who, weakened by fever and blood loss from vampire bat attacks, also had some type of larvae in the top of his nose, which in a few days burrowed into his head and killed him.

But the new hospitals were far from full, in spite of the fact that, mainly as a result of Charles Gadpaille's efforts in Jamaica, more and more labourers were arriving on the Isthmus. In May 1882, the British consul reported three thousand men at work along the line, and that 'The sanitary condition is very favourable. There are about forty-two cases of various types of fever, none of which are of an alarming nature. The average is fourteen cases in 1,000, which I consider a very feeble percentage, owing to the nature of the work the men are engaged in. I am happy to say that the natives of the West Indies stand the climate very well, and supply the Company with a good nucleus.' By this time the entire line had at last been cleared to a width of 300 metres, a task that had taken much more time and effort than had been anticipated.

The following month, June 1882, de Lesseps told the third annual shareholders' meeting of the details of the purchase of the PRR and asked for a bond issue to pay for this expenditure. He also announced that excavation work had now started at Gatún, Gamboa, Bas Obispo, Culebra, Gorgona and Paraíso. In twelve months' time, he breezily predicted, 5 million cubic metres would have been removed from Culebra. There was more good news. In February a contract had been signed with an American company, Huerne, Slaven & Company, to dig a channel to a depth of 2.5 metres between Colón and Gatún. The excavation was reckoned at 6 million cubic metres, which would take eighteen months from a start date of August 1882. This, de Lesseps explained, would bring American mechanical skill and might to the project, and bury for ever the fear that the United States was opposed to the canal. Shareholders were delighted and readily approved a bond issue to be held that September.

There were increasing numbers of Americans on the Isthmus. The process of parcelling out the work to contractors had continued through the year, with another American company taking on the dredging at the Pacific end. Machinery was purchased by the Company and then rented out to the contractors, who were to be paid per cubic metre excavated. But the arrangement was not to the liking of the *Agent Supérieur*, Armand Reclus, who complained about the overfavourable terms given by the Company and the bureaucratic chaos the approach was engendering. In fact, Reclus was losing heart. Since the return of the rains in May, frequent flooding of the works at Culebra had brought digging there to a standstill, and the hospitals were again filling up with fever patients and accident victims. Complaining of overwork, exhaustion and confused leadership from Paris, Reclus resigned in June 1882. He was given a job in the Paris office, but his time of influence over the canal, going back to the first Wyse expedition five years before, was now at an end.

In September, there was a further setback, when the Isthmus suffered an earthquake. Although there was only one fatality, the railway bridge over the Chagres was thrown out of line and the tracks damaged in many places. A Canadian doctor, Wolfred Nelson, who had been living on the Isthmus for six months and who would become a fierce critic of the French effort in Panama, cabled a report to the New York *Herald*. 'It all but produced an earthquake among M. De Lesseps' shareholders,' Nelson writes. 'He at once informed the world that there would be no more earthquakes on the Isthmus. Strange to say, despite the utterances of this celebrated man, the earthquakes kept on, to the unstringing of our nerves . . .' Cermoise describes the quake in less doom-ridden terms, despite tearing muscles in his legs jumping off a first-floor balcony to make his escape from a collapsing building. Cermoise would leave Panama a few months later, he says 'for family reasons', but in contrast to naysayers like Nelson he shrugs off the difficulties and frustrations of his time there. In fact, he was sad to be leaving the Isthmus, 'where,' he says, 'I have spent two of the best years of my youth'. Looking back while writing his account the following year, he missed the strangeness, novelty and the unexpectedness, 'and the friendship of good and loyal companions, with whom we have, together, played our part, however modest, in the most gigantic of all modern enterprises'.

The earthquake also failed to dampen the enthusiasm and optimism of de Lesseps and his *Bulletin*. In November, *Le Grand Français* announced that 'After two years' work . . . we are much farther advanced than we were at Suez after six years.' Of the 75 million cubic metres to be excavated, crowed the *Bulletin*, a quarter had been allocated to contractors. It was almost as if the work was already done.

In the US press, the French Company was dismissed as 'incompetent', and in Panama the *Star and Herald* warned that 'exaggerated statements' were causing 'doubt and distrust'. In France, however, confidence remained rock solid. The bond issue in September was a great success and massively oversubscribed. Although there were warning signs – the interest on the money had crept up from the share issue and yet more sweeteners were handed over to the financial institutions – it was a distinct knock-back for those predicting the imminent demise of de Lesseps' project. With 'applications for shares showering him from all quarters of France,' wrote the anti-canal New York *Tribune* soon after, 'he can now reckon with confidence upon the resources required for so vast a scheme. He can get the money . . . Englishmen and Americans may as well reconcile themselves to the situation.'

Then, at the end of the year, there occurred another setback on the Isthmus: the contractor Couvreux unexpectedly used a loophole in their contract to pull out of the project. It turned out that the veterans of Suez had found all their expertise worthless. As de Lépinay had predicted, Panama had nothing in common with Egypt. If anything the experience of Suez had actually hampered the efforts of Couvreux. The loss of Blanchet had hurt, too, with no one of similar stature from the company willing to go out to Panama. The break was amicable, Couvreux arguing that since smaller contractors were now in place, having a middleman between them and the Compagnie Universelle was a waste of money. The real reason for their defection emerged later, when the ashes of the Panama [illegible]ect were raked over in Paris ten years later: 'The truth is that during the trial period,' a government report reads, 'Couvreux and Hersent had been able to form a shrewd idea of the difficulties of the enterprise but were unwilling to undermine the [canal] company's credit by frank admission of the motive behind their retirement.'

With the works now stuttering after just two years, perhaps Ferdinand de

Lesseps should have taken personal charge on site, as he had at Suez. But that, he calculated, would have sent a disastrously alarmist signal to the markets. Appearance and confidence were all to a project living or dying on credit with the public investor. De Lesseps was needed in Paris. He was also distracted by the events at Suez, where the British had seized control, and by his ongoing eccentric scheme to flood the Sahara. Another explanation, put forward by de Lesseps' American detractors, was that age had finally caught up with the 'Great Engineer'. The Paris correspondent of the New York *Tribune* reported that de Lesseps on his return from his latest trip to Africa had 'aged a good deal . . . His handwriting, which was so clear and vigorous when he returned from Panama, is now a shaky scrawl.'

So instead of Ferdinand, his son, Charles de Lesseps, was sent to the Isthmus for a month-long visit. Charles, who possessed none of his father's verve or showmanship, would increasingly shoulder the burden of leadership of the project. With him was the first *Directeur Général* of the works, Jules Dingler (pronounced 'Danglay'), one of the most senior civil engineers in France. He had been appointed with a salary of the equivalent of $20,000, far more than anyone else in the French organisation was being paid. He looked unprepossessing – short, bald and round-shouldered – but his arrival at Panama, at the beginning of March 1883, would usher in the great heroic period of the French effort.

12

JULES DINGLER

By the end of the following year, Dingler's leadership had transformed the canal project. Even Wolfred Nelson described 1884 as a time when Panama was 'busy . . . and bright with hope'. 'The work moves steadily on,' reported the *Star and Herald* in November of that year. 'The progress which is being made is apparent to everyone who crosses the Isthmus.'

Dingler was passionately in love with the Great Idea of the canal, but he was experienced enough on large-scale engineering projects to see on his arrival that the effort on the Isthmus was drifting. His first move was to reorganise the Company's chaotic office in Panama City. From the Company's paperwork, a clean break is apparent dating from the arrival of the new *Directeur Général*. Armed with total authority on the Isthmus, unlike Reclus or Blanchet, Dingler set about tightening up payment procedures, demanding exact descriptions of articles requisitioned 'especially so where machine parts are required', and generally attacking the bugbears of waste and fraud. There was also a purge of the workforce, with many of the unsuitable adventurers and strays – whom Dingler referred to as 'idlers and traitors' – being sacked or moved from their comfortable offices out into the jungle. In the process the new leader made plenty of enemies, but also earned the respect of the majority of the workforce.

Next, he toured the line and put together the first truly detailed plan for the canal. This included a policy of creating much more gradual slopes for the trench, which, together with the idea of riveting the sides of the canal with vegetation, was designed to deal with the growing problem of landslides. The result was a large jump in the estimate of spoil to be excavated – 120 million cubic metres, 45 million more than judged necessary by the 1880 Technical Commission. In early autumn 1883, he returned to Paris to present his plan to de Lesseps and his board of advisers. Dingler got the go-ahead, although de Lesseps did not feel it necessary to revise either his costs or his scheduled completion date for the canal, which remained 1888.

When Dingler returned to the Isthmus, he brought with him his wife, son, daughter and her fiancé, as well as his collection of thoroughbred horses. The message of long-term commitment to the project was unmistakable, as was his belief that there was nothing to fear from Panama's climate. Dingler went further, announcing, 'I intend to show the world that only the drunk and the dissipated will die of Yellow Fever.' The family were an instant hit in Panama society. The *Star and Herald* reports a reception given by the *Directeur Général* in December 1883: 'Their rooms were crowded with many ladies, and a number of our most distinguished native and foreign residents. The handsome rooms were still further beautified with floral decorations and other adornments; the music was good, and gave a zest to dancing, whilst the cultivated hosts spread a charm through the scene of enjoyment.' The young Dinglers set about exploring the Isthmus, enjoying picnic and riding excursions.

Back in France, investors were also impressed with the new leadership. In October 1883 there was a second bond issue, which was again massively oversubscribed. 'This result has surprised all,' reported the Panama press, 'except those who know the popularity which Count de Lesseps enjoys, and the confidence investors feel in his projects.' Eleven months later there was a further bond issue. This time it didn't quite sell out, in spite of even more favourable terms. Nonetheless, French support remained solid.

By now, nearly 700 million francs had been raised. Dingler was spending it fast, giving out huge orders aiming to double or even triple the number of steam shovels, locomotives and other plant in operation on the Isthmus. Along with the new machinery came a host of new contractors. In May 1883 alone, Dingler signed seventeen contracts for excavation. Once back from France with his family, the *Directeur Général* divided the line of the canal into three divisions, each under the control of a single French engineer. The first covered Limón Bay and lower reaches of the Chagres; the second, the Upper Chagres and the hills between Matachín and Culebra; the third from Culebra to the approaches to the Bay of Panama.

The key contractor in the First Division was the American outfit Huerne, Slaven & Company, who had been taken on back in February 1882. The driving force of the company was H. B. Slaven, a Canadian-born drugstore owner from San Francisco. He had known nothing about excavation but, 'determined to have a finger in the canal pie', had, as Tracy Robinson put it, 'with an audacity akin to inspiration' put in, and had accepted, a bid to take on some of the work. Raising capital from a New York banker, he ordered the building of a series of huge, custom-made dredges. Although work on his section was supposed to have started in August 1882, it was not until April the following year that the first of these monsters had been completed in Philadelphia and, with great difficulty, towed to the Isthmus. The dredges were 36 metres long and 9 metres wide, square at both ends, with the seagoing capabilities of a large house. The first to arrive was destroyed by fire in Colón harbour, but the next – the *Comte de Lesseps* – arrived during the summer and was laboriously fitted out. A 20-metre wooden tower was assembled in the centre with a huge wheel on the top. Attached to this was a chain with a series of large steel buckets on a boom that was lowered over the side. Each of the buckets scooped up a cubic metre of soil from the bottom of the channel and hoisted it to the top of the tower, where it was emptied by jets of water into large pipes that extended 55 metres on either side, dumping the spoil clear of the working site. The whole contraption moved on huge legs, or 'spuds', 'by means of which,' says Robinson, 'she walked step by step into the material to be excavated'. The *Comte de Lesseps* started inland from the mudflats of Limón Bay in October 1883. Able to extract 5000 cubic metres a day, it made a huge impression. It seemed just the sort of fantastic machine that de Lesseps had promised would miraculously turn up.

In January 1884, an American naval officer, sent to estimate progress on the canal, reported back to Washington that 'this powerful and excellent machine' had already 'dug a passage 1075 metres long, 34 metres wide, and 4 metres deep; so that vessels of 13 feet draught may now, at this point, pass up the line of the canal for a distance of half a mile'. Another monster dredge arrived during the year, and in October, the *Star and Herald* reported that Colón and Gatún were soon to be in connection by water. 'So the visitor to Gatún . . . can easily satisfy himself that there is something serious and practical in this canal enterprise.'

Further up the line, work was undertaken on diversionary channels for the Chagres and digging up the alluvial silt of the river valley. This was managed by a Franco-Dutch firm, Artigue et Sonderegger, which had twenty smaller Belgian-made ladder dredges at work.

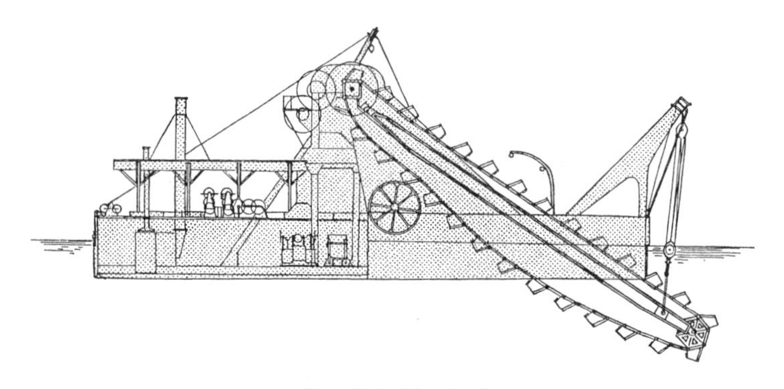

French ladder dredge

On the Pacific side of the Isthmus, much deepening work was needed in the Bay of Panama. This was started out by self-propelled dredges built by Lobnitz and Company of Renfrew, Scotland, which were sailed to Panama under their own steam. The first arrived in May 1883, having covered the five-thousand-mile journey from the Clyde in only a month. Such was the quality of their design and manufacture that they were kept on by the canal builders right up to 1914. Meanwhile, lighter dredges pushed up the valley of the Río Grande. It was steady progress, if less spectacular than that at Colón.

Inland, in *la grande tranchée*, it was a different story, with much more labour-intensive work being required. By the end of 1883 pilot trenches had been run the ten-mile length of the Continental Divide and a contract had been signed with Cutbill, de Longo, Watson and Van Hattum – usually referred to as the Anglo-Dutch Company. Work of excavation commenced with hand-pick and shovel, and the soil was removed in small iron cars, running on portable tramways. Once the trench was a few feet deep, a track was laid connecting with the main line of the Panama Railroad, and upon this steam excavators were brought up mounted on trucks. Most were US-built by Osgood or Otis. These machines commenced digging down in a series of stepped terraces, each about 5 metres wide and 5 metres deep, which was how far the excavators could reach. The spoil was loaded on to flat cars then dumped in a convenient nearby valley. 'From morning till night,' reported an American visitor, 'trains are moving about removing the excavations of the laborers and of the excavating machines, which latter do their work very well and very cheaply.'

Nonetheless, there remained a lot of hand excavation, and it was here that the majority of the canal labourers were concentrated. In mid-1883, the British consul reported some 1200 workers in this small space, mainly Jamaicans, Martiniquans and Italians. 'The Anglo-Saxon element prevails a great deal here in the way of officials and clerks,' he went on. The steam shovel operators and mechanics were also British or American.

So as well as ordering masses of new machinery, Dingler also determined to increase the number of labourers available to the contractors. There were ten thousand by September 1883, fifteen thousand by January the following year, and by the end of 1884, there were over twenty thousand on the payroll, making a total wage bill of some $40,000 a day. The majority of the workers were from Jamaica.

There was good money to be made on the Isthmus. Most workers earned about $1.50 a day, and as the pay was calculated on a piecework basis, extra labour would bring even more. This was a big deal for impoverished inhabitants of the West Indian islands, particularly in Jamaica. Soon large steamships were making the run from Kingston to Colón as frequently as every four days. Nonetheless, there were near riots at the docks as people fought each other to get a place on a ship. 'A stampede took place which is hardly possible to describe,' reported the Jamaican paper *Gall's News Letter* in early 1884 of the scene at Kingston docks. 'Men with trunks on their backs, women with little children tugging through the crowd, all trying to gain admission to the ship. In a few minutes the deck was crowded.'

Money being sent home to relations and the return of men from the Isthmus with their fortunes visibly transformed further fuelled Panama Fever on the island. 'Now and again,' complained a planter, 'you see a great swell with a watch and gold chain, a revolver pistol, red sash, big boots up to his knees, who swaggers about.' Such returnees soon became known as 'Colón Man', an almost mythical figure on the island, subject of many skits and verses, both admiring and gently ridiculing. 'One two three four/Colón Man a come', goes one. 'With him watch chain a knock him belly bam bam bam/Ask him for the time/and he look upon the sun/With him watch chain a knock him belly bam bam bam.'

As the 'Panama Craze' spread, even those few with stable and secure jobs joined the rush to Colón. 'The infatuation to go seems to have taken hold on the whole of them who are able to go,' reported the *Gleaner*. During the French construction period, some 84,000 made the journey to Colón from Kingston, at a time when the entire population of the island was under 600,000. Whole areas of the island became depopulated, and the demographics of those left behind was radically altered, leading to a decline in marriage and other unions, and a taking up by women and children of the work previously done by men. The birth rate fell, and women and became heads of families as never before. Children were frequently left to fend for themselves, as families were split up, often forever.

At the same time, the money coming back from Panama was invested in land, livestock and housing, leading to rising peasant proprietorship and ownership of goods, appliances and tools, in all contributing to an economic revival on the island at the end of the century. 'Colón Man' also brought home a new, less subservient attitude, his flash dress and accessories 'a flag of liberation'.

Official reaction to these huge changes was mixed. 'We are not of those who think it a calamity that so many of our people are going to the great Canal Work,' wrote one of the mouthpieces of the ruling plantocracy, the *Jamaica Witness*, in early 1884. 'It is a great enterprise this, the joining of two great oceans . . . We wish it God speed; and we feel rather proud that we can supply so large a portion of those who are to perform that necessary toil . . . Let them go by all means. It will give many a lesson in labour which they never had before – a hard day's work for good wages. Many will gain money, and return to acquire property; and they will be more men that they ever were, men who have felt their manhood, and who will more than ever prize their position and privileges as citizens of a great Empire.' But as early as mid-1883, Jamaican plantation interests in London were lobbying the Secretary of State for the Colonies to restrict 'the great outflow from the Colony of labourers, artisans, and respectable young men, who are properly described as the bone, sinew and hope of the country'.

The planters spread stories about the terrible risks of disease in Panama, and that the Company ran a brutal regime, but were unable to stem the tide. The *Witness* explains: 'the dangers and drawbacks of life on the Isthmus are counterbalanced by material advantages of an appreciable character'. As one of the West Indian work songs has it:

> Kill my partner
> Kill my partner
> Kill my partner
> Somebody's dying every day.
> I love you yes I do, you know it's true,
> And when you come to Panama how happy you will be,
> 'Cause money down in Panama like apples on a tree.

It was not just canal workers making the journey to what soon became known as 'New Jamaica'. There were also doctors, pastors, teachers, photographers, newspaper editors, railway and docks workers, clerks, domestic servants, small farmers, petty traders, 'scufflers' and criminals. Obeahmen and 'bush doctors' were much in demand to service the expatriates. The 1880s were the heyday of visionaries and healers such as 'The Prophet of Haddo', 'Shakespeare', and 'the Prophet Bedward of August Town'. Jamaican entertainers frequently went to Colón. 'Funny Murray', whose repertoire seems to have consisted of skits and verses in Jamaican dialect, performed frequently on the Isthmus and at the same time gathered material there for new skits such as 'Uncle Dober in Colón', which he put on in Jamaica. There were already well-established links between Colón and Jamaica, but these were further strengthened. Panama papers carried news from Jamaica including the results of the Kingston races. Jamaican washerwomen regularly travelled back and forth taking dirty linen from Panama back to the island for laundry.

On the Isthmus, the West Indians found themselves doubly vulnerable as foreigners working for a foreign company. There was no labour law in Colombia except for the 'freedom to work', and the workers' backgrounds mitigated against concerted labour action. Churches, community burial clubs and mutual support organisations were established, but there was little organisation in the workplace. Eight or nine strikes broke out on the Isthmus during the 1880s, yet these tended to be confined to the port or railway workers, whose stoppage could cause serious delays and backlogs for the steamer companies. The employers, however, found the Colombian authorities sympathetic to their side, and several strikes were broken up by soldiers. More than anything, though, the flood of new arrivals meant that there were always men to replace those who tried to improve their wages or conditions. Instead, to the frustration of the Company, the men would simply move to another area in order to get better pay, or to be with their friends. 'They have a way of shifting for themselves,' the *Star and Herald* reported in early 1884, 'and selecting their own masters and places of work . . . go[ing] about from section to section wherever they can best suit their own particular ideas as to wages and other circumstances.'

Most of their grievances were brought to the door of the British consulate, much of whose work involved interceding with the Company or the Colombian authorities on behalf of British subjects, or doling out emergency relief to those 'distressed'. Hugh Mallet had left for a post in Argentina in 1879, to be followed by a succession of short-lived appointees. For most of the decade, however, the job was carried out by Hugh's son Claude Mallet. Having arrived in Panama in 1877, aged seventeen, in January 1884 he became acting consul in Panama City at only twenty-four years old. He immediately got into a dispute with the long-serving vice-consul in Colón, Charles Crompton, whom he accused, in a letter to the Foreign Office, of drunkenness and racism. 'There have been many serious complaints,' he wrote, 'about the manner in which he treats those who seek at his hands assistance and advice, and redress for wrong done by the authorities, to British subjects. This is particularly true in the case of the Jamaica people, against whom he seems to have a most unreasoning and absurd prejudice, and as they compose the majority of the British colony in Colón it may be imagined what sort of reception the poor people receive at his hands . . . also, as frequently occurs, he yields to intemperate habits during office hours . . . complaints are made that whilst in this condition he has met applications for counsel and help with imprecations and blasphemy disgraceful to the position he holds.' Furthermore, the vice-consul had taken a job with the Canal Company, so had a divided interest as 'questions arising between the Company and the West Indian labourers are frequent and important'. The Foreign Office objected to the young upstart taking such a 'dictatorial tone' with such a long-serving diplomat, but when the charges were investigated on the arrival of a new consul, Mallet was found to be 'quite correct' and Crompton was dismissed.

Mallet had few resources and was stretched to the limit by the massive increase in workload that the huge immigration of British citizens brought, but carried out his work – much of which was unpaid – with determination and a lack of prejudice that was rare at the time. 'The practice in this consulate,' he wrote, 'has been to treat her Majesty's subjects . . . on precisely the same footing, whether they be white or black, or to whatever portion of Her Majesty's Dominions they may belong.' Three-quarters of his work was taken up with Jamaicans, and on several occasions he was officially thanked by the authorities in Kingston for help given to their nationals.

Some of the most demanding parts of his workload concerned the consequences of the great influx of 'aliens' into Panama's sensitive racial and social mix. The arrivals from Jamaica, who came from a radically different cultural background to Panama or greater Colombia, were seen as a threat and deeply resented. Antagonism was exacerbated by the hardships of the construction camp setting. As the number of Jamaicans grew, so too did instances of disturbances and bloodshed along the line of the canal. In March 1883, five Jamaicans and three Colombians were killed in fighting at Culebra, Obispo and Matachín, and the British consul requested armed protection for British subjects. But the attitude of the soldiery was just as antagonistic as that of the Colombian workers: 'I perceive that these men are partial in their protection and disorderly and brutal in their conduct,' the consul wrote.

Tension now simmered all along the line. 'Both sides are armed,' said the *Star and Herald* on 3 April 1883, 'both prepared, and alike expectant . . . the works between Matachin and Gorgona . . . are practically deserted.' More by accident than design – Colombians started refusing to work with Jamaicans – the two warring parties now lived and worked in separate areas, but friction was still building through the following year, threatening to spill out into open armed conflict.

For the moment, however, these labour troubles were not the greatest of Dingler's worries. In spite of the progress since he had taken over, the project was now beset with very serious difficulties. The harbour and warehouses at Colón could not cope with the volume of machinery and supplies being imported. Often, steamers would have to be unloaded by lighters, 'a ruinously expensive method,' wrote a visiting reporter from the New York *World*. 'The cost of coal is increased two thirds. It is worth £3 per ton in the harbour and £5 when landed.' Furthermore, the railway, in spite of a great expansion of rolling stock, did not have the capacity to move machinery inland as fast as it was arriving. So valuable equipment was left out on the waterfront at the mercy of the climate. 'Damage amounting to thousands of dollars daily is known to be going on,' reported a visiting journalist. The machinery itself came from many quarters – France, the United States, Belgium and elsewhere. It was constantly being modified and used in experimental, often ingenious, combinations, but much of it was found to be unequal to the task. A growing accumulation of discarded, inoperative equipment along the canal line testified to earlier mistakes. A later American engineer, piecing through the abandoned French plant, would describe it as 'of a character and complexity to defy description . . . some parts could only be classed as freaks. Apparently every crank who possessed influence was allowed to exploit his notions in the furnishing of machines to the company.' The experimentation and increasing number of small contractors combined to produce a host of contradictory specifications for spare parts, railroad track and truck gauge. At one time there were eleven different types of flatcar running on six different gauges. It was as far from a 'joined-up system' as can be imagined.

All the time, the Company was haemorrhaging money through a combination of mismanagement, extravagance and corruption. More than a hundred racehorses were imported from Europe and lavishly stabled at the Company's expense. There was also widespread pilfering. Inspectors sent to check on the contractors' excavation quantities were often bribed. One estimate was that the Company lost some 10 per cent of the work it paid for. The worst offenders were the Huerne, Slaven men, who were even accused of dredging soil from one side of their barges, for which they were paid, and then simply dumping it back into the water on the other side. The workers, also, became adept at exploiting the Company. 'There was no system or organisation,' reports a Nicaragua-born canal workman. 'A man can work on five different jobs a day, and when the week ended you collect for all five jobs. Their timekeeping system was poor.' The local Panamanians were also making a lot of money out of the Company, charging exorbitant rates for land the French needed, or bringing endless expensive legal cases against them. Like Reclus, Jules Dingler and his wife had to entertain a constant stream of visitors from Europe, Colombia and the United States during the dry season, and, in urgent need of an adequate house in which to entertain them, found no one willing to build something for less than $100,000. There was similar mass collusion over food supplies. Traders would board incoming ships carrying provisions, buy the entire cargo, and then fix the price of the goods on the Panama market.

As well as administrative difficulties, engineering problems were now beginning to pile up. In *la grande tranchée*, the rainy season brought continual slides which buried rails and machinery under thousands of cubic metres of sticky mud. The contractors in this section had to keep cutting the slopes back to flatten them, creating seemingly endless amounts of extra work. By the end of 1884 it had been decided that the gradient would have to be as gentle as one in four. This would have made the trench, had it been dug to sea level as planned, as much as three-quarters of a mile across in several places.

The spoil was removed in small dump cars to a convenient nearby valley, where a track would be laid on the brow of the hill. The cars were then tipped or laboriously emptied by hand, with the dirt thrown over the side. When a terrace had been formed, new track was laid on it and the process repeated. But the dump areas also became unstable, with terraces slipping away, destroying track and trains and leading to the whole system breaking down and the excavators lying idle.

'The rainy season, at last set in, is making up for lost time by pouring down oceans of water all over the Isthmus,' the *Star and Herald* reported at the end of May 1883. 'The effect on new embankments, fills &c., made by the Canal Company during the dry season, is not pleasant to contemplate. The work of months disappears in a day.' Almost exactly a year later, it was the same story: 'The heavy downpours of late are making short work of earth cuttings . . . A few hours of tropical rain caused the mighty Chagres to rise three feet. When it subsided the cut was found to be filled to within three feet of the top. The work of many days costing a great deal of money has disappeared as if by magic . . . This Chagres question is a mighty one.'

As had been anticipated back at the 1879 Congress in Paris, the problem of the Chagres was indeed among the most formidable faced by the French engineers. Dingler, in his grand plan, had stuck with the suggestion that to stop the river flooding the canal, a huge earthen dam be constructed at Gamboa, with another smaller dam twenty-five kilometres upriver. But this filled no one with confidence. There was no adequate rock formation at this site upon which to found such an enormous structure, and few believed that it would hold the pressure of the river at its most swollen. In addition, the basin behind the dam, in which it was hoped up to 6000 cubic metres of water would be held, had still not been adequately surveyed.

It was planned that the remaining flow of the river would be drained by a series of diversions running parallel to the line of the canal. But as the canal ran along the lowest points of the river valleys, the surface water of these diversions would be about 70 feet above that of the canal proper, requiring very strong guard banks. In effect, as a critic of the French plan pointed out, 'the water will have to be hung up on the sides of the mountains'. And just one of these channels would have to be thirteen miles long, with similar dimensions to the main canal. It was as if the canal had to be constructed two or three times. In all, every time the French engineers turned around, the task ahead seemed to have grown exponentially.

In all great construction projects the greatest cause of delay and financial loss – and the reason that considerable slack is worked into budgets – is generally termed 'unforeseen ground conditions'. Panama had these in spades. 'Fresh engineering difficulties present themselves,' wrote a British visitor in late 1883, 'and the magnitude of the work to be accomplished seems to increase.' It became clearer and clearer that, right from the start, de Lesseps had totally underestimated the task he was setting for his engineers. And as the problems mounted, in order to maintain the confidence of investors and sell new bond issues, de

Lesseps kept up a stream of promises and impossibly high targets. In June 1882 he had told the Company's annual general meeting that 5 million cubic metres would be excavated from Culebra in the next twelve months; the figure achieved in that period was only 660,000. The following year he reaffirmed his promise that the canal would open in 1888, and predicted a monthly excavation figure overall of 2 million cubic metres. But the force on the Isthmus was not exceeding a quarter of that. The more expectations were raised, the further de Lesseps had to fall. 'A day of reckoning is coming,' wrote the Montreal *Gazette* in August 1884.

Although the press in France remained on side, elsewhere criticism of the project, and of the exaggerated reports of the *Bulletin*, was mounting. In August 1884, the American *Engineer* magazine printed the report of a correspondent who had spent two months on the Isthmus, who estimated that it would need another twenty-four years and hundreds of millions of dollars more to finish the canal at the current rate of excavation, an analysis with which several US naval officers, sent to investigate progress, concurred. It was also now openly stated in Britain and the United States that the French press had been bribed to hide the truth from domestic investors.

In August 1884, the Montreal *Gazette*'s correspondent, back in Panama having visited the works six months earlier, reported that 'little substantial progress has been made . . . valuable plant remains unhoused, including locomotives, boilers &c'. The fault, as far as he was concerned, was with misplaced priorities: 'Time that might be used in building proper sheds is frittered away embellishing banks near houses, setting out tropical trees and plants to make the landscape attractive. '*C'est magnifique*,' he concluded, '*mais ce n'est pas le canal*.' In November the New York *Herald*, usually a fair judge of the project, accurately predicted, 'It is probable the present company will go into bankruptcy or liquidation within three years and the enterprise be taken up and completed by a new company or a government.'

On the Isthmus itself it was felt that real progress was being made, but the huge expenditure of capital had not gone unnoticed. Among many of the canal employees an air of heroic unreality had descended, as if infected by de Lesseps' fantastical pronouncements from Paris. One of the American 'inspectors' remarked on the 'tendency on the part of the canal officers to exaggerate everything that had been done by the company'. Others, though, testified that several of the Company's managers were privately saying that the project was catastrophically behind schedule. In July 1884, Claude Mallet reported back to London, 'It is generally believed here that the present Company can never finish the work, as the cost so far has greatly exceeded expectations.' For him, such a project could never be completed by private capital. Only a government could carry through such a task. The *Star and Herald* agreed, in the same month goading the US government to take on the job: 'It would be a pity,' it wrote, 'that a work such as this should be left partially completed as a monument of the folly and gullibility of Capital.' A government had to step in, and 'it would be well for Americans to remember that the government of France would have the most powerful motives to undertake it. There would be the natural desire to prevent the loss of French Capital, and the price of control and influence abroad is not a forgotten sentiment in France.'

A British observer, Admiral Bedford Pim, blamed the weaknesses of the original Wyse–Reclus surveys for the problems, and dismissed the sea-level plan as impossible. His only praise, after an extensive tour of the work, was for 'the gallant employees who have struggled manfully to carry out the wishes of their chief'. The New York *Herald*, reporting the pessimistic analysis of the latest US naval investigation, commented, 'Under such circumstances, there is something amounting to heroism in the persistence of the French Company at Panama.'

The French engineers and their workers were now facing more than just financial or engineering difficulties. Their very lives were at stake. In 1882, 126 people had died in the hospitals, mainly from yellow fever or malaria. The following year, as the workforce tripled, so did the number of deaths, to over four hundred.

But the official tally did not tell the whole story. Many died before they reached hospital. According to a lurid account by a New York *Tribune* correspondent, a number of workers ended up in unmarked graves. 'Death becomes a grim joke, burial a travesty,' he reported. 'An unconsidered laborer is buried under a hundred feet of earth – and very simply; rolled down an embankment, and twenty carloads of earth rolled after him.' Although the Company itself covered hospital expenses for its employees, the vast majority were on the payroll of contractors, who were charged a dollar a day for the care of their workers in hospital. It was alleged that some simply dismissed their men at the first sign of sickness rather than foot the bill. In addition, the hospitals themselves were feared and shunned, and with good reason. If you did not have malaria or yellow fever when you went in, you were likely to have it soon afterwards. In the absence of knowledge of the transmission of these diseases by mosquito there were no efforts to isolate known fever victims or to keep the insects from the wards. Furthermore, to protect the hospitals' much-admired gardens from leaf-eating ants, waterways had been constructed around the flowerbeds. Inside the hospital itself, water pans were placed under bedposts to keep off ants and other crawling pests. Both insect-fighting methods provided excellent and convenient breeding sites for mosquitoes carrying yellow fever or malaria. Doctor William Gorgas, who led the medical effort on the Isthmus during the American construction period, later wrote,

'Probably if the French had been trying to propagate Yellow Fever, they could not have provided conditions better adapted to the purpose.'

Gorgas reckoned that for every death in one of the hospitals, two occurred outside, which would put the 1883 toll at nearer 1300 than 419. This is conjecture, of course, and should be treated with caution. Nevertheless, there is fairly overwhelming anecdotal evidence to back up this claim. Charles Wilson, a half-Scottish, half-French sailor, was twenty-one when he arrived on the Isthmus in 1882 and started work for the Panama Railroad. Wilson was what became known as a 'tropical tramp'. He was born on board a ship and never belonged in any place except where there was money to be made. He found that working on the Isthmus earned him far more than the $20 a month he had been getting as a sailor. But there were, he says, 'thousands dying with yellow fever, malaria, and all kinds of diseases . . . everywhere, in the streets and under houses'. 'As for the men,' reported the Montreal *Gazette*, 'they die on the line and are buried, and no attention is paid to the matter. Two American carpenters are in an unnamed grave near Emperador . . .' Wolfred Nelson, the Canadian doctor, remembered an endless procession of funeral trains, reckoning that during the wet seasons of 1882 and 1883 'burials averaged from thirty to forty per day, and that for weeks together'.

The correspondence of the British consulate also draws a picture of illness growing into unmanageable proportions. The staff themselves were forever sickening and pleading to be allowed to leave the Isthmus to recover in Jamaica or back in England. Consul Edward March was invalided home in April 1882 after just a month in Panama, and six weeks later his replacement, Courtenay Bennett, reported that he had malaria, or 'miasmic affections' as his doctor described it. By June the following year his replacement was also ill and had to leave the Isthmus. The melancholy pattern, which was shared by all the other European consulships, was repeated for the rest of the 1880s, which meant that the young Claude Mallet was acting consul for much of the decade. Mallet reports a huge increase in the consul's workload as a result of having to sort out and return the effects of dead Jamaican labourers. He was also called on to help those abandoned by their employers. In September 1883 a Jamaican 'arrived in Panama by the afternoon train in a dying condition,' the *Star and Herald* reported. 'He had been put on board at one of the stations on the line; and, on arriving here, there was no one to receive or attend to him. He was incapable of thinking or acting for himself, and when he endeavoured to move, convulsions would seize him, during which he was restrained by force from injuring himself. Soon, however, these paroxysms passed, and he sank to the ground in the exhaustion of death. Mr Mallet, H.B.M.'s Acting Consul, was informed of the case and went to the station, where he found the dying man, had him placed in a carriage, and took him to the hospital.' On another occasion a thirty-two-year-old Liverpudlian,

Richard Oliver, was, Mallet reported 'found lying before the door of her Majesty's consulate in a dying state. I sent him to the hospital where he died a few days afterwards.' Often Mallet would have to pay for the care out of his own pocket and take his chances that a tight-fisted Foreign Office would refund the expense.

At the beginning of 1884 it was hoped that the advent of the dry season would again reduce the cases of fever, but it was clear that Panama was now in the grip of a major epidemic. Carelessly discarded spoil from the works had blocked water-courses and created permanent stagnant pools – ideal mosquito-breeding grounds – all over the Isthmus. On 21 January the *Directeur Général* Jules Dingler's daughter Louise, a pretty, dark-haired girl of about eighteen, died, in miserable agony, of yellow fever. The family was beyond grief. 'My poor husband is in a despair that is painful to see,' Madame Dingler wrote to Charles de Lesseps in Paris. 'My first desire was to flee as fast as possible and carry far from this murderous country those who are left to me. But my husband is a man of duty and tries to make me understand that his honour is to the trust you have placed in him and that he cannot fail in his task without failing himself. Our dear daughter was our pride and joy.' The day after her death Dingler was back at work at the usual time.

Louise Dingler's funeral was held in the cathedral. 'Seldom, if ever, during many years residence in Panama, have we witnessed such an affected and sympathetic assemblage as that which gathered yesterday,' the *Star and Herald* reported the next day. During her short time in Panama, the paper continued, 'no one had achieved higher social triumphs, or appeared destined to enjoy an Isthmian life so thoroughly as Miss Dingler, and seldom has such a thrill of regret passed through the community as that which was observable when her decease was announced'.

It did not stop there. A month later Dingler's young son, Jules, sickened with the same disease and died three days later. The *Star and Herald* reflected the horror and grief of the whole community: 'Mr Dingler was but 20 years of age, the picture of physical health and strength . . . Sympathy is weak, and words are powerless in such a cruel blow, to convey to the grief stricken parents the sense of loss and sorrow which these sad events have occasioned in the minds and hearts of all.' Soon after, Louise Dingler's young fiancé, who had come with the family from France, contracted the disease and also died.

The high-profile deaths in early 1884 caused a panic akin to that following the deaths of Etienne and Bionne back in 1881. Some three hundred French engineers applied to return home, which was refused. Nonetheless, men deserted from all parts of the line. For others, however, the spectre of death served to raise their work to higher, sublime levels of heroism. They were sacrificing themselves, one young engineer wrote, 'as surely as those who fought at Lodi or Marengo laid down their lives for France'. Although the teachers in the engineering schools in France were now quietly trying to deter their pupils from serving in Panama, idealistic young Frenchmen continued to journey to the Isthmus. One such was twenty-five-year-old Philippe Bunau-Varilla, who arrived that year and would become a key figure in the history of the canal. He had met Ferdinand de Lesseps in 1880 and ever since had been in the grip of the Great Idea of the canal. He was the apogee of graduates from the elite Ecole Polytechnique, France's top engineering college, where military uniforms were worn and the motto was '*La Patrie, les Sciences, et la Gloire*'. For Bunau-Varilla, the canal was 'the greatest conception the world has ever seen of the French genius'. 'The constant dangers' of yellow fever, he wrote in his autobiography, 'exalted the energy of those who were filled with a sincere love for the great task undertaken. To its irradiating influence was joined the heroic joy of self-sacrifice for the greatness of France.'

As Dingler pushed ahead with the work during the rest of 1884, achieving, in the circumstances, great progress, the death toll kept rising. In July, an outbreak of dysentery struck in Panama and Colón, filling the hospitals, while in the flood plains of the valleys, ideal breeding ground for mosquitoes, six out of seven Europeans working there during mid-1884 had contracted malaria or yellow fever. By the late summer of that year work was almost impossible in these sectors. One engineer told an American doctor that he had come over with a party of seventeen young French engineers. In a month all but one had died of yellow fever. At any one time, more than a third of the labourers were sick, making a mockery of Dingler's strenuous efforts to increase the workforce.

Shipping arriving in Colón harbour was also at risk. In September 1884, an outbreak of yellow fever saw 170 cases among steamer crews, with a mortality rate of over two-thirds. In October, a British bark, *William Fisher*, bringing coal to Colón, lost its captain, John Rigg, to 'Chagres Fever'. On the return voyage to Liverpool, half the crew then perished of the same disease.

The following month a floating hospital was set up in Colón harbour to deal with the overflow from the hospital, where the nurses, the Sisters under Marie Roulon, were also dropping like flies. By the end of the year all but three of the original twenty-four nurses were dead, and the annual figure for mortalities in the hospitals alone had tripled from the year before to nearly 1200. If Gorgas's estimate of the true level of fatalities is to be believed, this meant that more than ten people were dying every single day.

Many on the Isthmus put these deaths from yellow fever down to the 'abominable neglect of all sanitary measures' in the terminal cities. Where the French had jurisdiction, such as in Cristóbal, the new town built on reclaimed land to the side of Manzanillo Island, or in the settlements along the line, it was another story, but in Panama City and Colón the French were powerless. According to Charles Wilson, in Colón, 'There was green water, and all kinds of rubbish and rotting things in the center of the streets.' Certainly the lack of sewers and clean water would have contributed to the problems with dysentery, but, of course, they had nothing to do with yellow fever.

Others still subscribed to Dingler's original theory that 'only the drunk and the dissipated die of Yellow Fever'. The *Star and Herald* put the yellow fever deaths in Colón down to the 'host of idle loafers, who infect the town and load the air with their obscene and vulgar epithets at every hour of the day'. In April 1883 Wolfred Nelson was taken ill, but recovered, he says, 'thanks to abstemious habits'. 'Woe to the feeble person who doesn't know how to quench his thirst!' wrote a senior French engineer. 'He falls into drunkenness, and soon, aged, faded, with haggard eyes, shrunken, face drawn, yellow-skinned, he drags his broken spirit along in a body lacking in all vigour. And certainly he deserves it. It is fair that this shameful vice should be severely punished by nature.'

There was certainly no shortage of 'vice' in the terminal cities and along the canal line. In Panama, there were numerous bars, 'designed for nothing but hasty drinking', according to a French engineer, '. . . horrible dens that look and smell like the filthiest grog shops'. Colón, which retained its frontier-town atmosphere, and where the bars would always stay open later than in Panama City, was the worst, 'a veritable sink of iniquity', according to an American visitor. The single main street, running along the waterfront, he reported, 'was composed almost entirely of places for gambling, drinking and accompanying vices . . . and these diversions were in full progress day and night with such abandon as to make the town uninhabitable for decent persons'. Claude Mallet described the town as 'the hardest drinking and the most immoral place I have ever known'.

According to Tracy Robinson it was the fault of the 'spirit of venality and corruption [that] pervaded almost the entire French Company', which 'spread beyond the service itself, to debauch (there is no other word for it) the whole Isthmian community'. Wolfred Nelson blamed the myth that any human being in hot climates requires alcohol. 'Another point in this connection,' he continues. 'There is a general belief held by many intelligent people that a residence within hot countries has a marked tendency to increase the sexual instincts. Such is not the case. The real explanation is this. The majority are away from the refining influences of early culture and home life, – generally they are single men, – in a warm climate where all the conditions are supposed to produce general relaxation. There is a little society open to such men. If they become "one of the boys," – and the vast majority do, that is the end of it, and generally of them too, for this means late hours, gambling and other distractions, largely *pour passer le temps*. Such men readily become victims to disease.'

Doubtless, huge quantities of wine, champagne and *anisado* were consumed during the French years. On one road in Colón, where bars and prostitutes vied for trade, so many bottles were thrown out into the street that in time they formed a solid surface beneath the mud and when pavement laying crews arrived years later there was no need for them to put down a gravel base. Supplies were imported from France and sold at wholesale prices to Company employees, who would then sell them on to their friends. Claude Mallet reported that he 'could get claret for 6d a bottle . . . wines were so cheap that there was a habit of starting the day with a pint of champagne frappé "to kill the microbes"'. There was, of course, a sensible reason for this, as Mallet explains: 'No one dared drink the water that was sold in small measures from barrels.' In addition, as Nelson says, there were few other ways *pour passer le temps*.

In mid-1883 de Lesseps announced that he had ordered the opening of 'assembly rooms, provided with books, periodicals, and various indoor games' where employees could gather, but there is no evidence that this actually happened. Instead, in Panama City, employees had a choice between the 'horrible' local establishments or the Grand Hotel. A young French engineer describes the scene there: 'A great enormous hall with a stone floor was the bar-room.' In the centre were two huge billiard tables and beyond them a vast bar. 'In front of these rows of bottles with many coloured labels, most of the commercial business of Panama is transacted; – standing and imbibing cocktails, – always the eternal cocktail!' Across the hall was a little room crowded with people, 'where roulette was going on'. After each throw the croupier 'announced the number in three languages: "Treinta y seis, colorado! Thirty-six red! Trente-six, rouge!" Oh, this roulette, how much it has cost all grades of canal employees!'

'It was useless,' he goes on, 'to look for other pleasures. They were nowhere to be found. In this town there was neither theatre, concert no café, nothing but the hall of the Grand Hotel, to which one must always return.'

Furthermore, as one French visitor explained, 'passions run high owing to the constant proximity of death.' Another wrote that 'the Sword of Damocles hangs over everyone'. The threat of fever, he said, explained the fever of gambling that gripped the Isthmus. 'In this country,' wrote Henri Cermoise, 'death and *la fête* are perpetually hand in hand. Yellow fever threatens always and one is so unsure of tomorrow that one throws oneself into pleasures.'

The fast, frontier-town lifestyle contributed to a general lawlessness on the Isthmus. In addition, the money being poured into the great ditch had attracted to Panama numerous 'foreign men of dubious reputation', in effect desperate and vicious characters from all over Central America. The *Star and Herald* from this time is full of accounts of robberies and murders. In May 1883, for example, a man was stabbed to death during an argument, 'which as usual was occasioned by the vile rum which is sold so freely at all points on the line . . . Human life,' the paper concluded, 'is held at too cheap a rate on the Isthmus.' The situation worsened as the West Indians started arming themselves with revolvers to defend against machete attacks from their Colombian enemies. As always, Colón fared the worst, where the people were 'an agglomeration of all nations, and tribes and tongues, drawn from all lands and swayed by a thousand sentiments and impulses'. During the rainy season, when the work on the canal fell off, there were hundreds of unemployed in the town as the steamers continued to arrive from Kingston, and 'Fighting, drunkenness and the like are of everyday occurrences.' Charles Wilson was living in Colón's Washington Hotel. 'There were all kinds of people living in the town, and some of the worst kinds,' he writes. 'When you took a trip into it at night it was a question whether you would come out alive or dead.'

During 1884 there was widespread political instability in the province, exacerbated by the high inflation, food shortages and the general social unrest the canal project was causing. At one point, in October, there were two rival State Presidents, each with men under arms in Panama City. The following year, this would lead to full-blown civil war on the Isthmus, with serious consequences. Under such circumstances, the authorities were utterly incapable of policing the volatile streets of either of the main cities. Europeans and Americans increasingly looked to the foreign warships, frequently anchored in the bays of the terminal cities, for their protection. The small police force was ineffective and partisan, preferring to extort money from West Indians on the pretext of fines for vagrancy than solve any crimes. Dingler offered to contribute money to the establishment of a new three-hundred-strong police force, but this foundered on local objections.

There was to be one more personal tragedy for the *Directeur Général*. Around Christmas 1884 Madame Dingler started showing the unmistakable symptoms of the dreaded yellow fever. She died on 1 January 1885, completing the total destruction of his family. At last, Dingler's heroic steadfastness was beaten. His wife had frequently gone riding on one of two magnificent horses worth 25,000 francs, which had been a gift from Gadpaille in Jamaica. After her death, the director did not wish to encounter anyone else on the streets of Panama riding these horses, so he ordered the beasts to be killed. The staff refused to carry out the command. Finally they found a poor fellow who was given the role of executioner, but at the last moment his hand trembled and he could not finish the job. For hours the horses were heard, partially disembowelled, screaming in agony. In the end they were shot dead. This execution figures in the accounts of the Company and is billed as a hole of 33 metres paid at fifty piastres per cubic metre. Dingler was a broken man, deeply pitied by the canal workforce. He stayed on the job for another six months, but his wife's death ended his dynamic leadership of the canal project. He returned to France in June 1885 and, exhausted and heart-broken, was himself dead before the end of the year.

13

ANNUS HORRIBILIS

At least once a year US naval officers would tour the canal works and then write lengthy and detailed reports back to Congress on the progress, or lack of it, of the French project. Sometimes these can read as if the author had somehow become infected by the enthusiasm of their always generous hosts, or just by the sheer ambition of it all. But for the most part, with a crescendo as the years go past, they concluded that the current sea-level plan was running disastrously behind time and over budget. In the United States, de Lesseps' 'friends' – many, officially or not, on the payroll of the Company – continued to defend the canal against its attackers. Elsewhere, factories and workshops were kept busy supplying Panama's endless thirst for new and bigger equipment. But many shared the *New York Times*' view that the chance of the project ending in failure was 'not unlikely'. In the circumstances, commented the paper with not a little *Schadenfreude*, 'we can congratulate ourselves that it is chiefly foreign capital that will be swallowed up by it'.

Occasionally, a different voice is heard. 'Americans would much prefer that the American canal should be the work of Americans,' wrote a Commander Gorringe in the New York *Sun*. 'Evidently Americans had neither the courage nor the means to undertake it. The Frenchmen had; they have gone quietly to work; they ask us for nothing'. What's more, he continued, all their efforts would assist no country more than the United States, 'quite as much as British commerce and the British commercial marine were benefited by their work at Suez'. Not that respect for the French travails should hamper strategic good sense. It did not matter who did the work, he argued, because, 'When it is completed, if it becomes necessary or even important to our national welfare and safety that we should control it, there is no doubt that we shall take possession of the canal and the country through which it passes, with as little hesitation and trouble as the British recently took possession of Egypt and the Suez canal.'

The United States Navy, deterred from establishing naval bases on the Isthmus, was now sniffing around a bay on the north coast of the Dominican Republic, strategically placed to guard the passageways into the Caribbean basin and a canal wherever it was situated, 'which might,' the British Foreign Secretary was warned by one of his diplomats in December 1884, 'be so fortified as to become a second Gibraltar'.

The same month, not coincidentally, saw the climax of the early American efforts to build a canal, for and by themselves, in Nicaragua, which remained their favoured location for a waterway. Soon after the decision of the Paris Congress to plump for Panama, Ancieto Menocal, along with Admiral Daniel Ammen, ex-President Ulysses Grant and others had formed the Maritime Canal Company, to fulfil their vision of the breakthrough happening in Nicaragua, regardless of what the French were doing in Panama. In 1880 Menocal had negotiated a concession from the Nicaraguan government to build his canal. Late the following year, they succeeded in getting a bill introduced in the Senate for political and financial guarantees to be given to the company by the government. The bill was passed around interminable committees, until at last, in mid-1882, it was voted on, narrowly failing to win the necessary two-thirds margin, helped by lobbying from de Lesseps' men in Washington and a feeling, as the *New York Times* put it, that 'the time for guarantees and subsidies of bonds has gone by'. The great railway boom had been underwritten by the federal government, but now the power of the railroad barons was a national bugbear. 'Let them have protection and charters,' the paper went on, 'but let them persuade the capitalists of this country or of the world that they have a good thing and obtain their funds in a legitimate and business-like way. Therein at least they may take a lesson from de Lesseps.'

Ammen was furious that support had been denied, telling the British ambassador that he was going to 'abandon it as an enterprise backed by the United States' Government, and to seek the necessary capital in English and German markets for carrying out the work'. Rather than being an American canal for Americans, it would be neutral and free to all, multilaterally guaranteed.

Blaine's successor as Secretary of State, Frederick Frelinghuysen, was disappointed by the failure of the bill, for precisely this reason, muttering darkly to the British ambassador, 'there is reason to believe that direct overtures were made to the German Government by parties interested in the Menocal Concession'. But these efforts failed and the concession lapsed.

Frelinghuysen kept up negotiations with the British Foreign Office over altering the Clayton–Bulwer Treaty, albeit at a less shrill pitch than that of Blaine. But the Americans had little to offer in return for a move that would clearly benefit the United States at the expense of Britain, and got nowhere. Nevertheless, at the end of 1884 Frelinghuysen started fresh negotiations with Nicaragua for a canal treaty. A deal was signed on 1 December. Anticipating ratification, Ancieto Menocal prepared to depart once more for another survey in Nicaragua, this time not as an employee of a private company, but of the Secretary of the Navy.

On 10 December, the President, Chester Arthur, sent the treaty to the Senate with a strong message of recommendation. The work was too important to be left to private capital, he said. The Nicaragua canal would unite the country without recourse to the Railway Corporations, he went on, and deliver great benefits to US trade and shipping. European grain markets would be brought in reach of the Pacific Coast states, and China would be opened up to East Coast manufacturers, who would now find themselves 'midway between Europe and Asia'. By building the canal, he argued, the United States would make itself the centre of the world.

As the prospect loomed of two rival canals, the Mexican ambassador in Washington took the opportunity to pitch his homeland to the British ambassador there, Lionel Sackville-West. If the French and Americans were going to have their own canals, he told the ambassador, Britain should also have one, built at Tehuantepec. Sackville-West brushed this suggestion aside, but was keeping a close eye on the Frelinghuysen deal, which was in clear breach of the Clayton–Bulwer Treaty.

When it came to the vote, the 'much divided' Senate found itself in a quandary, as Sackville-West reported back to London. If it was rejected, it would 'be understood by European governments as a practical abandonment of the Isthmian policy, and entail humiliation. On the other hand its ratification would test the position of the British Ministry that the Clayton–Bulwer Treaty is still in force . . . but is the United States prepared for a controversy which might result in something more serious than diplomatic correspondence?' To overturn an international treaty was a very serious step, as was reflected by concerns in the country and press.

The treaty, which would have seen a US-government-funded Nicaragua Canal, was rejected, with 32 in favour and 23 against, narrowly missing by five votes the two-thirds majority needed for ratification. Party jealousies contributed to the defeat, but the Senate was also swayed by another argument in favour of delay: the canal would be a liability, fortified or not, until the United States had a competent navy to police the strategically vital waterway. A motion to reconsider the vote was introduced, but the inauguration of Grover Cleveland in March 1885 saw a change in policy and outlook. In his inaugural address Cleveland signalled a return to the traditional aversion to 'entangling alliances'. He was also opposed to government involvement in 'big business'. Shortly afterwards, he withdrew the treaty.

Thus America's canal effort passed back from government to be once again the responsibility of the 'folly and gullibility of Capital'. As hopes faded for Nicaragua, attention returned to Panama, where, whatever the setbacks, there were more men and machines at work than ever before. Suspicions of the French were mounting, with many believing that they were on the road to declaring a protectorate over a Panama state grateful to be detached from Colombian rule. In March 1885, the British minister in Bogotá had a conversation with his United States opposite number, who had recently been in Panama, in which the American 'described every Canal functionary as, in his opinion, a French Government Agent in disguise'. De Lesseps, he went on, was set on introducing French colonists to Panama, before annexing the whole of the Isthmus.

All the time, the political situation on the Isthmus was worsening. The scene was set for a show of military force by the United States.

The civil war in New Grenada in the late 1850s had ended with the triumph of the Liberal faction, and the drawing up, in 1863, of a new constitution that enfranchised the lower classes, reined in the influence of the Church on national affairs, and gave considerable autonomy to provinces within the country now to be called the United States of Colombia. This decentralisation suited Panama, which had always been more racially and ecclesiastically mixed, liberal and outward looking than distant Bogotá. But after a depression in the 1870s, caused in part by collapsing prices for Colombia's cash crops, there was a conservative backlash in 1884 led by the new President Doctor Rafael Núñez.

Núñez's efforts to reverse the Liberal reforms prompted civil war in mainland Colombia, which quickly spread to Panama, at a time when the French canal effort did not need new problems. In July 1884 an attempted Liberal coup in the province was suppressed with difficulty, and after further disturbances, martial law was declared and a military governor appointed.

Amid the growing anarchy, foreign consuls on the Isthmus hurriedly telegraphed for gunboats to be sent for their nationals' protection and on 18 January 1885 the US Navy's *Alliance* was brought up alongside Colón dock, trained her 3-inch guns on the town and stationed men on the wharf and around the Panama Railroad offices, with particular attention to safes and vaults. The demonstration of American power provoked the resignation of the mayor of the town in protest, but calmed the situation for a while.

Allegedly the landing of US troops had been at the request of the military governor, General Vila, but in Washington the Colombian minister shared his grave concerns with the British ambassador. Apparently, he had received a telegram from the government in Panama reporting that 'the sovereignty of the State was in jeopardy'. According to the Colombian, the danger 'arose from the intrigues carried out by the United States' Government to obtain control over the Isthmus'. The US, he said, 'were fearful of the establishment of a French colony . . . and would use the excuse of anarchy on the Isthmus to establish a filibustering colony'.

In Colombia, the revolutionary party was in the ascendant, with Cartagena on the Caribbean coast threatened by a powerful Liberal army. The Colombian President responded by requesting that Vila send a loyalist force from Panama for the defence of the city. Vila raised taxes further in the province and introduced conscription, but this merely led to another series of demonstrations against his regime. So Vila instead proceeded to the mainland with five hundred soldiers from the regular garrison. As it turned out, his action saved the city, and may have turned the course of the civil war, but it left only 250 loyalist troops in the province, most of whom were stationed in Colón. So it was in Panama City that Núñez's opponents first seized their chance.

On 23 March, acting British consul Mallet brought the Foreign Office up to date on events in Panama. 'My Lord, I have the honour to inform you,' the letter begins, in the customary way, 'that a Revolution broke out in this city on the morning of the 16th inst., . . .' 'At 2 a.m. on the 16th,' the *Star and Herald* reported a few days later, 'General Aizpuru gathered his men, 247 in number, at the garden of Paraíso, where after receiving some drinks to fortify them for the dangerous enterprise on which they were about to embark, they proceeded to enter the city.' The barracks and police station were speedily captured, but fierce fighting, in which some twenty people were killed, continued around the town hall, where the depleted Colombian garrison had taken refuge.

As soon as the diminished garrison departed from Colón on the railway to deal with this emergency, another Liberal leader, Pedro Prestan, seized his opportunity. With eighty men, he overpowered the small police force left guarding the Atlantic port and took control of Colón. Prestan, a mulatto and fervently anti-American, had a strong following among the poor and non-white in the town.

The Americans responded by landing troops from a warship, the *Galena*, to guard the railway and the wharf. Meanwhile, Prestan did his utmost to secure arms for his growing band of followers for the inevitable confrontation with government troops. After an uneasy truce of a few days, while in Panama City the government forces ousted Aizpuru, the hastily purchased weapons arrived in Limón Bay on a steamer from the United States.

The steamship line's agent in Colón was John Dow, an American with little love for Prestan or his followers. Dow refused to land the arms. Prestan was furious, and immediately arrested him, another Pacific Mail employee, the American consul and two officers from the *Galena*, who had been with their men at the railroad.

Prestan also threatened to fire on any of the crew of the US warships who disembarked and warned the US commander in the bay, 'Any aggression against us on the part of the US ships will imperil the lives not only of the hostages but also those of your fellow-countrymen living in Colón.' Prestan is quoted as saying to his staff, 'I wanted to prove to these people that their nationality and race does not protect them from my revolutionary authority. For the first time in the history of America a mulatto has dared put his hands on white US citizens, and this fills me with pride because I have vindicated by my act the dignity of the negro, outraged by the white man across the centuries.' The quote has the whiff of subsequent invention or embellishment, but the sentiment is accurate. The US consul, in fear for his life, urged Dow to deliver the arms. Dow consented, and the hostages were released. However, at this moment the captain of the *Galena* took possession of the shipment of weapons in the name of the government of the United States, and the hostages were quickly rearrested and taken to the rebel's position on Monkey Hill.

Here, Prestan's men were awaiting the rumoured return of government forces from Panama City. One hundred and sixty men were indeed on their way by train. They disembarked at Mindi and proceeded to attack the rebel positions at dawn on 31 March. Hopelessly outgunned, Prestan's men were driven back to Colón, where fierce combat continued for the next eight hours. At one point the fighting reached the walls of the British vice-consulate, where 'rebel bullets and cannon balls . . . completely riddled the building'. At four in the afternoon, with Prestan's forces on the brink of defeat, a fire broke out in the north of the city. Colón, almost entirely built of wood and lacking piped water or any sort of fire brigade, was soon an inferno, and in no time a rumour was circulating that it was Prestan himself who had ordered the fire to be started. Helped by a strong north-easterly wind, the fire burned for another twenty-four hours. Almost every building in the town was destroyed. Men were landed from the European warships in the bay to help save the foreign enclave of Cristóbal, happily separated from the main city by a shallow inlet, and soon the Americans and government troops were restoring order by shooting suspected looters. Within two days, the rebels had been destroyed or captured, although it seemed that Prestan himself had escaped.

In Panama City, Aizpuru had taken advantage of events at the other end of the railroad to launch a fresh attack. Fighting continued for eleven hours. 'The firing was hot and reckless in the extreme,' Mallet later reported. 'Thousands of cartridges were burned as the scarred and wrecked appearance of walls and interiors sufficiently prove.' By the end of the day, Aizpuru was victorious, and on 1 April declared himself the Military and Civil Chief of the city.

The bloody fighting and the tragic events in Colón had, however, given everyone a moment of reflection. Neither the government forces, now in control of what remained of Colón, nor Aizpuru in Panama City had sufficient force to overwhelm the other, and they signed an agreement to suspend hostilities for a month, 'to preserve the capital from criminal elements, and to give security to interoceanic traffic'. It seems an effort to prevent foreigners from having an excuse to intervene.

But US naval forces, including marines, were arriving in strength. The USS *Shenandoah* anchored off Panama City on 6 April. The *Alliance* returned on 8 April. Two days later the USS *Tennessee*, with an admiral aboard, appeared off Colón, together with a steamer of the Pacific Mail line packed with marines. More vessels arrived and by 15 April an entire marine brigade, along with two or three support battalions of blue jackets, was in place. Together with four field pieces, 170 men were ordered ashore on 8 April, and armoured railway cars to protect the transit were improvised with weapons and boiler plate from the ships. By 18 April, according to Mallet, there were some five hundred US troops based around the railway station in Colón, armed with a battery of Hotchkiss and Gatling guns and Dahlgron howitzers. In Panama City were a further three hundred men, with another two hundred scattered along the railway line. Offshore at either end of the line were half a dozen US warships with a further 1800 men and thirty more field guns. It was the largest overseas military expedition to be mounted by the United States between the Mexican War of 1846 and the war with Spain in 1898.

In mainland Colombia, Núñez's party was gaining ground against the revolutionists, and a loyalist force was being assembled at Buenaventura to depose Aizpuru in Panama. The general started erecting barricades in the city and preparing for a siege.

To the Americans, this was unacceptable. To protect foreign property and interests, troops were ordered out of their barracks and off their ships to take control of the city. Barricades were removed, and key positions in the city secured. All the bars and saloons in the town were closed down and Aizpuru, together with his senior officers, was arrested. 'The entry of the American marines into the city was a complete surprise for everyone and occasioned great excitement,' Mallet reported. 'The belief among natives was that the city was to be taken away from them; patriotic feelings were raised to a fever pitch, and threats were openly made that unless General Aizpuru was released and the American force withdrawn every foreigner would be assassinated, and the town reduced to ashes.' The US commander reassured the Panamanians that 'The presence [of the US force] is only temporary and simply to restore law and order. The idea of occupation or annexation of the Isthmus is one that has never occurred to the American mind.' On the evening of 25 April, Aizpuru was released, having promised to respect foreign nationals and not to fight within the city limits. The US force, now 1200 men with twelve howitzers, retreated back to the railway station and everyone waited for the arrival of the Colombian loyalist soldiers.

Two days later, their ships were seen anchoring off Taboga Island in the Bay of Panama. By now, Aizpuru had lost too many of his men to desertion to risk fighting and surrendered to Colonel Rafael Reyes, the loyalist leader, on 29 April. The next day, the Colombians landed and over the following week the American troops retired to their ships.

Aizpuru was later fined and exiled, and a witch-hunt was launched against his and Prestan's erstwhile supporters, with Jamaicans and Haitians singled out for special treatment. Many were shot out of hand, and others languished in jail without a trial for up to four months. Two were hanged on 6 May for starting the Colón fire, even though, while incarcerated on a US warship, they had helpfully signed testimonies which pointed the finger of blame at Prestan.

Prestan himself had escaped to the state of Bolívar but after the fire found himself friendless and soon fell into the hands of government troops, who returned him to Colón to face trial for arson. Held on 17 August, it was a military tribunal on which sat a motley collection of soldiers and locals, all enemies of the accused. Four witnesses were called for the prosecution, foreigners, none of whom actually saw the fire being started, although they testified that Prestan had threatened to burn the city at some point or another. In fact, it is unlikely that Prestan was responsible. He owned property in the town, had his wife and daughter living there, and had nothing to gain militarily from the act. But the verdict was never in doubt. None of the witnesses requested by Prestan in his own defence appeared, and the tribunal ordered that Prestan be hanged. The sentence was carried out the next day at noon, on a scaffold made out of railroad ties erected in one of Colón's main streets, a stone's throw from the entrance to the new canal. 'I saw no sign of fear,' writes Claude Mallet, who was a close witness to the execution. 'As he was dying he made no struggle and kept moving his arms as a sign of farewell to the crowd.'

The whole dramatic episode had many important repercussions. It had now been firmly established that the real power on the Isthmus was the US military, and with the defeat of the Liberals control of the country had been, it seemed, irrevocably handed over to the Conservative elite. In the *arrabals* and among the Colombian workers on the canal, the US intervention had fuelled fear and hatred of the 'Yankees'. Elsewhere, among the outward-looking elite and the foreign residents, the chaotic events had underlined the impotence and incompetence of the Colombian authorities as well as the malign influence of mainland politics. 'The State will never be free from such revolutionary nonsense,' wrote the *Star and Herald*, 'until it withdraws from the union and sets up a government of its own under the protection of the United States or the great nations of Europe . . . There is a strong and growing sentiment in favor of such a movement.'

The French man-of-war, with its accompanying marines, stationed in Colón harbour had, with the exception of helping fight the fire, played no part in the tumultuous events. Instead, even though it was a French company whose works and property were principally under threat, the force continued to maintain a strictly neutral stance, fearful of upsetting the United States' position, or doing anything that went against the sacrosanct Monroe Doctrine. In fact, the French military were under firm instructions to keep their distance from the Canal Company, and only to intervene if requested to do so by all the major consuls, something of which the Americans would never have been a part. But the brief war had a profound effect on the French canal effort. To carry out such a massive construction project in a stable political situation was enough; to achieve it in a state of anarchy and war was another thing altogether. It was just the first part in what would become an *annus horribilis* for the French effort.

For the benefit of share- and bondholders, and the ever-important confidence, the Company maintained in the *Bulletin* that they had lost nothing during the disturbance, but anyone on the Isthmus could see that this was patently untrue. Although Cristôbal had been spared, the fire had wrecked many large and valuable Company warehouses in Colón. In addition, offices, machinery, private residences and railroad plant had been destroyed or damaged. At a modest estimate, the loss to the Company was in the region of a million dollars.

There was to be a further ramification for the canal effort, and a nasty coda to the whole affair. On 3 May, the tensions simmering between the Jamaican and Colombian workers came to a grisly head. That night there was to be a circus performance near the work camp at Culebra. The men had just been paid. The local *alcade* (mayor) requested a picket of Colombian troops to keep order. Five men were sent, but these were, according to Mallet, part of Reyes' newly arrived force, who 'were ignorant of Isthmian affairs, and knew nothing of Jamaicans . . . and were animated only by a blind prejudice against all people who did not speak their language'. The soldiers tried to go through the camp to reach the site of the entertainment, but there was a rule that no armed men were allowed in the camps. The Jamaican watchmen, not knowing for sure if they were government soldiers or rebels, who were still roaming the countryside in small parties, disarmed them, on the orders of the camp chief, who said he didn't want a guard for the entertainment.

The men returned to their base at Emperador and reported what had happened. Their commanding officer was incensed and ordered out his whole force. On the way they were joined by a mob of Cartagenians armed with machetes and revolvers. The whole crowd seems to have been well-oiled, and there were raucous cries of '*Viva Colombia*'.

It was about two in the morning when they reached the labour camp. The offending watchmen were tracked down first. Arthur Webb, a Jamaican who had been on the Isthmus since 1882 and was in No. 4 barracks, saw what happened: 'I heard the watchman outside challenge some one who answered "Colombian". I opened the door, and saw four men around one of the watchmen chopping at him with machetes.' Webb took to his heels, was spotted, and fired at. The soldiers, some twenty-five in number, then attacked Webb's barracks, where it was believed another watchman had taken refuge.

At three in the morning, when volley after volley had crashed into the building or cut down the Jamaican workers as they tried to flee, the door was smashed down with machetes. Jamaican Samuel Anderson had taken shelter under his bunk. From there he saw 'a number of Jamaicans killed and hacked to pieces, their boxes and trunks were then broken open and robbed of their contents. Some Colombians then came into the barracks with kerosene oil and tried to set light to it. When I saw they intended to burn the barracks I left my hiding place. Several Colombians then seized hold of me, tied my hands, and struck me all over the head and body with the flat side of their machetes. I didn't resist and I was then made to accompany them to Emperador.' Another Jamaican was also taken prisoner, but they turned out to be the lucky ones. Arthur Webb returned to his barracks early the next morning 'and counted 23 Jamaicans lying killed, and hacked to pieces on the ground, and in their bunks. Some of them had their legs and arms chopped off, and many had their skulls split in pieces. Many of the dead appear to have been killed whilst attempting to dress.' He found that his possessions had been stolen, including $200 he had saved in the last three years. The floor of the barracks was awash with blood and the whole place scattered with the ransacked contents of the workers' trunks and boxes. Isaiah Kerr, a Jamaican who lived at nearby Las Cascadas, came to Culebra that morning as usual and 'on entering No. 4 Camp I saw my brother Augustus Kerr, who worked there, lying dead on the ground, his throat had been cut and one of his legs were gone. Many other Jamaicans were lying about dead and wounded.' He found that his brother's clothes and money had 'been taken away, and I could find nothing belonging to him'.

The Colombian authorities suggested that the Jamaicans had started the aggression by firing on the Colombians, but the witness statements, carefully gathered by the British diplomats, contradict this story. 'I have never before witnessed anything so horribly sickening as the scene of the butchery at the camp,' C. H. Burns, an American canal contractor, told Claude Mallet. 'Some of these unfortunate labourers lay upon their beds with only a night shirt on.' He saw no weapons among the mutilated bodies. 'It is not the first outrage upon Jamaicans, and all growing out of the prevailing hatred which the natives bear the "Chombo",' he said.

Samuel Anderson, taken prisoner that night, was confined in jail at Emperador. 'I remained in prison for nine days,' he said, 'four days of which I was kept with my feet in the stocks, and I was without food or water for 48 hours. On Monday 12th of May the judge at Emperador told me if I gave him fifteen dollars he would let me go. I had a watch in my possession, which I pledged for seven dollars which I gave to the judge, and he released me. On my return to the camp at Culebra I found all my clothes, and money had been stolen, and I am left without anything.'

A shocked and furious Claude Mallet demanded an investigation, but met only delay and prevarication. Important papers had, it seemed, gone missing. Raphael Reyes, now promoted to general, wrote to the *Star and Herald*, trying to excuse his men, but the paper did not believe him: 'In all these fights between Jamaicans and Colombians,' it said, 'the former are invariably represented as the aggressors, and as invariably are they beaten, demoralized and cut to pieces . . . it was a massacre, pure and simple.' The Panamanian government assured the Governor of Jamaica that his countrymen were safe, but Mallet told him that such promises were 'worthless', and are only made 'with a view of inducing Jamaica negroes to leave their homes and come to the Isthmus'. In fact, the massacre was symptomatic of a wider disregard for the rights of the imported workers: 'It must also be borne in mind,' Mallet wrote to the authorities in Kingston, 'that British subjects have suffered as much from constitutional as from the Revolutionary authorities. Alcaldes, prefects, judges and all in authority have paid little attention to the rights of the negro from Jamaica. The poor negro has been the legitimate prey of Executive and Judicial outrage of the gravest and most serious character. The records of this Consulate are made up largely with the story of their wrongs. The powerful Companies that bring them have taken little interest in their welfare and make no active efforts in their favour when they fall into the hands of the authorities.'

The fallout was both immediate and long-lasting. The Jamaicans fled the camp at Culebra, and within days no one who had worked there was on the Isthmus any longer. All along the line, Jamaicans abandoned the works to return to the safety of Kingston. Nor did the shocking events of that night lead to any change in the attitude of the locals to the Jamaicans. Those who remained were increasingly forced to arm themselves, and tensions rose further. Although new labourers did continue to arrive from Jamaica, the appeal on the island of 'Colón Man' was tarnished forever. Never again was the Company able to marshal such numbers of workers on the project as had been there at the beginning of 1885.

*

Even before the civil war, fire and subsequent events, confidence in the success of the canal was seeming more and more far-fetched. In early March Mallet had reported to London the visit of de Lesseps' second oldest son, Victor, along with others high up in the Paris Canal Company. They had professed themselves pleased with what they had seen, 'with the conviction,' Mallet wrote, 'that the enterprise will be ready for the world's commerce at the end of 1888. I may remark in passing,' he added, 'that there are few intelligent people outside of Canal circles, who share the sanguine expectations of these gentlemen . . . dissatisfaction and anxiety prevail'. Apart from anything else, the shock at the death of the last of Dingler's family was still being felt.

With the fire came a worsening of the bottleneck at Colón for the import of machinery and supplies, and the massacre at Culebra had led to labour shortages all along the line. April saw exceptionally heavy rains, which held up the work and exacerbated the problem of landslides. 'The Panama canal is in such a state that its ultimate completion is beyond question,' wrote the New York *Tribune* in May, 'but it appears equally certain that the present company can never complete it . . . In going over the canal route, one gets the impression that the work is practically stopped.'

But in some areas the sort of technical breakthroughs predicted by Ferdinand de Lesseps were occurring. Philippe Bunau-Varilla had been appointed Chief Engineer of the Pacific Division of the canal, even though he was only twenty-six. Through a study of the Bay of Panama he had accurately predicted that submarine trenches dredged there would stay free of mud and sand, thus dismissing a major worry in that sector. In recognition of this progress, in April 1885 he had been appointed head of the Atlantic Division as well. Here there was another breakthrough. The dredges of Huerne, Slaven & Company had been held up by hard rock at Mindi. When such material had been encountered at Suez, for instance between Bitter Lakes and the Red Sea, dams had been laboriously built, the area drained and excavation had continued 'in the dry'. But at Mindi, Bunau-Varilla, recalling an earlier experience in France, ordered a series of underwater holes to be drilled in the rock, a yard apart. In each was placed dynamite, which, when exploded, reduced the rock to paving-size slabs, which could then be dealt with by the dredges. Thus the cost of underwater excavation was reduced to that of cutting 'in the dry', and with even better methods and machines, Bunau-Varilla surmised, could be made yet cheaper. The realisation would lead to a clever suggestion to save the French canal.

It is difficult to get a handle on the extraordinary figure of Bunau-Varilla. There is little doubt that he was an engineer of genius, as well as having other talents, as would emerge later. In his own writings, however, he has an egomania verging on madness. It is reported he spoke – and he certainly wrote – not in sentences but in proclamations. Nevertheless, he was adept at making friends. Short, only 5 feet 4, he had perfect posture and a luxuriant dark red moustache. Many of those who met him found him an eccentric and slightly overwhelming figure. 'Mr Varilla's tremendous mental capacity becomes apparent when one looks at him,' wrote an American whom Bunau-Varilla would later befriend. 'His brain rises from an active, rather square face, but, as if to contain it, the sides of his head are much larger than the face.' 'His versatility was fantastic,' wrote another admirer. 'He had the energy of ten horses.' No one who met him ever doubted his fanatical devotion to the achievement of the canal, at whatever cost and by whatever means.

But time was already running out for de Lesseps' sea-level plan. By summer 1885 the excavation was falling seriously behind schedule. Menocal visited in August and reckoned that only 8 million out of the 120 million cubic metres needed had been excavated since the very start of the project. Furthermore, much of the money raised had been spent. In spite of de Lesseps' assurances that all the problems were surmountable, the Panama shares began to fall a little on the Bourse. The days of boom in the French financial markets had passed with the collapse of the Catholic Union Générale bank in late 1882. In place of the frantic speculation of the time of the launch of the Canal Company, traders on the Bourse were just as likely to be bear raiders, seeking through a variety of schemes to lower the price of the shares in such an enormous company in order to profit on the change in market price. Criticism continued in the newspapers. In London and New York the financial press was loud in its condemnation. In response to the undeniable reports of deaths from illness, a cartoon appeared in *Harper's Weekly* asking: 'Is Monsieur de Lesseps a Canal Digger, or a Grave Digger?' Even in France, doubts began to be aired.

At the opening of the Company's July 1885 AGM, the reality of the situation was beginning to undermine the faith of even the strongest believers in de Lesseps' Great Idea. Worries were expressed about the financial state of the Company, the falling rate of excavation, the sporadic labour troubles and the terrible rate of attrition from disease. De Lesseps countered with an inspired bout of oratory, speaking, without notes, for an hour, announcing that he would launch a lottery to raise the extra six hundred million francs more he now said were needed. Furthermore, he would personally visit the Isthmus to inaugurate the 'final stage of construction'. It says much for his magnetism he was still able to charm an overwhelming vote of confidence from his audience.

It was a lottery bond issue that had saved the Suez Canal. In 1867, after the failure of a bond issue, de Lesseps had issued 5 per cent lottery bonds, with four prize draws a year, each offering a 250,000-franc top prize. It had been a huge success and ensured the opening of the canal two years later. But under French law such an issue was restricted to undertakings of national importance and required a specific Act of Parliament. Straightaway, de Lesseps found the government of the Republic less helpful than had been the Imperial Senate. Although the Company organised, at great expense, a flood of petitions in favour of the legislation, the Minister of Works took no action until December, when he ordered a senior, and scrupulously honest government engineer, Armand Rousseau, to go to Panama to give judgement on the project. In Rousseau's hands, it seemed, was the future of the canal.

In the meantime, after a gap of two months after the departure of Dingler, a new *Directeur Général*, Maurice Hutin, had taken over in September. There had already been drift in the leadership of the project, but Hutin himself left Panama only a month later, struck down with yellow fever. He would survive and return to the canal story later, but again there was a vacuum at the top of the organisation on the Isthmus. Into the breach stepped twenty-seven-year-old Philippe Bunau-Varilla, as Acting Chief Engineer.

His duties at either end of the canal had already stretched the young engineer, leaving him time, he says, for only two or three hours' sleep a night, but he took on the new role with enthusiasm. His first initiatives were to improve recruitment and labour relations and to set a monthly excavation target of 1 million cubic metres for the first few months of 1886. With patriotic rhetoric, he urged on his French engineers and technicians.

Before the end of the year, however, disaster struck again on the Isthmus. On 2 December a violent storm lashed the Atlantic seaboard, with winds of up to a hundred miles an hour. Vessels crowded into the exposed harbour of Colón tried to escape out to sea, but at least ten were driven on to the shore. Bunau-Varilla rushed to the port, concerned above all about the safety of the newly built embankment of Cristóbal. At the harbour he was met by a terrible scene. Boats were smashed on the rocks or overturned, with their crews clinging to them like 'a human bunch of grapes'. The sea wall at Cristóbal survived, but more than fifty sailors were drowned. Meanwhile 'Rain poured in torrents,' the *Star and Herald* reported. 'The Chagres River has risen over twenty feet above its level.' The river was soon in huge flood, discharging twenty-five times its normal volume of water. In a stroke, much of the work and equipment was submerged.

The next day, with the sky again clear, Bunau-Varilla inspected his sector. 'The points where my locomotive passed on the previous day were now covered by fourteen feet of water,' he wrote, 'so I requisitioned three Indian canoes . . . as we paddled along through a channel apparently cut out of virgin forest all the workings were submerged and the tops of the telegraph poles were scarcely visible above the water.' After one canoe was damaged, the party had to crowd into two boats. 'The load was almost too much,' Bunau-Varilla continues, 'and the freeboard was not more than an inch above water. One of the engineers, a M. Philippe, said that he couldn't swim. I told him jokingly, "There is no danger. I could easily swim with you to the nearest trees." It was only then that I noticed the strangest phenomenon: the tops of the trees were not their usual green, but a distinct and ever-shifting black; as we drew nearer I saw they were covered with the most enormous and deadly spiders: tarantulas.'

The damage from the floods was straightaway noticed by Lieutenant William Kimball, who toured the works soon after to write the latest US Navy report on the construction. At Bohío Soldado he found 3 million cubic metres of sand deposited in the trench by the subsiding waters, with railtrack and spoil cars buried to the depth of 2 metres. Kimball's report provides a fascinating and even-handed snapshot of both progress on the works, and also, more generally, of life on the project at the turn of 1885–6. In some places he noticed progress from the findings of a report twelve months earlier. Work on the repair of the wharves had proceeded quickly. In *la grande tranchée* between Bas Obispo and Culebra 'some very good work has been done', even if a lot of it was by 'hand-drills, small blasts, and hand work'. At Matachín, the central machinery depot, almost entirely staffed by Americans from New Orleans, seemed efficient and well equipped, although it had cranes too light for the task. In general, the accommodation and infrastructure seemed to have improved, with the exception of the hospitals, which had proved inadequate for the number of sick.

Kimball noted with interest the construction of a dam on the Río Grande to enable dredges be floated further upriver – testament to Bunau-Varilla's idea to excavate as much as possible 'in the wet'. It was also hoped that the flooding of the marshes would 'improve the sanitary conditions near La Boca, which are at present very bad'. The water, it was believed, would prevent the 'miasma' disturbed by the excavation from affecting the workmen.

Elsewhere, the lack of overall progress from the year was more evident. At the site of the crucial dam at Gamboa, work had hardly started, and in the Paraíso section at the Panama City end of *la grande tranchée* he noticed severe problems with slides. 'The slips are not earth slides from the top of the bank,' he wrote, 'but rather movements of the whole hillside, which in some places carries one bank almost intact across the cut with the top surface unbroken, and with the vegetation undisturbed.' Looking up, Kimball could see both the railway line and a channel built to keep the Río Grande from the canal suspended on the hillside 30 and 50 metres respectively above the new ditch. 'There is a substratum of greasy clay all along the line,' he reported. 'It would seem as if both the deflection of the Río Grande and the deflection of the railroad must slide into the canal.'

All along the line there was the impression of spoil carelessly dumped, damaging the railroad embankment in one place, narrowing the Chagres in another, contributing to the flooding. The machinery he saw in action was 'considerable', but impressed him 'as neither large enough nor of the right kind'. The American Osgood and McNaughton steam shovels were doing a good job, as were the US and Scottish dredges, but they were too few in number. There was a lack of power drills and the French and Belgian ladder excavators were 'ineffective'. The miracle machines which had dug the sand out of the Suez Canal had come to grief. 'When working in soft earth, free from stones and roots,' wrote Kimball, 'there is no doubt that the chains-of-buckets machines have a greater capacity than those of the steam-shovel, American type; but these perfect conditions are not to be had.'

There was also a lot of plant idle, testament to failed experiments, and other factors. In one place five excavators were seen delayed for lack of spoil trains. This was due to an absence of proper switching arrangements to transport the earth to a dump only half a kilometre away. At Matachín a contractor had stopped work as he hoped to get a higher price for removing rock that had not shown up on borings. In general, the hundred or so small contractors seemed to be getting in each others' way, particularly over spoil removal. Furthermore, some of the companies, Kimball said, were 'irresponsible parties', who gave up when the going got hard or less profitable. Others had not the necessary financial resources for what they had taken on and, forced to borrow from Isthmian bankers at 2 per cent per month, soon went under. Either way, the result was delays, plant standing idle again, and the inevitable demands of new contractors for more favourable terms.

Stoppages were also caused by labour shortages, arising not just from fear of further political violence, but also from 'forethought of others, who decide to leave the Isthmus before they are killed by the climate . . . by the poor quality and high prices of provisions; by the exorbitant rates charged for small drafts by the small bankers, who control such business; [and] by the lack of sufficient guarantees for hospital attendance'. Furthermore, the 'men are in the habit of returning to their homes to spend what they accumulate, often leaving the works at the very time when from conditions of weather or arrangement of plant they can least conveniently be spared.' All this contributed to a turnover of workers of some 80 per cent a year.

Kimball also reported the tensions between the French and their US colleagues on the project. Apparently, the Americans were being accused of sabotaging excavators in order to stop the work, so their government could take control of the canal, 'and other wild statements'. Perhaps Bunau-Varilla's patriotic rhetoric had sharpened divisions. Certainly, Kimball noticed great dedication on the part of some of the French engineers. The Acting Chief Engineer, who showed Kimball 'unremitting and repeated courtesies', explained it thus: 'The contagion of my confidence in our success had taken hold of all my men,' said Bunau-Varilla. 'One man who fell was immediately replaced by another, and the battle went on.' 'It is an impressive fact that there is money value in the prestige of M. de Lesseps, the courage of the French and the determination to finish the canal,' noted Kimball, 'for otherwise the company would already have become bankrupt under the showing of 500m francs practically spent and not more than one tenth of the work accomplished.'

In spite of this gloomy, and, as it turned out, accurate, assessment of progress, Kimball still believed the canal would be built. 'That with sufficient expenditure of money, time, brains, energy, and human life, the canal can be finished is self-evident,' he concluded, although refusing to estimate 'the necessary quantity of all or any of them.' If the money from the lottery was forthcoming 'the canal will be so far advanced by the time the money for the new loan is expended that the necessity for finishing it will be apparent'. Put another way, the temptation to throw good money after bad would be irresistible. This new finance was the key factor. 'The Company has doubtless made some grave mistakes, but I am confident it has at its disposition all the necessary brains and energy,' wrote Kimball. 'As for human life, that is always cheap.'

It has been suggested that 1885 was the blackest of all for deaths from fever. We only have the figures for mortalities in the hospitals. For 1885, there was a similar official death toll as the year before, running at nearly a hundred a month, in spite of a slight lessening of the workforce. Records of burials in, for example, the foreign cemetery (of which Claude Mallet was the treasurer), suggest that these figures do not tell the whole story. On several occasions during the year Mallet was again called out to collect the body of dead British subjects from the streets of Colón or Panama City whom no one had been found to bury. There was a pattern of men employed on the canal or the railroad falling sick, being hospitalised at the contractor's expense, but then being discharged before fully fit. While looking unemployably ill, the patient would fall sick again shortly afterwards – basically have another malaria attack – and find himself unable to afford to go to hospital. Claude Mallet fought an ongoing battle with the Foreign Hospital to lower its costs of $2 a day, which he was obliged to pay for such 'distressed' Britons. Many of these were railway engineers and labourers displaced from Peru by the war there at the beginning of the decade. But the hospital now had more trying to get in than they had space. And, Mallet was told, 'under no circumstances would negroes be admitted'.

The Company still had its fair share of shocks. In October 1885, Bunau-Varilla was sent two new engineers to be Chiefs of Division. Both, he reports, were dead from yellow fever within two weeks. 'Many a man of them had been happy to

Spanish conquistador Vasco Núñez de Balboa discovers the South Sea and that Panama is a tantalisingly narrow isthmus

William Paterson, the Scottish promoter who declared that with possession of the Isthmus 'trade will increase trade, and money will beget money'

Mapping the route for the Panama railroad through thick jungle and swamp

Members of the American Selfridge expedition in the Darién jungle, 1870

Armand Reclus, the young French naval lieutenant who mapped the route of the French canal and led the de Lesseps effort in its early years

The hero of the Suez Canal, Ferdinand de Lesseps, depicted as Hercules pushing apart the continents of Africa and Asia

Ferdinand de Lesseps with his second wife and some of his many offspring

A triumphal arch, part of the lavish welcome given to de Lesseps when he descended on Panama at the beginning of 1880, fêted as the 'Presiding Genius of the Nineteenth Century'

Charles de Lesseps, who urged his father not to take on the challenge of Panama, but seeing that the old man had made up his mind, gave him his unconditional backing

Colón Harbour in 1884. The steamer was king, but much non-perishable freight was still carried by sailing vessel. The following year the town was destroyed by fire after a revolution on the Isthmus.

The beginning of the 'big ditch'. In spite of the hopes of the leadership, much of the French canal was dug by men rather than machines.

Jules Dingler, *Directeur Général* of the canal, 1883–85, who would pay a terrible price for his devotion to the endeavour in Panama

The execution of Pedro Prestan in Colón on 18 August 1885

A French ladder excavator. The machines that had triumphed at Suez proved unable to cope with the heavy clays of the Chagres valley.

Philippe Bunau-Varilla, French engineer, lobbyist and plotter extraordinaire

Bottle Alley in Colón. The small town had nearly 150 bars, with 40 in this one street alone.

Huerne, Slaven & Co. dredges at work. Visitors were hugely impressed by these monster machines, but the American contractors were among the most corrupt of all those working for the French Company.

The works in the Culebra Cut in 1888, with Gold Hill on the left

With the press running for cover, the directors of the Canal Company, led by Ferdinand de Lesseps, are brought to trial. Eiffel brings up the rear.

On the Isthmus, wreckage from the French effort was everywhere, with the jungle quickly returning

In a cartoon published six days afterwards, the *New York World* gives its impression of the 'Panama Revolution'

William Nelson Cromwell on one of his frequent journeys between Panama, New York and Paris

General Esteban Huertas in the regalia of the Commander-in-Chief of the Panama Army. Declared a 'Hero of the Republic' for his part in the 'revolution', he was quickly seen as a threat by the Conservative junta.

enlist,' he wrote of the still steady stream of new arrivals at Colón, 'but felt his heart sink at the sight of the warm, low and misty shores of the deadly Isthmus. Some bore on their faces the obvious mark of terror . . .' To Bunau-Varilla, and others, it was this very fear that made someone susceptible to fever.

The French consulate had a terrible year, losing three diplomats and two of their wives in the twelve months. Within five months of each other, two Italian consuls had died. Both the Spanish consul and his wife came down with yellow fever. The wife recovered from her delirium to find that her husband was already buried.

The British consulate saw their Colón vice-consul, the young Fred Leay, who had so meticulously taken the witness statements from the Culebra massacre, come down with 'Bilious Remittent Fever' and be invalided off the Isthmus. In January 1886, even Claude Mallet's hardy constitution was worn down, and he, too, contracted fever. He remembered lying in bed, and hearing one of the three doctors called for consultation say that he would not live to see daylight again. 'I had reached a state of semi-coma and did not care what happened,' he later wrote.

Fortunately a new consul, Colonel James Sadler, had at last arrived to relieve him. Mallet was allowed to retire for a short while to the healthier climes of Jamaica where he would recover, but he would therefore miss the two vital visits – of the government inspectors and of Ferdinand de Lesseps – the following month. His replacement, James Sadler, who had been through the Establishment treadmill of Eton, Oxford and the army, had little diplomatic experience outside of Europe. But it did not take him long to look around and assess the importance to Panama of the forthcoming inspections. Writing to the Foreign Secretary the Marquis of Salisbury on 27 January, he warned of the unpopularity of Núñez and the risk of further revolution on the Isthmus. Taxes and resentment were high. 'Crime is frequent,' he wrote, 'and remains unpunished from want of means to support the cost of imprisonment, though political offenders are treated severely.' Everything, he said, was riding on Rousseau's report. If it should be favourable and the lottery issue approved, 'the condition of the country may improve.' Otherwise, 'should the works cease, fresh misery and disturbance would certainly occur on the Isthmus.'

14

COLLAPSE

Rousseau arrived on 30 January 1886. With him, to relieve Bunau-Varilla, was a new *Directeur Général*, thirty-five-year-old Léon Boyer, along with his own hand-picked cadre of sixty engineers. Boyer was well known in France, having dabbled in politics, and also had a reputation as a brilliant civil engineer. He had a famous bridge to his name, and had been a holder of the Legion of Honour since the age of thirty. He was a very good catch for the Company. Also on the boat was Charles de Lesseps, in theory preparing a report of his own, but in reality keeping an eye on Rousseau and preparing for the arrival of his father.

They were met at Colón by Bunau-Varilla. Rousseau was accompanied by two other government-appointed experts. For two and a half weeks they toured the works, surprised, according to Bunau-Varilla, by the 'scientific order, the active discipline that prevailed everywhere'.

Rousseau left the Isthmus on 17 February. In the meantime, flags and bunting had been dusted off, streets cleaned, speeches and pageants rehearsed, and machinery, whether operative or not, whitewashed. *Le Grand Français*, the 'Presiding Genius of the Nineteenth Century', was about to descend once more on Panama.

On 31 January 1886, John Bigelow learnt that he had been invited, as a representative of the New York Chamber of Commerce, to accompany Ferdinand de Lesseps on his forthcoming visit to the works at Panama. Bigelow, lawyer, intellectual, former newspaper proprietor and US minister to France, had been at the lavish dinner for de Lesseps at Delomonico's back in March 1881. He was unsure whether or not to accept the invitation, and consulted friends and family. The first advice he heard was to have nothing to do with it, 'that it was probably a scheme to use my name as an ex-Minister to France to bolster the stock of a chimerical enterprise'. But then another friend urged him to 'embrace any opportunity of associating my name with such a magnificent enterprise'. Bigelow asked whether he could take his daughter Grace, for 'her company and assistance'. The reply came back in the positive – all would be at the Company's expense, including 'a satisfactory remuneration' for himself. 'This point, I confess, weakened my scruples about going,' he wrote in his private diary.

He accepted, but then spent a week worrying if he had made the right decision. Some big names, he learnt, had said no. He also worried about the illness on the Isthmus, and checked with a friend on the political situation to be reassured that the US Navy was there in force.

On 10 February he left with Grace on the steamer *Colón*, together with a journalist and other representatives from US chambers of commerce. A party had already left from France. Throughout their voyage, de Lesseps kept the fifty or so businessmen, engineers and diplomats entertained with 'witty lectures' and stories of his time in Egypt. 'He is indefatigable and inexhaustible!' exclaimed Gustave de Molinari, a journalist from the Paris *Economiste*. For the benefit of his readers in France, de Molinari spelt out the huge responsibility resting on the shoulders of *Le Grand Français*: 'The success of the Panama business does not (need I even mention) only interest the investors in the Company, it interests all of France. We live in a time where the power and the vitality of people is measured not only by the deployment of their military might and the success of their weapons, but also by their spirit of enterprise, the grandeur and utility of their works. If the French construction of the Panama Canal were to fail, the Americans (who are well disposed towards the idea) would take it upon themselves to buy out the project that we have given up on. As a result, with a lack of confidence in ourselves, the prestige of France would suffer for a long time in both countries. It would be like losing a battle. Let's hope that the battle is won!'

They reached Colón on 17 February 1886, just as Armand Rousseau was preparing to leave, his inspection completed. As before, de Lesseps, now eighty-one, had come at the best time of year and the weather was bright and dry. But Colón was still a wreck. A few houses had been rebuilt, and there were huge improvised market stalls. But in between, writes de Molinari, were 'vast open spaces left by the fire where stagnant rainwater, blackened beams, corrugated tin twisted by the heat, broken bottles and plates accumulate'. Mud and rubbish was everywhere, infested with toads, rats and snakes, and 'a myriad of mosquitoes breed in these low lying areas and spread through the mostly windowless houses and seek out their prey'.

John Bigelow's party arrived the next day, after twenty days at sea, and in the evening met de Lesseps for dinner on the USS *Tennessee* in the bay. Grace Bigelow was seated on de Lesseps' left. 'I was pleased to find that Grace and the old Baron got on admirably together,' wrote John Bigelow in his diary. 'Went to bed about eleven but was too heated and excited by the events of the day and evening to sleep well.'

For the next three days there was an exhausting schedule of visits to the nearby works and workshops, and trips on boats along the river and the completed section of the canal. Bigelow was suffering in the heat, changing his clothes twice a day, was horrified by the price of clean water and the filthy latrines, but was impressed with much of what he saw. At one chantier, he noted how the black labourers gave de Lesseps a warm reception: 'When we left they gave us repeated cheers which the old Baron returned with bows.' On board one British-built dredge, he noted approvingly its 'immense power'. New, even bigger machines were just around the corner, the visitors were constantly told.

Bigelow struck up a close and lasting friendship with Bunau-Varilla and also got on well with Charles de Lesseps, whom he called a 'very clear headed and capable man'. One evening, however, he was shocked to hear Charles confidentially and sadly prophesy that 'two or three years hence the United States would follow the example of England in the case of the Suez Canal, purchase an interest in [the Panama Canal] and take a share in its management'. In his diary Bigelow also noted private conversations he had with fellow Americans based on the Isthmus. Several who worked for the steamship companies told him that the Company would never finish the canal. Another told him that for the past nine years the corruption on the Canal and on the PRR had been 'shameless and that the Americans out there on the PRR rather excelled the French on the Canal'.

After three days, the party headed across the Isthmus. For the arrival of de Lesseps in Panama City a huge pageant had been arranged. Everywhere de Lesseps was acclaimed on his passage round the town by 'Vivas', shouts of 'Long life the Genius of the Nineteenth Century; long live the Man of Progress, the Grand old Frenchman.' Celebrations continued into the night with a torch-lit procession and, reported an exhausted de Molinari, 'banquets, dancing, lights, fireworks and who knows what else . . .'.

The party stayed for a further week, exploring the Pacific side of the works by day and dancing and banqueting by night. As before de Lesseps was 'indefatigable', restoring confidence everywhere he went. 'His stay is one continued fête . . .' wrote the new British consul James Sadler back to London. Had it been down to the people of Panama, he said, the lottery loan would be assured. De Lesseps confidently predicted that the year ahead would see 12 million cubic metres excavated; the next, 1887, would achieve twice that; and by 1888 there would be a monthly rate that would produce 36 million cubic metres for the year. With the same rate carried into the following year, he said, the canal would be completed by July 1889.

As the visitors prepared to return home, with no illness suffered, almost everyone declared themselves highly impressed. 'I am delighted to say that our expectations were exceeded,' de Molinari concluded. 'Even if the piercing of the Isthmus presents enormous difficulties, the effort made to conquer them is in proportion. Never has such a colossal work been undertaken, never have capital and science so united to deploy such a powerful machine to bring to an end the resistance of nature.'

John Bigelow was also overwhelmed by the grandeur of the project's ambition and size, writing that it had 'no parallel among private enterprises in all history'. He would remain a 'convert' to the cause of Panama, infected with its fever, for the rest of his life. But in his widely read report he also repeated many of the criticisms of Kimball – that plant was idle or discarded due to lack of leadership or system; that contractors were unreliable and corrupt; the challenges of the Chagres and the deep excavation in the Cut were not being met; that information from the company was 'often conflicting and rarely more than approximative'.

The fact that it was the dry season had not prevented him from noticing the 'insalubrity' of the Isthmus, particularly in the swamp area at the Atlantic end. 'You have a climate,' he wrote, 'where it may, without exaggeration, be said that – "Life dies and death lives".' Although, he said, 'human life is about the cheapest article to be purchased on the Isthmus', wages had steadily risen to keep attracting workers and mechanics to Panama, to a minimum of $1.75 silver a day, rising for skilled work to five times that.

Like Kimball, Bigelow believed that the fate of the canal would be decided by its finances. But he also agreed that 'too large a proportion of its cost has already been incurred to make a retreat as good polity as an advance'. In the meantime, the 'people of small means' who held the Panama stock would stick by de Lesseps, Bigelow believed, because success 'would rank among the half dozen largest contributions ever made to the permanent glory of France'.

Remarkably, he also anticipates what would be the American agenda in the next decade, a time that would see a radical change in the international balance of power and a transformation in US foreign policy. An open waterway would, Bigelow suggested, 'secure to the United States, forever, the incontestable advantage of position in the impending contest of the nations for the supremacy of the seas'. But there was a serious caveat: until the money was secured, and, crucially, the cost of the debt ascertained, it was impossible to say when and on what financial or political terms the work would be finished. 'And, for aught I see, this uncertainty must last until near the completion of the work,' he concluded, 'for nowhere in the world is the unexpected more certain to happen than on such a work at Panama. It is destined to be, from first to last, experimental . . .'

*

While the Company, indeed, the whole of France, awaited the verdict of Armand Rousseau, de Lesseps tried a new tactic to raise money – selling bonds on the Bourse rather than by private subscription. The experiment was not a success, with less than 40 per cent of the issue sold, even at an interest rate approaching 7 per cent. The decision on the lottery was now more important than ever.

Rousseau's report was submitted at the end of April 1886 to the new Minister of Public Works, Charles Baïhaut. 'I consider a cut through the isthmus . . . is a feasible undertaking,' Armand Rousseau wrote, 'and that it has now progressed so far that its abandonment would be unthinkable', a disaster not just for the shareholders, who were almost all French, but also 'for French influence throughout America'. If the Company failed, it would, he predicted, certainly be taken up by a foreign company, wanting to exploit the enormous sacrifices and the progress so far made. 'I believe that the government should . . . assist it,' he decreed. But before sending the lottery bill to the Chamber of Deputies, he warned, the government would have to content themselves that the Company was addressing the project's 'certain grave technical defects'. 'Important reductions and simplifications' were needed if the project was to be completed in anything like the time envisaged.

What he meant, though felt it was outside his remit to say, was that the sea-level plan was unworkable and had to be altered before it was too late. But, in combination with Rousseau's lofty ambiguity, others were now spelling it out. Jacquet, another government engineer who had accompanied de Lesseps to Panama, reported to the Cabinet that the sea-level plan was categorically impossible. It seemed likely that the bill would never make it to the Chamber unless this issue was resolved. On the Isthmus, too, in spite of continuing strong excavation figures for the early months of the year, Bunau-Varilla and others were exploring alternative visions to de Lesseps' 'Ocean Bosporus'. Léon Boyer, the new *Directeur Général*, took only a month to make up his mind that a canal *à niveau* was simply unachievable with the money and time at his disposal. Now, he urged de Lesseps, only the rapid adoption of a lock-canal plan could save the project.

But de Lesseps was not to be moved. He reluctantly agreed to certain time- or money-saving modifications – a reduction in the canal's depth and the scrubbing of plans for a large basin at Tavernilla in the middle of the waterway, for ships to cross each other in transit, thus increasing the crucial billable tonnage through the canal. Improvements to the terminal ports would, he promised, be reduced to a minimum. All these changes, de Lesseps insisted, could and would be reversed once money started to come in from tolls when a 'minimum canal' was open to traffic. After all, Suez was still being dredged and widened, nearly twenty years after its opening. But on the key issue – the conversion to locks – he refused to comply with the wishes of the government inspectors and the urgent appeals of his own senior engineers on the spot. The promise of an open, sea-level waterway, and its superior operating profits, had been the whole reason for choosing Panama in the first place. De Lesseps had from the very start nailed his colours to the mast by so energetically selling to the French public the simplicity and beauty of the idea of a canal *à niveau*. It would have been an embarrassing reversal if he gave in to the pressure.

By an unhappy coincidence, fever at a crucial moment robbed de Lesseps of two experts who might have changed his mind. In May, Boyer was suddenly prostrated, then dead. It was, the British consul reported, a particularly severe case of yellow fever. Bunau-Varilla, too, had 'been awakened suddenly' one morning soon after the end of de Lesseps' visit, by 'a violent vibration of my bed which I thought was a seismic movement'. It was 'the shakes', followed by a dose of fever, and before the end of April he had been invalided back to France. But even if the pair had been fit, it is uncertain whether de Lesseps would have listened to them. He didn't work by committee. At Suez, everyone had told him he was heading for ruin, but he had confounded his critics by never giving up. Panama, he now at last admitted, had proved many times more difficult than Suez. Yet to show weakness, he calculated, would surely be fatal to confidence.

The lottery issue had drawn the Company and the ever-shifting Third Republic political establishment together. It is hard to say who was the more hard done by. The Company found itself involved in a corrupt and fearful government, already mired in scandal. No fewer than twenty-six different cabinets served during the first eighteen years of the Republic, some lasting only a few days. The politicians, for their part, were now faced with juggling the interests of hundreds of thousands of French shareholders, as well as national pride, with the reality of the Monroe Doctrine, and local US military strength in Panama. The first reaction of the Cabinet to the Rousseau report was to seek to delay making a decision. But the report was leaked to the press at the end of May, causing a stir with a French public accustomed to the soft soap of the *Bulletin* and the pliant French press. Then, to everyone's surprise, Charles Baïhaut drafted a bill in favour of the application and presented it to the Chamber on 17 July. It later emerged that the politician had approached Charles de Lesseps and demanded 375,000 francs to perform the favour. The money had been paid and the bill moved onward, albeit slowly. The Chamber appointed a committee, which heard from Rousseau, de Lesseps and others. But still, as time ticked away for the effort on the Isthmus, nobody wanted to make a decision. On 8 July, the committee adjourned for the summer undecided. In the meantime, they asked, could they have a look at the Company's books and contracts?

De Lesseps was incensed. He simply could not wait that long for new money, nor risk the final verdict going against him. And to open up the books would be a clear admission of guilt. It was unthinkable. The following day he withdrew the application, telling shareholders, 'They are trying to shelve me – I refuse to be shelved . . . I work on, but not alone, assuredly, but with 350,000 Frenchmen sharing my patriotic confidence.'

The last chance to save the French canal had slipped away, through a combination of the dithering of the politicians and de Lesseps' stubborn refusal to give up the sacred sea-level plan. *The Times* warned of the political fallout in the likely event of the Company going under, predicting, 'The Republic will stagger under such a blow.' 'It indeed looks as though "The beginning of the end" is in sight,' wrote the New York *Tribune* of what it called the 'great imposture'. 'It is more and more a matter for congratulation that the American investor has never had faith in the Panama Canal scheme of M. de Lesseps, though of course sympathy will be felt for the unfortunate Frenchmen who have put their money in this gigantic sink-hole.'

At the end of July, de Lesseps faced the shareholders again, delivering once more an inspired bout of oratory and wishful thinking. 'In a year hence,' he announced, 'the machinery accumulated by the company will so assert its power that no doubt will remain of the success of the undertaking.' Now, the entire canal had been contracted for. Speculators, 'chiefly foreigners', had been attacking the Company's financial position, but the 'shareholders were too numerous and experienced to be deceived'. Once more, the assembled through rose to cheer de Lesseps to the rafters.

'M. de Lesseps in his reports coolly and intrepidly misrepresents almost everything upon which he touches,' reported the *Tribune* of the meeting. 'One thing, at any rate, is certain,' the paper concluded. 'M. de Lesseps's promise that he will complete the canal in the time or at the cost specified is mere boasting. He can no more do it than he can change Colombian rock to Egyptian sand.' Soon after, the paper started printing a series of lurid articles from an 'occasional correspondent' on the Isthmus, which painted a picture for its anti-canal readers of delusion and disease in Panama. 'Nothing is ever done by the canal company without a great amount of pomp, circumstance and red tape,' he wrote in August that year. But, he warned, 'of what one hears in Panama disregard one third, doubt one third, and disbelieve the other third . . . The air is as rife with deception as with miasma.' The canal employees were living in a pathetic, heroic dreamland. 'Where he, through canal spectacles, sees a dump,' the correspondent reported of one conversation with a French engineer, 'you see a virgin forest . . . Talk to them about [the work], and they lash themselves into a state of frenzied enthusiasm over their chieftain "*Le Grand Français*," . . . with a great rattling of glasses they drink long life to the company and damnation to the scoundrel stock-jobbers.' 'Every difficulty by which the company is beset requires two distinct efforts,' the writer concluded, 'it must be overcome and it must also be kept secret from the supporters of the project, who would otherwise be discouraged.'

Among the Anglo-Saxons, he reported, the talk was all of what would happen after the inevitable demise of the Company. 'Some Americans on the Isthmus express the belief that the work will be completed with American capital and under American control. The English say that they are sure, and always have been, that the work will ultimately become a British enterprise, and cite the precedent of the Suez Canal as a case in point.'

All over, emotions were heightened by the constant proximity of death by fever. Of the sixty hand-picked engineers Boyer had brought with him, nearly all were sick, demoralised or dead within a few months. The British consul, James Sadler, was invalided off the Isthmus in April, together with the son he had brought with him. Once more, Claude Mallet, returned from his convalescence in Jamaica, stepped into the breach. Since the advent of the rainy season, more than 80 per cent of the chief officials of the Company were out through sickness. Of thirty Italians who had arrived in a shipment twelve months earlier, only five now survived.

S. W. Plume was an American railway man, a veteran of South American projects. In 1886, after two years on the Isthmus, he was in charge of a gang of about a hundred workers, replacing rotten ties on the railroad. 'Every month or two I would lose a man, perhaps two men,' he told a US Senate committee some years later. 'I will explain it to you. If a man gets wet there with the rain he is sure to be sick the next morning . . . I never saw such a climate in all my life, and I have worked in the rice fields of South Carolina, and gracious only knows that is bad enough.'

The streets now saw a constant stream of funeral processions, and trains ran all the time to the cemetery at Monkey Hill. 'When I first went there,' said Plume, 'we used to run one train – perhaps it would be a car or two box cars – in the morning out of Colón, to Monkey Hill.' But by 1886, it was 'bury, bury, bury, running two, three, and four trains a day with dead Jamaican niggers all the time. I never saw anything like it. It did not make any difference whether they were black or white. They died like animals.' According to the *Tribune*'s 'occasional correspondent', some men, having expired from fever, were simply left out in the street to be pecked at by buzzards. In response, wrote the *Tribune* man, the Company's senior employees showed 'insane recklessness' and took up a 'habit of life . . . such as would result in wide-spread disease in any hot climate, even the most salubrious'. The day would start, he says, with a champagne cocktail laced with quinine for breakfast. Drinking of alcohol would continue all day, and in the evening, when it was too hot to read or play cards, it would be more of the same.

In spite of all this, the Company kept up a monthly excavation rate of a million cubic metres through mid-1886, according to Bunau-Varilla, who returned, after convalescence and now immune to yellow fever, late in the year. In the circumstances this was impressive, but anyone who wanted to could see that this was still not nearly enough to complete a sea-level canal within reasonable time. In the Culebra Cut, which had to be dug in places to a depth of over 300 feet, only an average of 12 feet had been removed, a paltry rate of just 3 feet a year. The Anglo-Dutch Company had been taken on in December 1884 and contracted to remove from Culebra 12 million cubic metres in four years, but after eighteen months had managed less than one million. As Bunau-Varilla wrote, they had proved to be 'a dismal failure'. 'During the dry season,' he explained, 'the works seemed to justify the best hopes. As soon as the first rains began, the dumps began to slide, the tracks were cut, and general subsiding of the ground inside the cut paralysed any movement of trains, and often overthrew the excavating machines.'

They were dismissed and a new contract was awarded to Artigue et Sonderegger. The new contractor's guiding light was Bunau-Varilla, whose brother was also high up in the same company. The new contract was part of a wider reorganisation. One outcome of de Lesseps' visit was that in mid-1886 the job of digging the canal was given to six large firms rather than the host of small contractors. This certainly helped reduce the suffocating bureaucracy of the Company and avoid the confusion and waste symptomatic of the 'small contractors' period. But it was expensive. Numerous firms had to be paid off, and the new work was inevitably contracted at a higher rate. Huerne, Slavens were kept on, with an even better deal. Artigue et Sonderegger's remuneration was so lavish that it led to the resignation of the Company's secretary in Paris.

There, the Company directors still hoped that the French government would rescue the project. After all, wasn't Charles Baïhaut, the Minister for Works, telling anyone who would listen that he believed in the canal, whatever the Rousseau report had said? In the meantime, the Company directors came up with a new tactic to raise funds.

August 1886 saw a new bond issue. The Company, seeking 250 million francs, offered 500,000 bonds for sale at 450 francs each, which, maturing in 1928, would be redeemable then at 1000 francs. Interest would be paid at 60 francs a year. It added up to a very good return of over 9 per cent. There was a further inducement that showed the Company sailing close to the wind. Twice a month, several bonds would be pulled out of a hat and redeemed at their full maturity value. It was as close to a lottery as the Company dared go without government sanction.

More than 100,000 subscribers came forward, and all but 10 per cent of the issue was sold out. It was a marked improvement on the issue of the previous April. The small peasant or artisan investor, what the *Tribune* called the 'Company's force, the whole secret of its power', was, it appeared, still backing de Lesseps. But the cost of the money, a key factor identified by Bigelow, was astronomical. After paying hefty fees to the bankers who organised the issue, the Company raised 195 million francs, for which it had undertaken to pay back 500 million, plus interest, over the next forty-two years. In the same month, in a carefully considered and well-informed editorial, *The Times* in London praised the courage and determination of the French effort, but analysed the money still needing to be spent against the long-term servicing of that debt through tolls as simply unworkable. 'It is magnificent,' the paper concluded, 'but it is not business.'

At last de Lesseps began to give way. In January 1887, he ordered his technical advisory committee in Paris to meet to consider the possibility of a lock canal. But he remained insistent on the original vision of an open waterway, demanding that all projects that construed permanent locks be excluded. Again, there was delay. A sub-commission was appointed which did not report until the autumn. The result was indecision on the Isthmus, as the latest US Navy inspector, Lieutenant Charles Rogers, reported after his visit in March 1887. The progress of the previous year had, he said, been 'creditable', virtually meeting Ferdinand de Lesseps' target of 12 million cubic metres. But he was doubtful whether the rate of excavation could be doubled in 1887, as planned. Moreover, he calculated that the Company had only enough money to continue for another three and a half months.

Nonetheless, Rogers, as many before him, was overwhelmingly impressed with the ambition of the scheme and the dedication of the project's leaders on the Isthmus. 'The most bitter opponents were our own countrymen and a few Englishmen or former employees of the canal who had been discharged or had some other grievance against the company,' he remarked at the end of his report. Such types 'were prone to exaggerated statements . . . or else malice'. 'The contractors are young, zealous, and energetic men,' said Rogers, 'the engineers are . . . both clever and capable, and no one can appreciate more than these men the difficulties that lie in their path. Instead of censure and detraction, they deserve the highest praise and respect . . . they wish well to an enterprise fraught with so much good for the human race, and they are doing their utmost under the circumstances to promote its success.'

Not all the Frenchmen on the Isthmus, of course, were quite so high-minded. At thirty-seven, Paul Gauguin had gone from riches to rags. His job as a broker had not survived the downturn of 1882, and he had since ruined his finances and his health through his taste for absinthe. In 1886, his long-suffering Dutch wife gave birth to their fifth child, and, with the situation desperate, he called on his wealthy sister, Marie, for help. She agreed to appeal to her husband, Juan N. Uribe, a rich Peruvian businessman who had offices and outlets all over and in particular, of late, in Panama. In the country where France was digging the canal, money was being made by the fistful and Uribe's fortune, it seemed, was benefiting from the project. Gauguin learnt that Uribe was setting up a brokerage office and bank and needed someone who knew finance and could be trusted to replace him when he took his holidays in Europe.

In March 1887, Gauguin resolved to try his luck on the Isthmus, writing to his wife, 'I will set off for America. I cannot continue to live here swamped by debts, a stultifying and lacklustre existence.' The plan was to settle in Panama, in the benign glow of the inspiring great work, and send for his family once he was established.

He arrived at Colón at the end of April, together with his friend, fellow artist Charles Laval. They were unimpressed with the town where, it seems, new hovels had sprung up since the fire, but nothing had been cleared away. In Panama City there was further disappointment. His brother-in-law was not running a bank but a general store, and not a very grand one either. There was certainly no work to be had for Gauguin. Finding that the price of land in Panama had so risen that it was unfeasible for him to settle there, Gauguin headed for the island of Taboga, which he hoped to find 'practically uninhabited, free and fertile . . . the fish and fruit can be had for nothing'. 'I'm taking my paints and brushes,' he wrote to his wife, 'and will, living like a native, immerse myself far from mankind.'

But he was in for another disappointment. Taboga had become something of a tourist trap. It was the favoured location for picnickers and day-trippers, and was dominated by the large sanatorium, to which well-favoured Company employees would retreat to be free, for a while, of the stultifying heat of the mainland. Well-organised guided tours crisscrossed the island. The 'native villages' had long ago wised up, and were now more than a little sham.

Gauguin returned to Panama, but the hotels were expensive, and his money was now running out. His friend Laval got work painting portraits of some of the better-off Company officials. Gauguin declined to do this, instead getting himself taken on as a labourer on the canal works. There he found that rumours were rife of the impending bankruptcy of the Company, and the reality of the job was far from the noble project he had envisaged. Gauguin was set to work with a pickaxe, chipping out holes for the dynamiters who would follow him. 'I have to dig from five-thirty in the morning to six in the evening under tropical sun and rain,' he wrote to his wife. 'At night I am devoured by mosquitoes.' The death toll wasn't that bad, he added sardonically, 'only 9 out of 12 of the negroes die while for the rest it is a mere half'.

Gauguin resolved to work only as long as it took to earn the fare off the Isthmus. But after a couple of weeks he fell foul of the new law and order regime of General Vila, who was back in Panama as Núñez's strongman, determined to stamp out dissent and resistance to firm rule from Bogotá. Gauguin was arrested for urinating in a street, which, he protested, was an open sewer anyway. He was imprisoned and then fined. Eventually back at work, he had just saved enough from his 600-franc a month pay for the fare to Martinique, when he was laid off, along with a raft of other workers, on orders from France.

Together with Laval, Gauguin left for Martinique on 8 June. On their arrival Laval came down with a fever, probably malaria. During one cycle of the disease, he became so depressed that Gauguin had to prevent him from committing suicide. Soon after, Gauguin, too, was ill. 'During my stay in Colón,' he wrote to his wife, 'I was poisoned by the malarious swamps of the canal and I had just enough strength to hold out on the journey, but soon as I reached Martinique I collapsed. In short, for the last month I have been with dysentery and marsh fever. At this moment my body is a skeleton and I can hardly whisper; after being so low I expected to die every night . . .' His stomach cramps and continued weight loss were probably due to amoebic dysentery, which had by then developed into what would today be diagnosed as hepatitis or an abscess of the liver.

The sacking of Gauguin was part of a general freezing of the Company's activities. It was more than indecision about the issue of locks. Money had pretty much run out, as Rogers had predicted back in March. Nothing more could be done until the outcome was known of de Lesseps' latest attempt to float a stock issue.

The July 1887 issue was along the same lines as the 'proxy-lottery' of eleven months earlier. This time, however, the take-up was only just over half. It was hardly enough to paper over the cracks. And again the money was ruinously expensive. By late autumn the Company's finances were once more in a perilous state, and speculators had forced the value of the original canal shares to a new low on the Bourse.

But at last, in October, a full two and half years after the Rousseau report and Léon Boyer's advice had been received, Ferdinand de Lesseps finally gave in to pressure to redirect the works towards the completion of a lock canal. What de Lesseps' commission recommended was based on experience on the Isthmus. Following the success of the excavation 'in the wet' at Mindi, after underwater blasting, several contractors had been experimenting with creating artificial lagoons along the line of the canal, and then assembling and launching water-borne dredges. These machines filled barges, which were then towed underneath a fixed-ladder excavator. This emptied them and lifted the spoil on to waiting trains or piped it out of the way. Thus water, rather than rail – vulnerable to rain and slides – would be used to carry out most of the moving of spoil. This led on to the new scheme. If pools could be created all along the route, separated by locks, then underwater excavation could continue with the canal open to traffic. As the various levels were lowered by dredging, the locks – five on either side of the Divide – could be gradually removed until the entire canal was at sea level. In the meantime, a working, and paying, lock canal would have been created, it was estimated by 1891, but, crucially, only as a means to the completion of what Bunau-Varilla called, 'the perfect, the final, project' of a canal *à niveau*.

Characteristically, Bunau-Varilla claims the entire credit for this plan. In fact, he had, in early 1887, built dams at either end of the Cut and experimented with excavating 'in the wet'. Whoever's idea it was, its strength did not lie in its engineering aspect alone. Indeed, the cost of the gradual transformation would still have been prohibitive. In addition, the constant supply of water to the summit level of the canal, essential to operate the locks, was dependent on vague, unsurveyed schemes to build tunnels or viaducts from higher up the Chagres River. But the beauty of the plan was the distance it went towards reconciling the reality in the Culebra Cut with selling the change of plan to de Lesseps and his army of supporters, who had for years been persuaded of the overwhelming superiority of a sea-level trans-Isthmian route.

There was still the question of finding the new six hundred million francs that this work had been estimated to cost. The following month de Lesseps reapplied to the government to run a lottery, outlining the temporary locks plan. At the same time, he announced that France's most brilliant engineer, Gustav Eiffel, had accepted the task of constructing the locks. Eiffel was newly famous as the creator of the giant iron structure just being started in Paris for the 1889 exposition. He was a good catch for the Company, although his name and expertise, it later emerged, came at an outrageous cost.

Nonetheless, Eiffel swung into action and by January 1888 his men were on the Isthmus starting the excavation of the lock basins as the giant iron parts began to be shipped from France. In the meantime, de Lesseps set out to sell the new plan to the stockholders, reassuring them that the original vision was postponed, rather than lost, and that by 1890 the canal would be sufficiently advanced for the passage of twenty ships a day.

From the government, however, there was an ominous silence. Then, after two months of deliberation, the Cabinet announced that they would not be submitting the necessary lottery bill to the Chamber of Deputies. Once more the

Company launched itself into lobbying, organising petitions, and, it later came out, outright bribery of politicians. So, in early March, nine Deputies of various political complexions introduced the bill the government had refused to back. Yet another commission was appointed to investigate.

De Lesseps could not afford to wait, and had to go with another bond issue. He promised that if the lottery was approved these new bonds could be converted to lottery bonds. But the issue in March 1888, the eighth in as many years, was the worst yet, with only a quarter being taken up. Clearly no more money was going to be forthcoming from this route. Soon after, de Lesseps was forced to borrow 30 million francs at a ruinous rate from his 'friends' at the Crédit Lyonnais and Société Générale banks to keep the Company afloat. The lottery was now the only hope.

It looked at first as if the commission would reject the bill, but after the surprising last-minute change of mind of Charles François Sans-Leroy, a hero of the Franco-Prussian War, it was approved by a margin of six to five. The debate in the Chamber, which was packed with canal supporters, on several occasions degenerated into a brawl. Nonetheless, the bill was approved by a wide margin on 28 April, and was rubber-stamped by the Senate on 8 June, though the Company was compelled to state in its loan prospectus that the granting of permission implied no government guarantee or responsibility.

Shares in the Company, which had fallen to a low of 250 francs in December, now soared. On the Isthmus, Bunau-Varilla believed that nothing could now stand in the way of the successful completion of 'his' plan. In the Culebra Cut, he now only had to lower the floor to 140 feet above sea level, rather than thirty below. He was confident he could do this in three years, and now had nearly three thousand men on the site, working around the clock with the assistance of recently installed floodlights. The actual results of his company, Artigue et Sonderegger, do not back up Bunau-Varilla's boasts. In addition, in spite of the harsh regimes of Vila and his Conservative successors, tension and violence among the workers was again on the increase, made worse as the Company was forced to look further afield for recruits, bringing new communities, such as Africans and Puerto Ricans, into the volatile racial mix. Yet whatever the chances of success with the lottery money secured, it was certain that should the issue fail, then all would be lost.

The lottery bill authorised the Company to borrow a further 720 million francs, 600 for the completion of the work and the rest for investment in French government securities to guarantee payment on the bonds and to furnish the cash prizes. Two million bonds bearing 4 per cent interest were offered for 360 francs each, redeemable in ninety-nine years for 400 francs. Bi-monthly draws promised maximum wins of nearly 700,000 francs. Against their better judgment, the

Company was persuaded by their bankers to offer the entire sale in one go. It would start on 20 June and run for six days.

The Company threw everything it had at efforts to promote the sale, spending, it came out later, over 7 million francs on 'publicity' and over 3 million on 'patronage'. No one in France was unaware of what was at stake.

On the morning of the start of the sale, someone put a hoax notice out by wire that Ferdinand de Lesseps had died. There was an instant denial, but the damage was done. Two days later speculators dumped Panama shares on the Bourse, causing a sharp fall in its value. By the end of the six days, although 350,000 people had subscribed, three times the 1880 figure, less than half of the bonds had been sold. Money, it seemed, was exhausted. Once the sum had been put aside for interest and the prize fund, the Company had gathered in only a little more than 100 million francs, a sixth of what it required.

Still, de Lesseps refused to accept defeat. On 1 August, addressing the annual general meeting, he urged his troops on to one final, patriotic effort. 'All France,' he announced, 'is joined in the completion of the Panama Canal. Actually more than 600,000 of our compatriots are directly interested in the rapid success of the enterprise. If each of them will take two lottery bonds or get them sold, the canal is made!'

Then de Lesseps and his son Charles set off on a gruelling tour of twenty-six French cities, with Charles now doing most of the speaking, while his father's presence still guaranteed huge turnouts. 'Spontaneous' local committees were organised by the Company to recruit new investors.

The remaining bonds went on sale on 29 November, with a final exhortation from de Lesseps: 'I appeal to all Frenchmen,' he said. 'I appeal to all my colleagues whose fortunes are threatened . . . Your fates are in your own hands. Decide!' By this time, the price had been cut to just 320 francs, with generous terms for paying in instalments. It was decided that unless 400,000 were sold the subscription would have to be annulled.

Soon after the opening of the sale, bear raiders made another attack. By 8 December, lottery bonds on the Bourse were selling for 40 francs less than de Lesseps was asking. The final day of the sale had been set for 12 December. The day before, the American journalist Emily Crawford visited the Company's headquarters at 46 Rue Caumartin. The hall of the building was packed with investors, 'flushed and excited, but willing to stake their last penny on the hope of retrieving their fortunes . . . They were like desperate gamblers,' Crawford would report in the New York *Tribune*, 'whose hopes rise highest when their losses have been greatest.' The crowd grew in number and noise, until suddenly at 4.00 p.m., it was hushed by the appearance of Ferdinand de Lesseps. The old man clambered on to a counter in the corner of the room and cried out: 'My friends, the subscription is safe! Our adversaries are confounded! We have no need for the help of financiers! You have saved yourselves by your own exertion! The canal is made!' So overcome that he was weeping, de Lesseps joined the crush in the room, shaking the hands of his exultant, cheering investors.

No details were given, but soon rumours were circulating that 800,000 bonds had been sold. The next day saw the same chaotic scene in the Company headquarters, and then, late in the afternoon, Charles de Lesseps appeared. How had the subscription gone? everyone asked. 'The subscriptions reached a total of 180,000 bonds,' Charles began in a low voice. 'This being below the minimum fixed by M. de Lesseps, we will commence returning the deposits tomorrow. You see, I am telling you exactly how things are.'

There was a shocked, dazed silence. How had the picture changed so radically overnight? someone asked. 'My father is younger in spirit that I,' Charles replied. 'His remarks were made on the strength of a hopeful report . . . the result is bankruptcy or the winding up of the company.'

In desperation Ferdinand de Lesseps placed before the Chamber of Deputies a bill authorising the Compagnie Universelle to suspend payments of all debts and interest for three months, while he attempted to float a new company. On 15 December, the Chamber threw out his bill by 256 votes to 181.

Ten minutes after the vote, a reporter called at de Lesseps' house with the news of the bill's rejection and his company's liquidation. The old man turned pale. '*C'est impossible*', he whispered. '*C'est indigue*'.

15

SCANDAL

The crash of the Compagnie Universelle was the biggest of the nineteenth century, and the greatest since the markets began. It is a key moment in the history of finance capitalism, and opens a fascinating window on the forces of optimism, corruption and unreality that swept through Victorian financial life. But it echoes, too, many of the concerns of the modern markets – public confidence, rumour and disgrace.

The liquidation wiped out the hard-earned savings of 800,000 private investors. A staggering thousand million francs ($280 million) had been expended, and the Company had liabilities of nearly three times that sum. De Lesseps made one last, fruitless effort to save his Company, with a spectacularly unsuccessful bond issue in January 1889, but the next month an official receiver was appointed. *Le Grand Français* never recovered from the blow, retiring to his country house at Chesnaye. His great will finally broken, he quickly lost awareness of the world around him, preferring to stare out into the garden, or sit musing by the fire. By the summer of 1889, he was often confined to his bed.

While the investors struggled to comprehend what had happened, the liquidator, having ensured the barest minimum of maintenance on the Isthmus, started going through the books. Such was their complexity that the first man on the job was exhausted and ill within a year. His replacement also found it too much and soon resigned.

When the news of the liquidation arrived in Panama, it was, according to Tracy Robinson, 'like a stroke of paralysis'. Foreign consuls had been predicting fierce riots, due to the 'unsavoury' nature of most of the workers, and gunboats had arrived offshore from France, Britain and the United States. There were disturbances, but on the whole a sense of dumb shock prevailed, as contractors and workers alike laid down their tools, the giant machines were shut down and peace returned to the jungle for the first time in seven years.

All along the line thousands of labourers found themselves suddenly thrown out of work. Shops closed as merchants pulled down their shutters and relocated, the prostitutes and professional gamblers set off for more inviting pastures, the railway closed down stations, and the banks suddenly stopped honouring cheques from the Company. Rent rates and land values collapsed just as immediately.

Within no time at all, the workforce had fallen from 14,000 to just 800 involved in basic maintenance. Many of those laid off left the Isthmus, either for home or for other employment in the region. Two months later, a British visitor reported seeing 'a few labourers . . . retained, just sufficient to show that the works are not altogether abandoned . . . Everything here is quiet, and a great commercial depression exists.' But many of the workers, however, found they had insufficient funds for the rail fare to Colón, let alone the steamer back to Kingston. Slowly, they made their way on foot to Colón, where they congregated in desperate groups.

'There are hundreds [of destitute Jamaicans] absolutely starving,' reported the *Star and Herald* in early April, 'who have not tasted food for days . . . Despair is taking possession of the people.' Many sought the help of Claude Mallet, who telegraphed the Jamaican government to send steamers to pick the men up. Receiving no reply, he was forced to feed 1500 out of his own pocket. Eventually the Government of Jamaica agreed to start bringing their thousands of destitute countrymen home. In all, 7244 were repatriated. Some six thousand Jamaican men, women and children were left behind when the fixed period for repatriation ended, and most took up residence in shanties along the line, growing bananas and other crops for subsistence. Some also stayed behind as they refused to believe that the project could really be abandoned. It seemed impossible that so much work and sacrifice could have been in vain.

In France the immediate aftermath of the great crash, when attention was diverted by the great exposition of the centennial of the French Revolution, saw a similar refusal to accept the end. Surely the French government would come to the aid of the scheme? But stern warnings from the new US President Benjamin Harrison not to interfere ended what little inclination to do so that might have existed in the French Cabinet. As this realisation percolated through the country, the recriminations started.

Throughout the year following the bankruptcy, petitions from bondholders seeking redress from the government poured into the Chamber of Deputies. At the same time, rumours of wrongdoing in the Company and in the government gathered momentum. In June 1890, the Deputies voted to refer the business to the Minister of Justice, who seemed unwilling to proceed further as he saw insufficient grounds for an indictment and argued that the receiver should be given an open field.

The delay prompted the publication of *La Dernière Battaille* by Edouard Drumont. A professional anti-Semite, Drumont had made his name in 1886 with the publication of *La France Juive*, a thousand-page attack on the evils of finance capitalism. Drumont was a devout Roman Catholic and blamed the fall of the Union Générale in 1882, by which thousands of Catholics had been ruined, on 'The Syndicate', a sinister Jewish and foreign-backed conspiracy of financial interests bent on the corruption of judges, ministers and journalists. The same villains had organised the Franco-Prussian War, Drumont alleged. The book had been a runaway success, despite the previous marked lack of anti-Semitism in France. Drumont's new book had in its sights the greatest ever venture in finance capitalism – de Lesseps' Canal Company. Thirty thousand had died in Panama, he wrote in *La Dernière Battaille*, while the bosses had lived in luxury; 'champagne had flowed in torrents', as all the time the Isthmus was becoming 'an immense boneyard'. At fault was *Le Grand Français* ('an anti-phrase in reality') himself, a fraud, a cheat and a liar. Addressing de Lesseps directly, he asked, 'What have you done with the money?' The book was a bestseller, and launched a fever of recriminations against the Panama project.

As the voracious Paris press jumped on the story, the Minister of Justice was forced, in June 1891, to appoint an examining magistrate to investigate the leaders of the scheme. The first step was to summon the eighty-six-year-old de Lesseps, his son Charles and another senior Canal Company director. Ferdinand de Lesseps, against the advice of his doctor, roused himself, donned his uniform of a Grand Officer of the Legion of Honour, and went to meet the investigator. According to Charles, 'He had apparently recovered all his strength; he remained three quarters of an hour . . . and when he left his face radiated charm and energy as it always did under difficulties. But the reaction soon set in; it was frightful.' Back home, the old man took to his bed and hardly spoke to anyone for three weeks, except to say to his wife, 'What a terrible nightmare I have had. I had imagined I was summoned before the examining magistrate. It was atrocious.'

The magistrate's report, delivered in May 1892, accused the Company of 'dissipating' funds 'in a manner . . . more consistent with the personal views and interests of the administrators and directors . . . than with the true interests of the company.' Secondly, they were culpable of repeated 'false announcements of progress', which concealed the 'true situation' and misled investors. The files were handed over to the Public Prosecutor, who took his time sifting through the evidence, while police raided the offices of ex-canal officials and engineers, including Gustave Eiffel.

The delay gave the anti-Semite Drumont a chance to renew his attack on the canal, and the whole corrupt system, which, he alleged, had led to the disaster. In September his new newspaper, *La Libre Parole*, started a series of articles headed 'The Secrets of Panama' under the by-line 'Micros'. This turned out to be an ex-employee of the Company with a grudge. The revelations were sensational: more than twenty Deputies, he said, had taken bribes to support the lottery bill. The writer named the Company's chief financial adviser, Baron Jacques de Reinach, conveniently for Drumont a Jew of German birth, as the distributor of the bribes. It was alleged that de Reinach had used an intermediary, Léopold-Emile Arton, another Jew, to approach the Deputies.

Arton had disappeared, a fugitive from justice for another crime of corruption, so de Reinach was the next to be questioned by the ongoing investigation. Yes, he conceded, he had passed Arton 1 million francs, 'for publicity', and had personally distributed 3 million francs to the press. But, he said, he had never bribed any elected official.

The Public Prosecutor had baulked at a criminal investigation, unable to see the great Ferdinand de Lesseps facing a prison sentence, but at a special Cabinet meeting on 15 November, the premier, Emile Loubet, handed the matter over to the new Minister for Justice, Louis Ricard, who, as it happens, bore a long-standing grudge against de Lesseps. 'My decision is firm,' Ricard announced. 'I have already sent to the prosecuting attorney the order to bring criminal proceeding against de Lesseps, Eiffel, and others.' The charge was fraud and maladministration. Included on the summons was Baron de Reinach.

But early on 21 November, just as the papers were about to be served, the news exploded that de Reinach had been found dead. His family doctor diagnosed 'cerebral congestion' but the rumours had it as suicide or murder, or even that he had escaped over the border leaving behind a coffin full of stones. Later that day, Jules Delahaye, an outspoken Deputy and Drumont sympathiser, charged in front of a packed Chamber that 3 million francs had been distributed throughout the house by the Canal Company, and that no fewer than 150 members had taken bribes to assist the Company in its dealings with the government. 'There are only two kinds of Deputy,' he announced in a loud voice. 'Those who took the money, and those who did not.' 'The names, the names!' the Chamber, in violent uproar, demanded. 'If you want names, you will vote an inquiry,' Delahaye shouted over the din.

The government had no choice but to order a formal parliamentary investigation into the allegations of 'Micros' and Delahaye, and in the meantime the Deputies were demanding an autopsy on the hastily buried de Reinach to find out the truth about his death. The Cabinet hoped by refusing this to quench the flames of the Panama scandal. They were unsuccessful and were forced to resign within a week.

Now 'the Panama Affair' had outgrown its original focus and became a general stick with which to beat one's political opponents. Out in the country, too, the word '*Panamiste*' had a wider meaning as synonymous with corruption anywhere. '*Quel Panama!*' meant 'What a mess!'. On 6 January *La Libre Parole* organised a large anti-Semitic rally in the centre of Paris, where the speakers proclaimed the Panama disaster the fault of the Jews, who, they said, were now laughing at France's misfortune. While most of the crowd cheered, a small number protested. A fight broke out, which degenerated into a riot. Paris was now jumpy. A bomb in a police station was blamed on anarchists, whose numbers were swelling as disillusion with the government grew. Then, amidst talk of a royalist coup, military units were put on the alert. Foreign consuls described France as on the brink of a revolution.

Such was the backdrop for the sensational trial of de Lesseps father and son, Eiffel and two other Company directors, which started on 10 January 1893. Ferdinand de Lesseps was excused from attending because of his rapidly fading health. The defendants were well represented. Neither Ferdinand nor Charles de Lesseps had made any real money from the canal, their lawyer pointed out. Clearly, neither had benefited from any fraud. As for maladministration or misleading the investors, didn't every great project run vastly over budget? The Marseilles Canal was supposed to cost 13 million francs, but was not finished before an expenditure of forty-five; the Manchester Ship Canal, too, had massively exceeded its budget. If the de Lesseps had sinned, the defence argued, it was only through 'excessive optimism'.

But the public clamour for a scapegoat had made the Advocate General determined to get a conviction, describing the attempt to build the canal as 'the greatest fraud of modern times'. Ferdinand de Lesseps and his son were both indicted for fraud and maladministration and given five-year prison sentences. Charles de Lesseps, his face in his hands, wept as the judgment was handed down. It was taken for granted that his father, who, reportedly, was not even aware of the trial taking place, would not be fit to go to prison. Two other directors got two years and a fine. Eiffel, found guilty of making a 7-million-franc profit for work he had barely started, was sentenced to two years' imprisonment and a fine of 25,000 francs.

Four months later the sentences were quashed on a technicality by the Supreme Court, but by then the situation had worsened for Charles de Lesseps. Although the belated autopsy on Baron de Reinach had produced no new evidence – his entrails were too badly decomposed to test accurately for poison – the Parliamentary Investigation, carried out in a febrile atmosphere of allegation and counter-allegation, had turned up one sensation after another.

The Chamber of Deputies Committee did genuinely try to answer the question posed by Drumont back in 1890: what happened to the money? Much, it was immediately apparent, had gone to the large contractors taken on in late 1886. Few of them had achieved a fraction of their contracted total excavation, but some had somehow, nonetheless, cleared huge profits. One company made over 20 million francs for having excavated less than three and a half million cubic metres out of its contracted total of 29 million. Bunau-Varilla's Artigue et Sonderegger had pocketed a profit of over 11 million francs, in spite of having manifestly failed in its allotted tasks.

The various banks and syndicates that handled and occasionally underwrote the bond issues had also made eye-wateringly huge profits from their business with the Company. Some 10 per cent of the Company's total receipts, over 100 million francs, had disappeared against the cost of the flotations. Everyone, it seemed, had had their hand out.

Part of this cost for the flotations was money paid to the press, and not all of it for advertising space. Twelve million francs had been distributed between 1880 and 1888, it emerged, with the largest 'subsidy' going to *Le Petit Journal*, which had stayed loyal to the Canal Company right till the end. An astonishing 2575 magazines and periodicals had received cash payments to plug the canal venture, including, bizarrely, *Marriage Journal, The Poetic World* and *Foresters' Echo*. Some small journals, it appeared, had been established purely to benefit from the 'Panama Cheque'. The report was damning of the Fourth Estate: 'With extraordinary imprudence,' it said, 'the press published the most deceptive articles, inspired confidence, and asserted that the canal would be completed even after the enterprise had become a hopeless wreck . . . It failed to perform its duty towards the public.' Two shame-faced newspaper editors even returned their subsidies, or part of them, to the liquidator.

And when the Company became involved with the government over the lottery request, it had found a whole new raft of even greedier and more dangerous parasites. When it moved on to the accusations of government corruption, the inquiry met a storm in the Chamber. Insults were hurled across the floor, Deputies challenged each other to duels, and on several occasions members had to be forcibly removed, kicking and screaming, from the Chamber. The ministers of Finance and Justice both came under suspicion and resigned. Then the investigators came across a banker who had, at de Reinach's request, changed a single cheque issued by the Canal Company for 3,390,000 francs into twenty-six separate cheques for varying amounts made out to bearer. The stubs were confiscated from the bank and on all but one were written the initials or the first few letters of the recipient's name. Amongt them were a string of distinguished Senators and Deputies, including Charles Sans-Leroy, who at the last minute had changed his deciding vote in favour of the lottery bill. The evidence caused a sensation and a new word, a '*chéquard*' – cheque taker – entered the French language.

When it came to the prosecutions, both the Chamber and the Senate had to pass special legislation to remove the immunity of their members. The list of the accused reads like a who's who of French political life. Among the Senators and Deputies were one ex-Premier, two ex-Ministers of Justice, one ex-Minister of Commerce and one ex-Minister of Fine Arts.

The corruption trial opened in March 1893, by which time the accused had been reduced to Charles de Lesseps and the Company secretary, on trial for giving bribes, and five Deputies and one Senator for receiving them. This included Sans-Leroy and Charles Baïhaut, who had forwarded the original lottery bill to the Chamber back in 1885.

Charles de Lesseps, in prison since December, seemed a different figure from that of the first trial. His 'face was drawn and his skin yellowed', and his earlier courteous manner had been replaced by a quiet outrage. He was not guilty of bribery, he argued, but a victim of extortion. The Minister for Works, Charles Baïhaut, had demanded 1 million francs to keep the lottery bill alive, he testified. The 375,000 francs was just the first instalment. When the bill was subsequently rejected, Charles de Lesseps had refused to pay any more. Since the failure of the very first share issue in August 1879, the Company had found it necessary to pay for press and political support, and for the goodwill of the Bourse. 'They seemed to rise up from the pavement,' Charles said of the *chéquards*. 'We had to deal with their threats, their libels, and their broken promises.'

Charles Baïhaut, uniquely, tearfully confessed to having received money from Charles de Lesseps for his support of the lottery bill. But Sans-Leroy insisted that although he had suddenly paid off debts to the tune of 200,000 francs, the money had come from elsewhere. Likewise, the other politicians denied any wrong-doing, and claimed the money they had received from de Reinach was for other business matters.

For the verdict on 21 March 1893, the aroma of whitewash was heavy in the air. All of the Deputies but Baïhaut, who 'had been stupid enough to confess', were acquitted of the charge of receiving bribes. Nonetheless, Charles and an intermediary he had used were still found guilty of bribery and were sentenced to one year's imprisonment each. Baïhaut was given five years' solitary confinement in the notorious Etampes prison, a huge fine, and was ordered to repay the bribe. Should he be unable to manage this, Charles de Lesseps became liable.

There had been one noticeable absentee from the trial. The two biggest of the de Reinach cheques, for a million francs each, were to Cornelius Herz. An original member of the Tür syndicate, the US-born Herz was supposedly an electrical engineer and inventor who had many friends in high places. He had fled to England shortly before the trial, where he remained out of reach, too ill to be extradited. It emerged that he was in fact an accomplished con man, who, for reasons that remain unknown, was blackmailing de Reinach for huge sums from the Canal Company coffers.

Herz soon became the centre of the mystery surrounding de Reinach's death, and accusations grew ever more lurid. Herz was a British spy, it was alleged. Then, in the Chamber, a Deputy called Paul Déroulède accused the grand man of French politics, the Radical leader Georges Clemenceau, of being Herz's accomplice in treachery. On a foggy afternoon at the St Ouen racetrack, just north of Paris, the inevitable duel was fought. Clemenceau was a crack shot, and before proceeding to the meeting Déroulède bade farewell to his friends, embracing them repeatedly and dividing among them a few locks of his hair. Clemenceau, who the day before had practised his shooting at a gunsmith's in Paris, more quietly reassured his tearful staff that they had nothing to fear.

The carriages of the two duellists were followed to the racecourse by a pack of pressmen, and soon a large crowd had gathered to see the action. It was to be at twenty-five paces, with each man allowed to shoot three times. At five past three, Déroulède, bare-headed, and Clemenceau, in a black hat, took their places. Both missed with their first shots. They fired again, and still neither was hit. At this point, the men's seconds and witnesses rushed forward, urging the antagonists to consider their honour satisfied. But they were shooed away, and the men fired again. This time Déroulède reached anxiously for his coat-tail, but was himself unharmed. Clemenceau, also, was unwounded, but back in the Paris the mud of Panama had stuck. He lost his seat and would endure a decade of political eclipse.

Herz never returned to France to face trial and took his secrets to the grave. The fact that he was Jewish, however, added further fuel to the fire lit by Drumont. Although the outbreak of anti-Semitism in France, which would lead four years later to the terrible Dreyfus Affair, was part of a larger Europe-wide pattern, the Panama Scandal, or, more exactly, Drumont's version of it, had provided the spark that started the fire.

A few months after his second trial, Charles de Lesseps was allowed to leave prison to visit his father on a day release. Accompanied by two policemen, Charles made the journey to Chesnaye. His father greeted him with 'Ah, there you are Charles . . . Has anything new happened in Paris?' The old man never commented on the ever-present policemen, and after a long walk with his father in the woods near the house, Charles was returned by his guards to his prison cell.

Soon after, Charles became ill and was in hospital when released in September 1893. But within a couple of months he was forced to flee to London, having become liable for Charles Baïhaut's fine. By this time, Ferdinand de Lesseps' mental state was mercifully such that he knew little of what was going on, and he remained sequestered at home within the family circle. He died in December 1894, a few days after his eighty-ninth birthday. However many people had lined their pockets at the expense of the shareholders, de Lesseps had not been among them. His family were so poor that the funeral expenses had to be met by the board of directors of the Suez Canal.

The waves of scandal did nothing to help the receiver in his battle to rescue something from the mess on behalf of the investors. The option of abandoning the project altogether was seriously considered, but the impossibility of dividing up the remaining assets among the legion of creditors, of all descriptions, and the almost negligible return it would have meant, persuaded the liquidator that the best option was to keep the concern alive – either to press on and build the canal, or sell it to the highest bidder. When the liquidator's representatives toured the work in 1890, they found a great deal of plant still in good condition. Fifty million cubic metres of earth and rock had been moved, they reported, equivalent to two-thirds of the total excavation of Suez. Their estimate of the value of the work done, the plant and the Panama Railroad was 450 million francs ($90 million). A sea-level canal was ruled out of the question, but a lock canal, it was estimated, could be built for 900 million francs, in eight or nine years. As the concession from the Colombian government was about to run out, Lieutenant Wyse was contacted and sent to Bogotá to ask for an extension.

Negotiations dragged on for months, the Colombians under the false impression that Wyse was backed by the bottomless coffers of the French state. In addition, US diplomats in Bogotá did all they could to block an extension of the concession. The deadlock was only broken by urgent appeals from Panama to conclude a deal. This was approved on 26 December 1890. A ten-year extension was allowed, provided work started before February 1893. But, while the scandal of recriminations raged in Paris, nothing was achieved about beginning work and the start date of the concession had to be extended to October 1894. The new deal allowed ten years for the completion of the canal.

If this were not to lapse as well, the liquidator had to get a new company up and running promptly, for which he estimated he needed 60 million francs start-up capital. There was no way that the public were going to be persuaded to fund the new venture, so the liquidator, and the bondholders' representative, the *mandataire*, together came up with an ingenious scheme whereby those contractors deemed to have made unacceptable profits out of the de Lesseps Company were forced to contribute to the financing of the new venture. In return, the threat of legal action would be lifted. Eiffel was the first to be targeted, in spite of his acquittal by the Supreme Court and growing celebrity following the success of his iron tower in Paris. At first, Eiffel determined to resist the pressure, moving his accounts to Belgium, and then to another country. But much of the Panama mud had stuck, and in February 1894, still proclaiming his innocence, Eiffel, 'desirous of coming to the aid of the liquidation and of the old company', handed over 10 million francs. Couvreux, the original contractor who had abandoned the project at the end of 1882, was next, accused by the liquidator of setting a dishonestly low price for the work before the launch of de Lesseps' company. Couvreux disclaimed any responsibility for the disaster on the Isthmus, but, as 'a friendly gesture', contributed half a million francs. One by one, the other contractors who were still solvent were forced to dig in their pockets; Bunau-Varilla's company, accused of corruption and kickbacks, invested just over 2 million francs. Only the American company Huerne, Slavens, 'one of the enterprises which made some of the most scandalously excessive profits out of Panama', escaped, having their books in the United States, safely out of reach of the liquidator. In all cases, 'penalty shareholders' were forbidden involvement in the running of the new company.

Then the *mandataire* went after the bankers who had creamed some 75 million francs off the Panama cake. The first one privately approached to invest in the New Company or face prosecution held out, but lost his case in the courts and was forced to contribute anyway. After that, most of the sixty-three approached agreed to pay up straightaway. To go after the corrupt press, however, proved too complicated and time-consuming.

The administrators thought they should give the French public a chance to have part of the Company and issued 600,000 shares in September 1894, but fewer than seven thousand individuals came forward, and only 35,000 shares were taken up, raising a paltry three and a half million francs. But together with the forced contributions, the liquidator now had nearly forty-three and a half million francs. The rest of the sixty million was made up by the liquidator, with money belonging to the creditors of the Old Company, and the Compagnie Nouvelle du Canal de Panama was formally incorporated on 20 October 1894, only eleven days before the new concession ran out. The same month, engineers working for the New Company arrived on the Isthmus.

What they found there was an open canal from Colón to Bohío Soldado, but at varying depths. Incursions by the Chagres, now in many places following through the excavated canal, had caused intermittent damage. The diggings were all covered in thick vegetation but this had helped prevent slides by holding the banks together. Termites had consumed everything non-metallic left out in the open, and Eiffel's great steel lock gates, found dumped on the beach near Colón, were ruined. Nonetheless, much workable machinery had been sufficiently well maintained to be put back into action.

The whole effort of the New Company is characterised by caution and parsimony. A vague plan had been adopted to continue with a lock canal, with a dam at Bohío, but in the absence of any firm blueprint, work, which resumed in October 1894, was concentrated in the Culebra Cut, which would need to be lowered whatever the final scheme decided on.

The New Company took advantage of the initial glut of labour on the Isthmus among those left stranded by the collapse of the Old Company to set a wage scale far lower than had been the norm the decade before. But this led to strikes in 1894 and 1895, when the Company found itself unable to attract sufficient numbers to the Isthmus. The governments of the British West Indies were hostile to recruitment on their patches as a result of the cost of having to repatriate their nationals after the failure of the Old Company. Eventually an arrangement, suggested by Claude Mallet, was agreed whereby the Company, or, more usually the emigrant himself, was required to deposit twenty-five shillings in the Jamaican treasury to be used if necessary for repatriation. The worker would also receive compensation and passage home in the instance of injury or disease. Under the terms of this deal, some three thousand men journeyed from Kingston to Colón.

The New Company enjoyed some good fortune. With much smaller importation of non-immune workers, the disease rates were nothing like the decade before, and at twenty-five deaths per thousand annually actually compare pretty well with the American period to come. But as an operation it was a shadow of the Old Company. The files show its letterhead to be reused Old Company paper, with the new name written in by hand. At its peak in 1897, there were only four thousand workers on the canal, excavating in a year what the de Lesseps Company had managed in a month. By 1898, half of the original capital had been expended, and there was still no sign of the French public putting their hands in their pockets.

Efforts were made to attract investors, while keeping the canal a French project. To make it more sellable, a detailed plan – for a lock canal with a dam at Bohío – was put together by yet another international commission of experts. Bunau-Varilla, acting independently and considered something of a liability by the New Company directors, even managed to catch the interest of the Tsar of Russia after a fortuitous meeting with a Russian prince on a train to Moscow. But nothing came of it. No private company had the enormous resources needed for such a work, and no foreign government would dare defy the United States by setting up in their backyard. At the end of 1898, the directors had only one choice. Reluctantly, clutching their latest plans, they headed to Washington where they were received by President William McKinley. Bunau-Varilla describes himself as horrified by the 'passivity' of the New Company. Patriots like him saw the abandonment of Panama to the Americans as the last, sorry episode in a long decline of French influence in mainland North America, which at one time, before the Seven Years War, had been overwhelming. The Company, said Bunau-Varilla in a typically martial turn of phrase, 'laid down its arms without fighting'. 'Today,' he would later write, 'we behold in foreign hands the great work our minds conceived, and which our gold and our blood brought forth from the domain of the impossible.'

According to Bunau-Varilla, the Americans received the offer with 'scepticism'. 'It would never have come to anything,' he says, 'had I not at that moment begun my campaign.' How the Americans came at last to take over the Panama Canal, in a US-controlled zone of a newly independent country, is as controversial and murky a story as any from the de Lesseps years.

PART THREE

The American Triumph

'What is Costaguana? It is a bottomless pit of ten-per-cent loans and other fool investments. European capital has been flung into it with both hands for years. Not ours, though. We in this country know just about enough to stay indoors when it rains. We can sit and watch. Of course, some day we will step in. We are bound to. But there's no hurry. Time itself has got to wait on the greatest country in the whole of God's universe . . . We shall run the world's business whether the world likes it or not'

– US financier Holroyd, Joseph Conrad, *Nostromo*, 1904

16

HEROES AND VILLAINS – THE 'BATTLE OF THE ROUTES'

The scepticism of the American leadership in late 1898 had nothing to do with the principle of a United States government-owned and controlled canal in Central America, or the 'entangling alliances' that it would involve. In fact, the 1890s had seen the US become more outward looking and expansionist than ever before. Important thinkers like Alfred Mahan, whose *The Influence of Sea Power Upon History* was published in 1890, were arguing that a canal, guarded by an expanded navy, was necessary not only as a means of commercial expansion for the now pre-eminent US economy, but as a conduit for sea power (what Mahan saw as the truest judge of a nation's greatness) and a means of spreading 'superior Anglo-Saxon civilisation' over the region. The war with Spain in 1898, motivated in part by fears of German ambitions in the region (Admiral Tirpitz had declared an interest in German control of a Panama Canal), had seen the acquisition by the United States of the Philippines, Puerto Rico, Hawaii, and, effectively, Cuba, and the need for a short sea route to link these new possessions, as well as the two coasts of mainland United States, seemed more urgent than ever. Before the war, McKinley had set up a commission under Rear Admiral John G. Walker to study the canal question, but it was an incident during the fighting that effectively ended the long-standing objection of Congress to taking on the work as a government-funded project. At the beginning of the conflict, a battleship, the *Oregon*, stationed in San Francisco, was ordered to proceed at once to the Atlantic, a 15,000-mile course around the Horn. Sixty-seven days later, its heroic progress followed daily by the press, the vessel arrived to join in the decisive Battle of Santiago Bay. The demonstration of the military significance of a short cut through an Isthmian canal could have been made to order.

'The construction of such a maritime highway,' proclaimed McKinley at the end of 1898, 'is now more than ever indispensable.' The President also instructed his Secretary of State, John Hay, to restart negotiations with Britain to rid the United States of the restrictions of the Clayton–Bulwer Treaty.

Already, there was legislation for an American canal in process in Congress. In June 1898, when it became known that the Isthmian Canal Commission under Admiral Walker intended to recommend the construction of a waterway at Nicaragua, Alabama Senator John Tyler Morgan introduced a bill allowing for the building of a fortified Nicaragua canal by the US government. Morgan, who chaired the Senate Committee on Interoceanic Canals, had been a colonel in the Confederate Army during the Civil War, and believed that a canal through Nicaragua would return to the South the prominence it had lost since the war and make the ports of Mobile and Galveston thriving hubs of trade. By the late 1880s the construction of a Nicaragua canal had become an obsession, and he had many times argued unsuccessfully for Congressional support for the private company that actually started work on a Nicaragua canal for a short time in late 1889 before going bust. Like de Lesseps, Selfridge, Menocal and Ammen, Morgan was gripped not only by the Great Idea of an Isthmian canal, but by a clear view of a waterway of a particular type, in this case a lock canal at Nicaragua. His obduracy would have serious consequences. Morgan's bill came up for debate in January 1889, when the ink on the Treaty of Paris with Spain was barely dry. It passed through the Senate with ease, and was presented to the House of Representatives.

Although Walker's Commission had visited the French works at Panama, such as they were in early 1898, at no point had the Panama route been considered a serious option. The scandals in Paris, the well-publicised attrition from disease and seemingly insuperable engineering and political problems had given Panama a distinct odour of failure. There were a few voices raised for Panama, including that of John Bigelow, who, despite his largely pessimistic report in 1886, had become infected by the project. But they were few and far between and had no representation in either House in Washington. Nicaragua, on the other hand, was seen as a clean slate free of the taint of poisonous European influence. As the New York *Herald* wrote, 'The Nicaragua canal is a purely national affair, conceived by Americans, sustained by Americans, and if later on constructed, operated by Americans according to American ideas, and for American needs. In one word, it is a national enterprise.' All seemed set fair for the Nicaragua route, which would make the New Company's property worthless. It appeared it would take a mighty battle and a miracle of persuasion to change the nation's preference.

William Nelson Cromwell and Philippe Bunau-Varilla were two of the most skilful lobbyists ever to work the corridors of power in Washington. It was largely as a result of their efforts that American engineers would, in 1904, arrive to restart work on a canal not, as everyone expected, in Nicaragua, but in a newly independent and US-controlled Panama. Their contribution would also see the Panama Canal at its rebirth mired, as before, in controversy, scandal and recriminations.

Cromwell was one of history's great fixes. From a modest Brooklyn family, he was, like Bunau-Varilla, short of stature and fatherless, and he shared the Frenchman's aggression and determination. By the age of thirty-three he had risen to become the guiding light of one of Wall Street's pre-eminent corporate law firms. Sullivan and Cromwell, as it became known, was a new type of company, born out of the railroad boom, offering all sorts of services from finance and accounting to press relations and political lobbying. Above all, it was ferociously well connected. Its clients included the huge railroad companies and the nation's most trusted banks, including JP Morgan; and, from 1896, to represent their interests in the United States, the Compagnie Nouvelle du Canal de Panama.

It was a good choice by the directors in Paris. Cromwell knew the Isthmus, having been general counsel, a director and a shareholder in the Panama Railroad for three years. But, above all, he was a superb operator. Affable, with disarming manners, he was known as 'The Fox' because of his extraordinary cunning. 'He is one of the readiest talkers in town. No life insurance agent could beat him,' wrote the New York *World*, which would lead the investigations into the 'Scandal of Panama'. '[He] has an intellect that works like a flash of lightning, and . . . swings about with the agility of an acrobat . . . He talks fast, and when he wishes to, never to the point.' As well as his contacts and his lawyer's talent for obfuscation, Cromwell had the crucial ability to master such a complicated brief as that presented by the French. He did not come cheap: he would bill the New Company some $800,000 for his services. But when the fee came up for arbitration in 1907, Cromwell was able to argue with justification that his services had 'involved almost every branch of professional activity – engineering, law, legislation, finance, diplomacy, administration and direction'.

As a 'penalty' shareholder in the New Company, Philippe Bunau-Varilla was barred from direct involvement with the running of the venture, but he had far from given up on his canal dream and the final accomplishment of 'the Great Idea of Panama'. Apart from anything else, he had over a million francs of his own money forcibly invested in the New Company. After the collapse of the de Lesseps setup, Bunau-Varilla had unsuccessfully stood for office in France, and then with his brother taken over a newspaper, *Le Matin*, as an alternative outlet for his campaigning for the completion of the French canal. In addition, on the prompting of his friend John Bigelow in New York, in 1892 Bunau-Varilla published a book outlining his plans for finishing the canal at Panama. Bigelow saw that it was distributed widely among opinion-formers in Washington.

John Bigelow also provided introductions to influential Americans living in or passing through Paris. Even if Bigelow could not provide a link, Bunau-Varilla made himself the master of 'chance encounters' and thereby managed to pitch for Panama to many US citizens who, once converted, would be vital to his later efforts. Frank Pavey, an influential New York lawyer who would soon be working for Bunau-Varilla, described (for the benefit of a later Congressional investigation into the whole 'Panama Scandal') meeting the Frenchman in Paris. Pavey's attitude prior to the encounter was typical of his countrymen, that 'There was a hole in Panama into which a lot of French money had been sunk, and that no canal would ever be possible there.' But Bunau-Varilla gave him the full evangelical treatment. 'He never let go of an American victim when he got one in that library until he thought he had converted him,' said Pavey, 'and the first time I dined in his house I stayed until 2 o'clock the next morning, listening to his picturesque and fascinating argument in favor of Panama and against Nicaragua . . . [he] made a special effort to convert me to the cause of Panama, which I am frank to confess he did.'

A desire to claim credit for the great achievement of the canal was a weakness that Cromwell and Bunau-Varilla cleverly exploited among their enemies. But it also dominates their own accounts of their involvement in the events that led to the start of the American Panama Canal. In the Frenchman's published writings there is one hero of the story, namely himself, a new, hyper-patriotic *Grand Français* who steps into the breach to steer events and protagonists towards the saving of 'the noble conception of French Genius through its adoption by America'. Cromwell's version emerged when his 65,000-word justification for his enormous fees at the arbitration court in 1907 was handed to the press. The leak caused a sensation, for Cromwell, naturally putting the best shine on services rendered, claimed to have decisively influenced US government decisions in favour of the Panama Canal to a breathtaking extent.

But Bunau-Varilla, like de Lesseps before him, was not one to share the limelight, calling the claims of the man he disparagingly called 'the lawyer Cromwell' 'a tissue of erroneous and misleading assertions'. Cromwell, in turn, would play down the contribution of Bunau-Varilla and even sought to discredit his motives for campaigning for a Panama Canal. Both men talked down their mutual co-operation. Each wanted to be, and subsequently saw himself as, uniquely, the hero who made the waterway a reality. They also shared an obsession with the canal. Bunau-Varilla's was well established, but for Cromwell, too, the longer he was involved the more it became greater than just another lucrative job. As an American journalist would later write: 'Once you have touched Panama, you never lose the infection. Some call it canalitis.'

But to others, one or both of the men were the villains, rather than the heroes, of the piece. To his enemies, such as the Nicaragua lobby led by Senator Morgan, and those who objected to the United States' shady involvement in the Panama Revolution, Cromwell's undoubted influence and interest, combined with the taint of a new, runaway Wall Street, made him a perfect scapegoat. He was portrayed as a corrupter of American public life. A Congressional investigation was told that Cromwell was 'the revolutionist who promoted and made possible the revolution on the Isthmus of Panama'. He was, the investigation's leader suggested, one of the most dangerous men the United States had spawned for a long time. Almost worst of all, he was 'one of the most accomplished lobbyists this country had ever produced'. The *World* concurred, writing that Cromwell's 'masterful mind, whetted on the grindstone of corporation cunning, conceived and carried out the rape of the Isthmus'.

However, to Panamanians the 'rapist' was the 'traitor' Bunau-Varilla, who, as shall be seen, blackmailed the infant Republic into acceding to a deal with the United States that was patently unfair. The Frenchman, the self-anointed heir of Ferdinand de Lesseps, would stop at nothing to see the Panama Canal built.

It is Cromwell, however, who takes centre stage in the early parts of the story. When he was contracted by the New Company in 1896, he told his employers that 'no one in the United States doubted that the Panama Canal in itself was an impossibility . . . Public opinion demanded the Nicaragua Canal.' In turn, the directors in Paris for the moment kept Cromwell on a tight leash, still hoping that the last resort of selling out to the Americans might be avoided. When, by 1898, this looked impossible, they acceded to Cromwell's pleas to allow him to press energetically the case for a sale to the United States government.

Cromwell straightaway set up a special press bureau for the production and dissemination of anti-Nicaragua and pro-Panama propaganda, at the same time lobbying engineering societies, shipping interests and influential politicians. 'We must make our plans with Napoleonic strategy,' he told his French clients. For Cromwell this meant being 'ubiquitous and ever present' on Capitol Hill, as one of his enemies would later complain. It was Cromwell, naturally, who set up and presided over the meeting of the New Company directors with President McKinley in early December of that year.

But the President was not impressed, in part because of the clause in the Wyse concession that forbade sale of the canal works to a foreign government. On 5 December 1898 he sent his message to Congress supporting Morgan's Nicaragua Bill.

When this measure sailed through the Senate, Cromwell had to concentrate all his resources on the House of Representatives to prevent Morgan's legislation becoming law. If the bill passed, then all his client's assets would be worthless. The job of getting the Nicaragua Bill through the House of Representatives fell to Iowa's William P. Hepburn. Both the Republican Hepburn and the Democrat Morgan considered the Nicaragua option inevitable and were manoeuvring to secure the honour of the legislation for their party and themselves. Someone – what a later investigator called 'mysterious influences' – played upon Hepburn's vanity by getting him to introduce a bill of his own, rather than just sponsoring Senator Morgan's in the Lower House. This complicated the passage of the bill, and Hepburn was persuaded to accept an amendment that called for a new study to look at all the feasible routes for a canal. Effectively, the bill was killed, and a new Commission was ordered, again under the direction of Admiral Walker, to look once more at the best route for a canal 'under the control, management and ownership of the United States'. To the fury of Morgan, who saw the powerful transcontinental railroad interests behind the 'delaying measure', it was the first crack in the Nicaragua edifice, and a great victory for the Panama lobby.

Characteristically, Bunau-Varilla claimed the credit, saying that the field had been reopened through the efforts of his carefully cultivated American friends. But in this instance, the 'mysterious influences' were almost certainly Cromwell, who somehow got himself invited before the committee studying the bill and argued 'for hours on the most profound study of the technical sides of the question'.

Cromwell did not rest on his laurels, however, immediately doing his utmost to influence the selection of the new Walker Commission. In this he was only partly successful, failing to block the appointment of several experts who had already pronounced for Nicaragua on the previous commission. But he did manage to arrange that the Commission's first port of call would be Paris, where all the talk would be of Panama, rather than Nicaragua. Cromwell left ahead of the Walker party, sailing for France on 9 August 1899 'to prepare and direct the presentation of the case'.

In Paris, the nine eminent engineers and military bigwigs of the Walker Commission were subjected to a barrage of plans, maps and figures by the New Company, but also, because of Cromwell's efforts, elaborately wined and dined. One lunch included six courses and four different wines. Bunau-Varilla popped up, too (that Cromwell did not meet and consult with him at this time if not before is impossible to believe). Supposedly thanks to another introduction from John Bigelow, now eighty, Bunau-Varilla had dinner with three of the Commission engineers, including the eminent George S. Morison, whom he subjected to the full Panama treatment. All were converted, or, as Bunau-Varilla puts it, 'the scales had fallen from their eyes'.

While Bunau-Varilla provided the high notes, Cromwell got down to business on his return to Washington. The key man to get on side, Cromwell decided, was Mark Alonzo Hanna, who had been the chairman of the Republican National

Committee during the 1896 election. Hanna was very close to McKinley and was considered the most powerful man in Washington. He also had a long interest in canals, had recently been tasked by the President with getting on top of the Isthmian canal question, and had joined Morgan's Senate Committee on Inter-oceanic Canals. Cromwell got an introduction from Hanna's banker who was also a client of Sullivan and Cromwell. The meeting opened with Cromwell slapping down on Hanna's desk a $60,000 donation to the Party funds compliments of the Compagnie Nouvelle. It was an outrageous move – not only was it a vast amount of money, but also no one in Paris had authorised such a payment on their behalf.

Cromwell's efforts seemed to be paying off when the Republican Party Convention in June 1900 changed its call for a 'Nicaraguan' canal to an 'Isthmian' one, but on other fronts Panama was stalling. In spite of the lavish hospitality, the Walker Commission had returned from their Paris trip disappointed by the New Company's inability to state either a firm price for their venture, or that they had the legal right to sell to the United States government. Maurice Hutin, who had briefly been *Directeur Général* of the de Lesseps Company back in 1885 before being invalided off the Isthmus with yellow fever, was now in charge of the New Company. In April Walker had again asked him to name a price. Hutin did not reply for ten weeks, but on 26 April took the precaution of buying for 5 million francs in gold an extension to the Colombian concession up to 1910. But he still stalled on Walker's demands. Hutin's long-standing and traumatic involvement with the French canal project led him to hope that the United States would somehow be taken on as a partner, rather than simply buying the French out. The result was that Walker, in a preliminary report issued in November 1900, indicated that because of 'all the difficulties of obtaining the necessary rights . . . on the Panama route' Nicaragua presented 'the most practicable and feasible route'. The Panama lobby was in crisis.

A few days later, according to his own account, Bunau-Varilla received an invitation to speak to the Cincinnati Chamber of Commerce, from a US businessman he had met and converted to the Great Idea in Paris earlier in the year. 'The bugle-note had been heard,' Bunau-Varilla wrote, and set sail straightaway for the United States, arriving at the beginning of January 1901. By a coincidence or not, it was exactly the time it would have taken for Cromwell to summon him after the preliminary report by the Walker Commission.

Bunau-Varilla's whirlwind three-month tour of the United States, reminiscent of those undertaken by Ferdinand de Lesseps in the 1880s, began in Cincinnati on the evening of 16 January 1901, in a large hall decked with the flags of France and the United States. It was an unqualified success. The Frenchman seemed a strange, exotic creature, with his theatrically impeccable manners, grandiloquent gestures, large head and moustache waxed to two fine points, but his passion for Panama was plain. It was the 'intensity of conviction which inspired all your utternaces', one of the guests wrote to him, that gave what he said so much impact. 'I love a man,' the American went on, 'who loves a great cause.'

In his own account of the tour, Bunau-Varilla maintains that it was 'Fate' that ensured he met the key US decision makers. 'Every time I was in need of a man he appeared,' he writes. In truth, Bunau-Varilla left little to chance. Several of his converted American friends were working for him at his expense, setting up meetings and opening doors. After Cincinnati, he headed to a business club in Cleveland, supposedly thanks to a 'chance encounter' with a friend of a businessman there he had met on the boat over. Again he spoke about the advantages of the Panama over the Nicaragua route – the railroad, the superior harbours, the shorter length and lower cost. He also warmed to a new theme that had made a considerable impact during his first speech, the suggestion that, unlike Panama, the Nicaragua route was bestridden by volcanoes. The Cleveland audience was particularly important as it included key friends of McKinley and Hanna.

At every meeting, Bunau-Varilla, or his friends working behind the scenes, came away with a new invitation. After Cleveland, he headed for Boston, then Chicago. In New York, he dined again with the Walker Commission's George Morison, and then addressed the city's Chamber of Commerce, courtesy of fixing by his old friend John Bigelow. Those present included J. P. Morgan, John D. Rockefeller and Andrew Carnegie. All the time, Bunau-Varilla was liberally spraying around gifts of cigars, flowers and theatre tickets, giving dinners and writing countless letters seeking introductions or just pushing the case for Panama.

In New York, Bunau-Varilla had 13,000 copies of a leaflet, 'Nicaragua or Panama?', printed and mailed to every congressman and business leader, the editors of four thousand newspapers and magazines, and key members of boards of trade and chambers of commerce, as well as the ever-growing list of contacts he had made himself. Bigelow arranged for copies to be sent to Admiral Walker and Secretary of State John Hay, who had once been the old man's aide in the Paris embassy.

Now Bunau-Varilla considered heading for home, but there then occurred another of the 'chance encounters' that he had made his speciality. Reading between the lines of his own account, it was a setup. Bunau-Varilla was staying in the Waldorf-Astoria, as were several key Washington politicians. 'Towards midnight,' Bunau-Varilla writes, 'as I was about to go out for a breath of fresh air before retiring, I met a party of people in evening dress entering the Waldorf-Astoria. My surprise was great when I saw at the head of them Colonel Herrick [a contact from Cleveland] with a lady on his arm, and behind them a short, stout gentleman who limped slightly. His characteristic face, so frequently reproduced in the newspapers, was familiar to me.'

It was Mark Hanna, identified by Cromwell as the key man to get on side. Herrick feigned surprise as he made the introductions, and Bunau-Varilla came away with a pressing invitation to call on the senator in Washington. Not content with this priceless coup, Bunau-Varilla continued to loiter in the lobby of the hotel until another friend, this time one of the 'converted' Cincinnati businessmen, happened to come past in the company of the US Comptroller of Currency. Through him, the Frenchman secured an interview with McKinley himself.

In no time, Bunau-Varilla was in Washington, to 'attack the political fortress'. He had a number of meetings with Hanna, which, apparently, culminated in the Senator for Ohio pronouncing, 'Mr Bunau-Varilla, you have convinced me.' The interview with the President was briefer. Bunau-Varilla did not wish to 'inflict' a lecture on him. Besides, he knew 'that the opinion of Senator Hanna would be his [McKinley's] own.'

Before the Frenchman's return to Paris, there was to be one more interview in Washington, arranged with enormous difficulty by Bunau-Varilla's friends. One evening in early April, Bunau-Varilla headed to the 'enemy headquarters', the Washington home of the Alabama Senator, John Tyler Morgan, the foremost exponent of the Nicaragua route. It was an icy meeting. According to the Frenchman, Morgan had an 'almost demented state of the mind . . . after twenty years' uninterrupted efforts for Nicaragua' which 'prompted him to see conspirators everywhere'. Morgan, 'trembling with passion' and just as fanatically devoted to the Nicaragua route as Bunau-Varilla was to that in Panama, did all the talking, something the loquacious Frenchman found impossible to deal with. At last Bunau-Varilla interjected, 'But the volcanoes of Nicaragua . . .' Morgan was incensed. Rising to his feet, he spat, 'You would not put one dollar of your own money in this absurd project – in this rotten project – of Panama!' Bunau-Varilla, too, stood up and raised his hand as if to strike his enemy. At the last moment he checked himself, announcing, 'You have just inflicted on me, sir, a gratuitous and cruel insult. But I am under your roof, and it is impossible for me to show you my resentment without violating, as you do, the laws of hospitality.' With that, Bunau-Varilla swivelled and marched out of the door. On 11 April he sailed for Paris, shaken by the confrontation with Morgan, but confident that he had made a significant dent in US public and official opinion in favour of Panama.

How Bunau-Varilla came to be summoned to the United States, and who paid for the lavish trip, remains a mystery. But certainly Cromwell had not been idle in the meantime. On the news that the provisional Walker report would favour

Nicaragua, he was on to the Colombians, warning them that the future of their canal was in jeopardy. Bogotá responded by sending a senior politician and close friend of the Colombian President to Washington to press the case for Panama. The envoy met Hay on 13 March 1901, and early exploratory negotiations between the Colombian envoy, Walker, and Cromwell representing the New Company, went well. Apart from anything else, the Colombian's presence in Washington indicated to Cromwell and his clients that Bogotá was happy for the New Company to sell out to the US, despite the terms of the Wyse concession forbidding its handing over to a foreign government.

But there were potentially fatal problems and distractions. The Colombians were convinced that the Nicaragua option was a red herring designed to get better terms for the US, and that Panama was the only serious option from a technical point of view. The United States, for its part, was still involved with the complicated negotiations, initiated in 1898, to change or abrogate the Clayton–Bulwer Treaty with Great Britain, which continued to stand in the way of any unilateral action by the US on the canal issue. Worst of all was that the New Company still refused to set a price for its assets – demanding instead independent arbitration to settle the issue – or to confirm that it had the right to sell. Utterly frustrated, Walker went to Cromwell in July 1901 to demand that the directors in Paris name a price. Cromwell, equally annoyed, pressed Hutin for an answer in the most direct terms.

For the New Company directors, enough was enough. Not only had Cromwell made free and easy with their money, often in questionable ways, but now he was adopting an unacceptably aggressive tone. In July 1901, he was sacked.

So while the other distractions continued through the summer, the New Company now had no representation in the US. Naturally, Walker's Commission had to consider the political aspects of the choice of canal location as well as the purely engineering issues, and it did not look good for Panama. On the Isthmus itself, hopeful rumours swirled about, but to those in the know it was clear that the preliminary verdict in favour of Nicaragua was not about to be reversed when the Commission submitted their final report in November 1901.

Then, on 6 September, there happened a combination of two of the recurring events of the end of the nineteenth century – expositions and anarchist violence. President McKinley was attending the Pan-American Exposition in Buffalo, New York, when a lone anarchist called Leon Czolgosz fired two shots from a .32-calibre revolver into the President's upper body. McKinley died eight nights later.

On the same day, 14 September 1901, a new Chief Executive was inaugurated, the former vice-president, Theodore Roosevelt. And with his arrival, everything would change for the canal.

*

In his first address to Congress, the new President promised an American-built and controlled trans-Isthmian canal. 'No single great material work which remains to be undertaken on this continent,' he declared, 'is of such consequence to the American people.'

Roosevelt, who was actually descended on his mother's side from one of the survivors of the Scottish 'Darien Disaster', had already, while New York State Governor, intervened in the canal debate, or, more specifically, the negotiations with Britain for the abrogation or alteration of the Clayton–Bulwer Treaty. Hay's discussions with the British had been slow and laborious, but in February 1900 he had at last signed an agreement with British ambassador Sir Julian Pauncefote that abrogated the restraining treaty that had prevented the US from building a canal on its own. But its replacement, which forbade fortification, stipulated that the waterway should be 'free and open in time of war as in time of peace, to vessels of commerce and of war of all nations', and looked to an international guarantee, found no favour among the Roosevelt circle. While some press commentators applauded the lack of aggression inherent in the neutrality clause, which gave the US's neighbours less 'cause for suspicion', Roosevelt, a great admirer of Mahan's theory of the importance of naval power, wrote directly to Hay complaining that the 'international guarantee' would flout the Monroe Doctrine and took direct issue with the ban on fortification. 'If that canal is open to the warships of an enemy,' he wrote, 'it is a menace to us in time of war; it is an added burden, an additional strategic point to be guarded by our fleet. If fortified by us, it becomes one of the most potent sources of our possible sea strength.'

Hay was shocked by these criticisms and sniffily told Roosevelt that such matters of Great Power diplomacy were outside the remit of a mere State Governor. But opposition in the Senate, led by Morgan and Roosevelt's friend John Cabot Lodge, forced Hay back to the negotiating table. Happily, he found Great Britain in an obliging mood. Embroiled in a costly and internationally unpopular struggle with the South African Boers, and worried about Russian expansionism towards India and the German naval programme, the British were keen to nurture an informal détente with the United States. Senior British politicians believed, like Arthur Balfour, that a US-controlled canal would 'strengthen our position enormously and . . . with England at Suez and the U.S. at Panama we should hold the world in a pretty strong grip'. Soon after the revised treaty was signed, Britain started reducing her costly garrisons and naval squadrons in the Caribbean. The British were beginning to learn how to use American power to their advantage.

With the signing of the second Hay–Pauncefote Treaty on 18 November 1901, which implicitly allowed the fortification of an American canal, the long rise of the US to local pre-eminence was complete. As the President of Colombia noted, the treaty 'ruptured the dikes placed against so-called American imperialism . . . It changed the face of the question and made the situation for the [Colombian] Government obscure, delicate and complex: action and inaction equally presented great problems and reason for anxiety.' Just how 'delicate and complex' would soon be illustrated.

The signing of the treaty cleared a significant obstacle from the path towards an American canal. Its location, however, still looked like Nicaragua. The Panama lobby had been wrong-footed by the death of McKinley – Roosevelt had not even been considered as a target for their propaganda – and it was believed that the new President would be far less influenced by the carefully 'converted' Mark Hanna. In October, Hutin in Paris at last gave Walker an estimate of the value of the New Company's property, set at just over $109 million, albeit open to arbitration. The next month Walker's Commission reported.

Bunau-Varilla and Cromwell had done their work well – the engineers clearly preferred Panama from a technical point of view: the route was much shorter, would need fewer locks, and was hindered by less curvature; a ship could pass through in twelve hours rather than thirty-three; it would also be cheaper to build and maintain. Walker estimated a Panama canal would cost just over $144 million, against $190 million for Nicaragua. But when the price of the French Company was factored in, Panama became much more expensive. Therefore the Commission, to the particular dismay of engineer George Morison, plumped for Nicaragua.

The following month the US signed a canal convention with the Nicaraguan government, and on 9 January 1902 the House of Representatives overwhelmingly backed a new bill from Congressman Hepburn appropriating $180 million for the construction of a Nicaraguan canal and sent it on to the Senate. In Washington the Panama venture was now being described as 'a worthless ditch'.

In Paris, these developments caused a panic. In December there had been angry scenes at a shareholders' meeting, leading to the police being summoned. Attacks on the board of directors were led by Philippe Bunau-Varilla, who urged the shareholders to sell to the US at whatever price they could get or see their investment become entirely lost. Hutin was forced to resign and was replaced by Maurice Bô, president of the Crédit Lyonnais bank, who, as a penalised shareholder, was not strictly allowed direct involvement with the Company. But he was a friend of Bunau-Varilla and Cromwell, and was happy to take the drastic step now required. On 4 January Bô wired Walker a revised price of $40 million, the Commission's own valuation of the New Company's property.

This news changed everything and set the scene for a decisive intervention by the new President. On 10 December Roosevelt had received a letter from his fellow Harvard graduate George Morison, outlining the engineer's reasons for disagreeing with the majority of the Walker Commission's preference for Nicaragua. Roosevelt may have also been influenced by Mark Hanna, who remained a senior figure in the Party, or by the fear that the unfinished Panama Canal would be completed by a European power if the US pressed on in Nicaragua. Whatever the reason, Roosevelt now had a fixed idea: he wanted the Panama route. As soon as the news came in from Paris of the revised price, the President summoned the Walker Commission and instructed them to change their verdict. Several protested, but as the price had been the sticking point for Panama, on 18 January they complied with Roosevelt's demands and published a revised report recommending Panama. An amendment was drafted which authorised the President to purchase the French Company's Panama property and concessions for $40 million; to acquire from Colombia perpetual control of a Canal Zone at least six miles wide; and to build a Panama canal. If a clear title or a satisfactory agreement with Colombia could not be reached 'within a reasonable time', then the President was authorised to proceed with the Nicaragua route. John Coit Spooner, a past master at steering difficult legislation through the Senate, was chosen to introduce the amendment, and thus to face the full fury of Morgan and the Nicaragua party.

But the Panama lobby would be on hand to help. On 27 January, Maurice Bô reinstated Cromwell as the New Company's US representative, albeit with an order to stick to 'legitimate means'. The same day Cromwell met with Bunau-Varilla (they both claim this is the first time they worked together), who had rushed to Washington to prepare for what would be the climax in the Senate of the long 'Battle of the Routes'.

The astonishing, Roosevelt-led turnaround caused great confusion in the United States press, long accustomed to the American preference for a Nicaragua canal. Much of the concern about Panama was the prospect of dealing with Bogotá. 'The Colombians . . . have negro blood enough to make them lazy, and Spanish blood sufficient to make them mean,' declared *Harper's Weekly*. Somewhat prophetically, the New York *World* commented, 'Talk about buying a lawsuit – the purchase of the Panama Canal would be buying a revolution.'

Since the beginning of the year, the Hepburn Bill had been in the hands of Morgan's Senate Committee on Interoceanic Canals. After extensive cross-examinination of all available experts, the Committee, in spite of the Walker Commission's decision, voted seven to four in favour of Nicaragua. But Hanna, with Cromwell's help, produced a minority report in favour of Panama. The date of 4 June was set for the start of the debate in the Senate over which of these would be adopted.

In the meantime, Bunau-Varilla was as busy as ever, writing to newspapers and politicians, producing pamphlets and pouring pro-Panama rhetoric into the ears of anyone who would listen. Then the Panama lobby had a stroke of luck. On 8 May, the volcano Mount Pelée on the Caribbean island of Martinique exploded with devastating effect. The town of St Pierre in its shadow was utterly destroyed, and more than thirty thousand people were killed. Although Martinique was nowhere near either route, volcanoes were suddenly on everyone's mind. Then, just a week later, the news came in that a volcano in Nicaragua itself had erupted. Friends of Panama in the press had a field day.

Senator Morgan opened the debate in the Senate with a spirited counter-attack on the 'volcano scare', brandishing a letter from the Foreign Minister of Nicaragua (who had somehow been persuaded to deny the eruption had taken place), and pointing out that Panama had itself recently experienced an earthquake. But the main thrust of his argument against Panama was political. Its people were 'mixed and turbulent'; it was chronically unstable; to build a canal there, the US would have to take the country by force, an action, predicted the Alabama Senator, that would 'poison the minds of people against us in every Spanish-American republic in the Western Hemisphere, and set their teeth on edge against us'.

The next day, Senator Mark Hanna replied in favour of his minority report. It was to be the greatest speech of his career. Shunning rhetorical flourishes, he spoke in a slow, businesslike way, illustrating his points with an impressive array of visual tools, including a huge map showing active volcanoes in Nicaragua. All this had been prepared for him by Cromwell and Bunau-Varilla. The Panama route was shorter, he pointed out, had less curvature, better ports, a railway, fewer locks and 'was a beaten track in civilization'. Furthermore, the engineers wanted Panama, and 'there are now done a great many things which fifty years ago were unheard of, never dreamed of, never thought possible, as a product of human intelligence and ingenuity in engineering. It has become a byword today that in the hands of a skilful engineer nothing is impossible.'

The speech, over two days, certainly changed votes, although the Panama lobby was not home yet. Pro-Nicaragua senators suggested that the whole effort of the Hanna party was to delay any canal in order to serve the interests of the transcontinental railroads. There still persisted, too, a feeling that Panama was irrevocably stained by corruption and what was seen as the vice of the French years. 'It is the certainty of moral defilement,' declared Senator John H. Mitchell of Oregon. 'Panama cannot be touched with safety by American people.'

The volcano argument was also foundering. The Nicaraguans were sticking by their story that there had been no recent eruption in their country, and the whole scare was starting to be seen as an invention of the Panama lobby. On 6

June, a cartoon appeared in the influential Washington *Star*, showing Hanna slapping imaginary volcanoes on to a map of Nicaragua aided by a comical Frenchman, Bunau-Varilla, and James J. Hill, head of the Great Northern Railroad. As Bunau-Varilla wrote, 'If the vote were to be taken under this impression Panama was done for ever . . . Fortunately I had a sudden inspiration.'

Over the next few days Bunau-Varilla scoured the philatelists of the capital looking for a certain 1900 one-centavo Nicaraguan stamp, which he had come across the year before. In the foreground of the stamp is pictured a busy wharf while in the background rises the magnificent bulk of Mount Momotombo. In an artistic flourish the illustrator had added smoke to the top of the volcano, which was actually more than a hundred miles from the proposed Nicaragua canal. Just before the vote, every senator was sent this 'evidence' of the dangers of the Nicaragua route.

It was almost the last shot, but this went to Senator Morgan, who used his final speech before the vote on the minority report to launch a bitter attack against his enemy Cromwell. The 'direct, constant, and offensive intrusion of the Panama Canal Company' into the workings of the US government was, he said, 'humiliating' and 'repulsive'. 'I can not neglect Mr Cromwell,' he said. 'I trace this man back . . . to the beginning of this whole business. He has not failed to appear anywhere in this whole affair.' The contagion of Panama, 'death's nursery', had through its agent Cromwell poisoned everything it touched.

Everyone knew that the vote on 19 June would be close, and the press and the country, which had followed the fourteen-day debate closely, waited with bated breath. When the result came it could hardly have been narrower, with Hanna's minority report in favour of Panama winning by just eight votes. 'The battle was won,' wrote Bunau-Varilla. 'Truth at last triumphed.'

After this, the passage of the Spooner amendment was a formality, and the House of Representatives was persuaded to back the Act as well. In part, the pro-Nicaragua faction assumed that either the French title would prove defective or an agreement with Colombia would not be forthcoming and they would get their preferred option after all. If they had lost a battle, they had not yet lost the war.

17

'I TOOK THE ISTHMUS'

The first potential stumbling block was dealt with quickly. The US Attorney General sailed to Paris and exhaustively went through the available books and contracts before pronouncing that the deal with the New Company was legitimate. But on the second issue, of coming to a deal with Colombia, it was not to be so simple.

Since October 1899 the country had been in the throes of its longest and most devastating civil war since independence eighty years before. The 'War of a Thousand Days', fought between Conservative and Liberal factions, would claim the lives of between 150,000 and 250,000 Colombians.

With Liberal armies threatening Bogotá itself, the administration, led by the elderly Conservative José Manuel Marroquín, was showing signs of confusion, and was having great difficulty staying in contact with its representatives in Washington. Marroquín tried to delay the signing of any deal with the US until his position was more secure, but amid calls from the Nicaragua party that the 'reasonable time', as stipulated by the Spooner Act, had run out, a treaty was signed with the Colombian legation's secretary, Tomás Herrán, on 22 January 1903 (the previous envoy had resigned in disgust at what he saw as the bullying tactics of the US administration). Herrán himself had been concerned that further delays would lead Roosevelt, whom he described as 'impetuous [with a] violent disposition', simply to seize the Isthmus.

The terms of the treaty were that in return for an annuity of $250,000, with a $10 million gold one-off payment, the US would receive a six-mile-wide Canal Zone on a hundred-year-lease renewable at the sole option of the United States. Although Colombian sovereignty was specifically recognised, this was something of a fig leaf: the US was to be allowed to establish its own courts within the proposed Zone and, in an emergency, to land its forces without Colombia's consent to protect the canal.

Morgan did his very best to wreck the treaty in the Senate, proposing a number of amendments, including anti-Catholic measures, which he knew would make the deal unacceptable to Colombia. Along the way, he described the treaty as a compact with 'a crowd of French jail-birds, cleverly advised by a New York railroad wrecker . . . and a depraved, priest-ridden people'. The Alabama Senator successfully filibustered into March, but overplayed his hand. Roosevelt pushed, Morgan broke down and the treaty was ratified without amendments on 17 March. Now the ball was firmly in Colombia's court.

In mid-March 1903, with the civil war at last over, and as the new Colombian Congress was being elected, the Panama *Star and Herald* commented, 'Few of the members who will assemble in Bogotá, competent observers say, have ever seen the ocean . . . They are comparatively indifferent to the advantages of the project, while feeling great pride in their soil and sovereignty, and a corresponding fear of the gradual absorption of their territory by the United States. These things count against ratification.'

Soon after, Claude Mallet returned to the Isthmus, having served for two years in Bogotá. It was not good to be back. In 1892, Mallet had married Matilde de Obarrio, a Panamanian twelve years his junior. The following year their first child, a girl, had been born, but had suffered from diphtheria. Their son, born in England a year later, contracted meningitis as an infant, and Mallet decided that Panama was no place to bring up his children. Thus his family life was conducted by letter – on average one every two days from Panama to Brighton or Croydon – for much of the following decade. It appears to have been a marriage tailor-made to confound national stereotypes. 'What a difference there is in our nature,' he wrote during one of their occasional long-distance spats. 'I hot blooded, brimful of love and affection. You, cold, severe, undemonstrative such as I thought only my countrywomen embodied in their characteristics.' In all events, Panama carried endless reminders of his distant wife and had lost much of its appeal. Furthermore, there was 'a great deal of illness' in Panama, and yellow fever in Colón – 'I have heard of four cases (two deaths) since I arrived on Thursday.' More than anything, there was great depression about the chances of the canal treaty going through. 'Religion here has taken an extraordinary hold upon the people,' he wrote on 1 June. 'A few years ago such a scene [a procession of girls carrying an effigy of the virgin] would not have been permitted. The Jesuits are getting in their work and unless the canal is made we shall lapse back to what the place was fifty years ago.'

Panama's Senator José Agustín Arango, who also worked as a lawyer for the Panama Railroad, believed that the result of the forthcoming debate in Bogotá was a foregone conclusion and refused to attend the Senate opening at the beginning of June 1903. Instead, he believed that the best hope for a canal to bring much-needed prosperity back to his homeland was through secession. By May a small revolutionary group was active, centred on Arango's sons and sons-in-law, all young men educated in the US. Soon after, Federico Boyd, the son of the founding editor of the *Star and Herald*, and Doctor Manuel Amador Guerrero, the Railroad's seventy-year-old, frail-looking head physician, were brought on board. The group met secretly at Amador's house, or at the Panama electric light plant.

Petitions were sent from Panama to Bogotá both for and against the treaty. Liberals were opposed to the 'selling' of Panama to the United States. Cromwell helped organise pressure in favour of the treaty. At the beginning of June, Ricardo Arias, a rich landowner and also a member of the Arango conspiracy, wrote in an open letter to his countrymen that the treaty 'for the Isthmus of Panama is a question of life or death . . . The gravest responsibility will rest on whomsoever, letting slip this opportunity, the last offered us, may sink this land into eternal ruin.' A petition sent from Panama to the Colombian President Marroquín warned that rejection would 'give rise to unpatriotic feelings'.

In fact, Marroquín was in a near-impossible situation. In the United States he was seen and portrayed as an all-powerful dictator, but this was far from the case. His power was extremely fluid, varying from issue to issue, and he had made enemies across the political spectrum. It was imperative to his political survival that he did not alienate any of his fragile support.

Indeed, the canal question carried political high explosive. In Colombia, sovereignty was of prime importance, the chief symbol of national permanence and unity in a land of disordered change. In fact, the constitution specifically forbade the transfer to another power of the sovereignty of any part of the country. It was one thing to give a concession to a private company, quite another to hand one to the voracious power to the north, which had already demonstrated its aggression in Cuba and the Philippines. 'Not an atom of our sovereignty nor a stone of our territory,' wrote the newspaper *El Correo Naçional*, should be given up, even if it meant 'renounc[ing] the honour of a canal across Panama'.

Anti-Marroquín newspapers, of which there were many, attacked the treaty as a way of damaging the President. Herrán had sold out, it was stated; the deal was an example of Yankee imperialism; there was still hope of a European country, Britain or Germany, riding to the rescue to build the canal; Morgan's comments in the US Senate debate about 'depraved, priest-ridden people' were printed, causing widespread resentment.

Under the aggressive leadership of Roosevelt, the United States had been throwing its weight around in the region. At Guantánamo Bay in Cuba, ideally situated to guard the Windward Passage into the Caribbean and thence to the Isthmus, a US naval base had been established in February 1903. At a speech in Chicago in April Roosevelt declared that 'our nation has insisted that because of its primacy in strength among the nations of the Western hemisphere it has certain duties and responsibilities which oblige it to take a leading part thereon'. What would become known as the Roosevelt Corollary to the Monroe Doctrine emerged as a policy soon after Roosevelt entered the White House. Not only was the United States committed to excluding European powers from the hemisphere, but it was also taking on the role of 'international police power' intervening in cases of 'chronic wrongdoing' or 'incompetence'.

To Colombians, this posture was both frightening and insulting. It is 'a warning to our countries,' wrote the Bogotá paper *El Porvenir*. 'It is the conviction of his irresistible superiority and vigour that makes the Yankee, from Mr. Roosevelt to the rag-picker, treat the turbulent republics of Latin America with haughtiness and contempt.' The authority to intervene, said the paper, was 'derived from nobody knows where . . . as though the great nation had received from some universal power the mission to put in order those who live in disorder!'.

Against this background of distrust and fear, public sentiment on the Hay–Herrán Treaty quickly changed, the US minister reported back on 15 April, 'from approbation to suspicion and from suspicion to decided opposition'. The minister, Arthur Beupré, still believed that Marroquín had the power to force the measure through Congress, but that an open vote would see it rejected.

But this would not be put to the test. Marroquín had no intention of acting so vigorously on the canal question. He saw himself in an unwinnable position. 'History will say of me,' he had written the previous year, 'that I ruined the Isthmus and all Colombia, by not permitting the opening of the Panama Canal, or that I permitted it to be done, scandalously injuring the rights of my country.' The way out, as he saw it, was to hand over the responsibility for the decision to Congress, a step the constitution demanded anyway.

But the Colombian President was also personally ambivalent about the canal, which, if built, would open up his country as never before. Like US Secretary of State John Hay, he was a novelist. In his 1897 book *Entre Primos*, he used a cultural confrontation between an effete Englishman and an idealised, hard-working Colombian to show the frippery of the outside world and the superiority of the insular Colombian character. During the civil war he had risked ruining his country in order to protect it from the demands of the Liberals – railroads, foreign influence and capital. In many ways the canal represented the greatest threat of all to everything he held dear – the sheltered, Catholic genteel age of nineteenth-century Bogotá, untrammelled by technology, modernism or Protestant capitalism.

Discussions of the treaty continued through the spring. In April, the Colombians again indicated that, even if they gave in on the sovereignty question, an even greater sticking point was the issue of Colombia's right to a proportion of the money to be paid to the New Company. On this issue, though, Cromwell had engineered his great coup. 'We pointed out,' the lawyer later wrote, 'that Colombia had already pledged herself morally to consent, and that her consent should be imposed on her as being demanded by international good faith.' Even Hay asked Cromwell whether, perhaps, some $5 million or so could not be paid over from the $40 million, but Cromwell succeeded in persuading him that this would be tantamount to giving in to blackmail. Cromwell's influence, on behalf of his client, right at the centre of the US government, is astonishing.

Then the Colombians hinted at another possible way out of the impasse. If the 1900 extension to the concession, organised by Hutin in the midst of the Colombian civil war, were declared illegal, then they could simply let the original term of the deal with the New Company expire in October 1904, and then sell the lot to the Americans for $25 million. But Cromwell need not have worried. Roosevelt and Hay were appalled by this threat, which confirmed their opinion of the Colombians as shifty and grasping. Hay had strong views on property rights, calling the Colombians 'greedy little anthropoids'.

The Americans decided that a firm hand was needed. On 9 June, eleven days before the debate in the Colombian Senate was due to start, a serious threat was issued from Hay's office: 'If Colombia should now reject the treaty or unduly delay its ratification, the friendly understanding between the two countries would be so seriously compromised that action might be taken by the Congress next winter which every friend of Colombia would regret.' All efforts by Marroquin to reduce the humiliation of the deal were now met by a firm rejoinder: any amendments or other delays would be 'tantamount to a rejection of the treaty'.

Behind this brow-beating tone was the determination of the President, Theodore Roosevelt. With elections looming in 1904, he was talking up the grandeur and national pride that the construction of the canal would bring to his country. It was, he told an audience in Chicago, the 'greatest material feat of the twentieth century – greater than any similar feat in any preceding century'. Of course, it 'should be done by no foreign nation, but by ourselves'.

The Panama lobby was also keeping up the pressure. On 13 June, Bunau-Varilla, at huge expense, cabled Marroquín. 'The only party that can now build the Panama Canal is the United States,' he wrote. 'Neither European governments nor private financiers would dare to fight either against the Monroe Doctrine or the American Treasury for building Panama Canal.' Failure to ratify, he warned, would lead to either the 'construction of Nicaragua Canal and absolute loss to Colombia of the incalculable advantages resulting from construction on her territory the great artery of universal commerce, or the construction of the Panama Canal after secession and declaration of independence of the Isthmus of Panama under protection of the United States as has happened with Cuba.'

Cromwell was busy, too. On 12 June he had paid a public visit to the White House, and the next day a story appeared in a New York newspaper, which turned out to have come from Roger Farnham, Cromwell's press agent. 'President Roosevelt is determined to have the Panama Canal Route,' the piece read, saying that a combination of 'the greed of the Colombian government' and the 'frenzy over the alleged relinquishment of sovereignty' made defeat of the measure 'probable' in the Colombian Senate. But, the article continued, 'the State of Panama will secede if the Colombian Congress fails to ratify the canal treaty'. Supposedly, Farnham even told the paper's editors the date of the 'revolution' – 3 November, when US newspapers would be full of returns from the mid-term elections.

Indeed, the treaty never stood a chance in the Colombian Senate. The debate started on 20 June, and was dominated by attacks on Marroquín that had little to do with the canal. Two weeks later General Rafael Reyes, back in Bogotá after a period of exile and a firm supporter of the treaty, asked Beupré for an additional $5 million from the US and $10 million out of the $40 million for the French Company to break the deadlock. Hay replied that the US Senate would not approve it. 'Any amendment whatever or unnecessary delay in the ratification of the treaty would greatly imperil its consummation,' he told Beupré. A few days later Roosevelt wrote to Hay backing up this firm stand. 'Make it as strong as you can . . . Those contemptible little creatures in Bogota ought to understand how much they are jeopardizing things and imperilling their own future.' In fact, Hay, apart from his improperly close relationship with Cromwell, had his hands tied. The close result of the vote in the Senate on the Hanna minority report meant that any deviation from the strict terms of the Spooner Act could see the treaty fail to make it through the Senate. Thus the intransigence of Morgan and the Nicaragua party doomed the treaty as much as any opposition in Colombia.

The rejection from Bogotá, when it came on 12 August, was overwhelming, with twenty-four voting against, with three abstentions. Even Marroquín's son voted against the measure. In the US, the vote was seen as an attempt to extort more money out of the US or the French Company. Patience with Bogotá, never extensive, was now at an end. It was time for a new plan.

*

Roosevelt and Hay now weighed the options open to them. The first was to persevere with Colombia and hope that the treaty might be ratified the following year. The second was to push ahead with the Nicaragua option, as the Spooner Act directed. Or, the whole question could be handed over to Congress to decide. The fourth was somehow to proceed with the Panama route without recourse to Bogotá.

The first option was quickly written off. The President was not inclined to continue negotiations with what he now called 'the foolish and homicidal corruptionists in Bogota'. But to turn to Nicaragua (or see this option chosen by Congress) would for Roosevelt not only represent a personal defeat, but also be 'against the advice of the great majority of competent engineers', as the President declared. Furthermore, there was a growing consensus that the great new warships under construction in US yards as part of Roosevelt's naval expansion would struggle with Nicaragua's narrow and winding rivers. Roosevelt was set on Panama. In September the French minister in Washington, Jusserand, sent a despatch to his government reporting of the President that, 'I know, for having heard him say so, how intensely he wants it [the canal at Panama]; he will neglect nothing that may enable his country to perfect this work and be the master thereof.'

So nothing was to be ruled out, including seizing the Isthmus by force. In March 1903 US spies had been sent to Panama to obtain information to assist military operations there. Days after Colombia's rejection of the treaty, a paper had been forwarded to Roosevelt by Hay's deputy, Francis B. Loomis, which seemed to offer a fig leaf of respectability for such a move. Written by an expert in international law, Professor John Bassett Moore, the paper argued that under the justification of 'universal public utility' Colombia had no right to stand in the way of an improvement that would benefit the entire world.

Such a move carried great political risks, both domestically and internationally. But there was another option. Only two days after the rejection of the treaty (but before the news reached the US), Senator Shelby Cullom gave a press conference on the canal question. Cullom was the chairman of the Senate Foreign Relations Committee, and had just been conferring with Roosevelt at the President's summer residence at Oyster Bay. Roosevelt was prepared for bad news, said Cullom, but was still determined on a Panama canal. When asked how this would be possible if the treaty failed to be ratified in Bogotá, Cullom replied, 'We might make another treaty, not with Colombia, but with Panama.' A month later Hay wrote to Roosevelt dismissing the possibility of making a 'satisfactory treaty with Colombia', but going on, 'It is altogether likely there will be an insurrection on the Isthmus against that government of folly and graft that now rules in Bogota . . . Something we shall be forced to do in case of a serious insurrectionary movement in Panama, [is] to keep the transit clear. Our intervention should not be haphazard, nor, this time, should it be to the profit . . . as heretofore, of Bogota.'

In Panama the independence plot had gathered momentum and important friends. Following Arango and Amador, more Panama Railroad employees had been brought on board, including Herbert Prescott, assistant superintendent of the Railroad (whose brother, another plotter, was married to Amador's niece), and James R. Beers, freight agent and port captain for the Pacific terminus of the PRR. Both were United States citizens and the US consul Arthur Grudger also joined the group. In July 1903 Cromwell, *de facto* head of the Railroad, had summoned Beers to New York, at about the time of his planted story in the *World* about Panama secession. Beers' meeting with Cromwell went well. The lawyer gave Beers a cable code, and warned him to keep secret the involvement of the PRR, as it could forfeit its concession from Colombia. Cromwell also suggested a date for the revolution, 3 November.

On 26 August Amador was sent by the plotters to New York. They were aware that without US help any move by Panama towards independence could easily be crushed by Colombian forces. Amador sailed on board the *Seguranca*, a Panama Railroad and Steamship Company steamer. He had few funds for his trip, but managed to win a goodly sum playing poker during the voyage. Also on board, on unrelated business, was José Gabriel Duque, the Cuban-American owner of the Panama *Star and Herald* and the lottery, head of the fire brigade, and reputedly now the richest man on the Isthmus. Duque, an American citizen, knew about the plot, but was not part of the revolutionary junta. He later claimed that much of the money won by Amador was from him.

The boat arrived in New York on 1 September, and Amador saw Cromwell the next day, receiving 'a thousand offers in the direction of assisting the revolution'. But it had been José Gabriel Duque who had been met off the boat by Roger Farnham and taken straight to the office of Cromwell and Sullivan at 41 Wall Street. For Cromwell, Duque had two distinct advantages over Amador: he was rich; and he had no awkward connection with the Railroad. The lawyer assured Duque that there was no chance of Colombia coming to a deal, and that if Duque lent the revolution $100,000 on Cromwell's security, the lawyer would arrange for him to become the first President of an independent Panama. Of course, Cromwell went on, Duque should go to see Hay, and picking up a phone on his desk he organised it there and then. The following evening Duque was on an overnight train to Washington (to avoid having to register in a hotel), and met Hay at ten o'clock the next day. The Secretary of State all but told him that the US would support the revolution: 'The United States would build the Panama Canal and did not propose to permit Colombia's standing in the way,' Hay pronounced. If the revolutionaries took Panama City and Colón, he went on, American warships would prevent Colombian troops from landing in order to keep fighting away from the all-precious transit.

No sooner had he left Hay's office than Duque was on his way to see his old friend Tomas Herrán at the Colombian legation. Perhaps because of some slight from the junta in Panama, from the influence of his wife, a fiercely patriotic Colombian, or because he still hoped to shock Bogotá into ratifying the treaty, Duque told Herrán everything. The next day, 3 September, the Colombian minister cabled home, 'Revolutionary agents of Panama [are] here. Yesterday the editor of La Estrella de Panama had a long conference with the Secretary of State . . . There is the probability of revolution with American help.' Herrán also set detectives on Amador's trail and fired off a warning to Cromwell that the Compagnie Nouvelle and the Railroad would lose their concessions – everything they were hoping to sell for $40 million – if they supported revolutionary activity.

This had Cromwell running scared. The next time Amador went to his office, he was 'out'. Amador said he would wait, but still the lawyer refused to appear. Eventually Cromwell burst out of his office and physically removed the Panamanian doctor from his premises. Soon after, Cromwell made arrangements to leave the country on business. He knew his card was marked and that someone else would have to take up the challenge of engineering the revolution.

Amador was confused and downhearted, cabling a single-word message – '*Desanimado*' ['Discouraged'] – to his co-conspirators in Panama and prepared to sail on the next ship. But then Amador heard that, should he remain in New York a little longer, he would receive help 'from another quarter'.

Philippe Bunau-Varilla later claimed that his voyage to the United States at the beginning of September 1903 was motivated by the illness of his thirteen-year-old son, who was staying with John Bigelow. In fact, the Frenchman was up to his neck in Cromwell's plot. At the beginning of the month he had written an article for *Le Matin*, predicting revolution on the Isthmus and naming that same date – 3 November. He arrived in New York on 22 September – exactly the time it would have taken if Cromwell had summoned him straight after Herrán's warning.

Bunau-Varilla met Amador two days later and found the doctor in a state of fear and indignation. 'All is lost,' said Amador. 'At any moment the conspiracy may be discovered and my friends judged, sentenced to death, and their property confiscated.' The Frenchman reassured him that he, Bunau-Varilla, would handle everything. Just over a week later, Bunau-Varilla, through Francis Loomis, one of the many Americans he had cultivated as they passed through Paris, secured a meeting with Roosevelt, ostensibly to discuss *Le Matin*'s role in the Dreyfus Affair. Of course, the conversation turned to Panama. Bunau-Varilla announced that there was a revolution coming. The President was naturally unable to give overt support, but the Frenchman picked up what was left unsaid. Roosevelt later wrote to John Bigelow of the meeting: 'I have no doubt that he was able to make a very accurate guess, and to advise his people accordingly. In fact, he would have been a very dull man had he been unable to make such a guess.'

A week later Bunau-Varilla met Hay who agreed that an insurrection was imminent and let him know that US naval units were already standing by to dash to the Isthmus 'to keep the transit open'. When Bunau-Varilla got together with Amador again shortly before the doctor's return to Panama, he assured him that US help would be forthcoming for the revolution, as long as it happened on 3 November. The money – $100,000 – needed to bribe the Colombian garrison would come from the Frenchman's own resources, on the condition that Amador agreed to make Bunau-Varilla Minister Plenipotentiary in Washington for the new republic.

Bunau-Varilla now took total charge. As he wrote, 'I held all the threads of a revolution on the Isthmus.' The weekend before Amador's departure he spent at the Bigelows' house at Highland Falls writing a declaration of independence, military plans and a new cipher code – Amador was 'Smith', Bunau-Varilla 'Jones' – while his wife and Grace Bigelow sewed together a new flag made out of silk purchased by Bunau-Varilla at Macy's. The whole 'revolution kit' was wrapped in the flag and presented to Amador when he left on 20 October.

Back in Panama, Amador found his co-conspirators unhappy about Bunau-Varilla's demand to be made Minister Plenipotentiary, the flag (it was much too similar to the Stars and Stripes), the small amount of money promised and the lack of firm proof of US military assistance. Where was the signed agreement from Hay or Roosevelt? Who exactly was this Bunau-Varilla, and what authority did he have to offer promises of help? The plotters, for the most part wealthy landowners or professionals, had much to lose. The 'revolution' experienced its first serious wobble.

Worse was to come. While Amador had been away in the United States, the junta had been working to bring into the conspiracy key players on the Isthmus. The mayor of Panama City, who happened to be the brother of Amador's young wife, María de la Ossa, was successfully recruited, as was the deputy head of the police force. General Esteban Huertas, the young commander of the local garrison, and married to a Panamanian, seemed sympathetic although so far uncommitted, but his second-in-command, when approached, indignantly threatened to reveal the plot. To get rid of him the State Governor José de Obaldía, who lived with Amador and was unofficially in on the conspiracy, had invented an invasion in the north of Panama by Nicaraguan troops, and despatched a force under his command to investigate. But Obaldía, to cover his back, also cabled Bogotá on

25 October about the invasion scare. Three days later, Obaldía heard that the Colombian authorities, acting with uncharacteristic haste, had readied a force at Cartagena under the supreme commander of the army, General Tovar, to proceed to Panama to assist in repelling the supposed invasion.

The news caused renewed panic among the conspirators, who demanded that Amador should rapidly provide proof of American support and the veracity of Bunau-Varilla's promises or the whole project would be abandoned. The next day, 29 October, Amador sent the following cable to New York: 'Fate News Bad Powerful Tiger Smith', which translated as 'For Bunau-Varilla. More than two hundred Colombian troops arriving on the Atlantic side within five days.' 'Urge vapor colon' Amador went on, abandoning the code. Obviously he hoped that Bunau-Varilla, on his own authority, could order a US Navy steamer to the Caribbean side of the Isthmus.

Of course, Bunau-Varilla had no such power, but he did have friends in the right places. The same day he rushed to Washington. 'It was a test to which I was being submitted,' he later wrote. 'If I succeeded in this task the Canal was saved. If I failed it was lost.' His aim was to make the US government understand that 'its duty was to send immediately a cruiser in anticipation of probable events, rather than to wait for their explosion' as it had done in 1885 during the Prestan uprising. In Washington Bunau-Varilla saw his friend Loomis, who was standing in for the Secretary of State while Hay was away on holiday. Loomis agreed that the situation was 'really fraught with peril for the city of Colón' and gave the Frenchman to believe that a steamer would be despatched straightaway.

Bunau-Varilla had been watching the reports of US Navy ships in the newspapers. He knew that the *Nashville* was at Kingston, and, according to his account, guessed that this would be the vessel sent to Panama. Calculating the speed of the craft and the distance to be covered, he estimated when the gunboat would arrive at Colón. The next day he cabled 'Smith' in Panama saying a US warship would be with them in two and a half days.

Bunau-Varilla's confident tone gave the conspirators new heart. In a frenzy of activity, the flag was redesigned, a new declaration of independence was penned and Duque and his fire brigade of some three hundred young men were recruited and armed. Herbert Prescott brought his boss, Colonel James Shaler, into the plot and, realising the importance of the railway – the only way across the Isthmus – they arranged for all the line's rolling stock to be moved to the Panama side. Shaler, a tall, white-haired seventy-seven-year-old, was a popular figure in Panama, and would later be made a Hero of the Republic.

As Bunau-Varilla had predicted, the *Nashville* appeared in Colón harbour late in the afternoon of 2 November. The ship's captain, Commander John Hubbard, however, was not yet suspecting anything out of the ordinary. His orders were simply to consult with the US consul and report back on goings-on on the Isthmus. Nor were Colombian loyalists suspicious of the arrival of the two-stacked gunboat – the *Nashville* had been at Colón just two weeks earlier. But to the conspirators here was irrefutable proof that Bunau-Varilla and the Americans were going to deliver on their promises.

At around midnight on the same day, a Colombian gunboat, the *Cartagena*, also arrived in the harbour. On board were three generals and about five hundred *tiradores*, or expert marksmen. The next morning Hubbard went on board to be informed by General Tovar that he was landing his men. Hubbard was determined to play it by the book. He had as yet received no orders to prevent the disembarkation, and there was so far no disturbance on shore to merit his intervention. Thus, shortly after first light on 3 November, Generals Tovar, Amaya and Castro, followed by Colonel Eliseo Torres, the next most senior officer, resplendent in uniforms of yellow, blue and gold, glittering with medals and braid, stepped ashore on to the wooden wharf at Colón, closely followed by the rest of their men.

It was a bitter blow for the conspirators. Not only was the Colombian force formidable, but the Americans had singularly failed to prevent their landing, as had been promised. Fresh panic swept the group, and even Amador considered calling the whole thing off. However, Señora Amador was made of sterner stuff, rallying the plotters and quickly devising a trap to neutralise the Colombians.

The generals were met by local dignitaries and reassured that all was well to the north and that they should re-embark straightaway. But something made Tovar suspicious, and he demanded to be taken to Panama City. Enter Shaler, to play his part to perfection. Unfortunately, said the Railroad's superintendent, there were at the moment insufficient cars to transport the troops. However, there was a special luxury carriage available which could ferry the generals and their aides across to Panama. The Colombians protested, but were reassured that their men would be on the very next train.

Once on board the car, Amaya suddenly became jumpy and announced that he was going to stay with the men, but at that moment Shaler pulled the signal cord, jumped off the train, and waved cheerfully at the generals as they steamed out of the station. Soon after, Hubbard received orders from Washington, sent the day before but delayed, instructing him to prevent the landing of any armed force, or its use of the railroad. He therefore ordered Shaler not to transport the Colombian troops at Colón, thus giving the superintendent another excuse to buy time for the plotters at Panama City.

As soon as Amador heard from Prescott that the generals were on the way, he appealed once more to Huertas. 'If you will aid us,' he said, 'we shall reach immortality in the history of the new republic.' If he didn't, the elderly doctor warned, Huertas would surely be relieved and sent to some violent interior province of Colombia, far from his friends and family in Panama. At last Huertas agreed to be part of the uprising, his decision helped by the offer of $50 for each of his men, and $65,000 for himself.

The generals' luxury train arrived at Panama at 11.30, to be met by General Huertas, a military band playing patriotic songs and crowds of children waving Colombian flags. As Tovar later said in his defence, 'There was nothing that did not show the greatest cordiality and give me the most complete assurance that peace reigned throughout the department.' After a procession through the city, the Colombians were taken to a hotel to have a siesta.

Meanwhile Amador and Duque prepared for a mass meeting to take place in the city at 5.00 p.m. with the fire brigade poised to arrest those who might resist the uprising and ready to distribute rifles. But rumours were everywhere, and at 1.30 p.m. the generals were awoken to be told that a demonstration was going to take place. Then a note arrived from a local Panamanian loyal to Colombia warning Tovar to trust no one.

The general roused himself and demanded to know why his men had still not arrived. While Shaler continued to invent excuses as to why the men could not be transported, Huertas took the generals to lunch. All the time, their suspicion was mounting. After lunch, having again ordered the Governor, Obaldía, to organise the immediate despatch of their men, the generals proceeded to the barracks to carry out an inspection. By 5.00 p.m., Tovar had heard reports of a mob gathering and making its way towards them. Huertas suggested that a patrol be sent out and Tovar agreed. But as the men detailed for the patrol proceeded out of the barracks, as if to pass in front of the generals seated on a bench near the sea wall, they split into two columns, one marching in front of the seated men and one behind. On a command the men wheeled round and stopped, their fixed bayonets pointing towards the astonished Colombian top brass, who were told that they were now under arrest.

Tovar charged at one of the soldiers but was immediately hemmed in by bayonets. Castro also made a run for it, but was quickly recaptured having been found hiding in a toilet stall. Captain Salazar led the prisoners away to the jailhouse to cries from the growing crowd of '*Viva el Istmo libre!*' '*Viva Huertas!*' '*Viva el Presidente Amador!*'

In order to maintain the fiction of his non-involvement, Obaldía was arrested. Then Ehrman sent a message to Washington detailing the successful uprising, and at around 6.00 p.m. the leaders of the revolutionary junta proceeded to Cathedral Plaza to be acclaimed by an enthusiastic crowd. Now only the small matter of the five hundred heavily armed soldiers at Colón stood between Panama and independence.

*

Colonel Eliseo Torres, the commander of the Colombian force at Colón, had heard nothing of the goings-on in Panama City, but was becoming increasingly aggressive about Shaler's constant refusal to transport his men. Then, early on 4 November, he received a letter from Hubbard informing him that the railroad was closed to all troops. At lunchtime the same day, Torres was approached by Porfirio Meléndez, the junta's man in Colón, and told, over a drink at the Astor Hotel on Front Street, about the arrest of the generals and the uprising in Panama City. Meléndez then offered the colonel a bribe if he would remove his men. At first Torres refused to believe the news, but then he flew into a rage at the treachery of the Panamanians and their American friends, threatening to burn Colón to the ground and kill all American citizens in the town if the generals were not released.

Hubbard immediately readied his tiny force on the *Nashville* and started evacuating American and British women and children on to boats in the harbour, while their menfolk were herded into one of the stone buildings belonging to the Railroad. Some forty US sailors and marines were landed to defend the building, which was soon surrounded by Torres' greatly superior force. At this, Hubbard moved the *Nashville* close to the wharf, causing the *Cartagena* quickly to slip away, leaving her troops stranded. The American gunboat then trained her armament on the Colombians, and a tense stand off ensued.

But twenty-four hours later, when told that a US force of five thousand men was on the way to the Isthmus, and satisfied with his brief defiance, Torres agreed to leave for the payment of $8000. The money for the bribe had to be borrowed from the safe of the US-owned Panama Railroad. There was not enough, however, also to pay for the passage on a steamer, so more money had to be obtained from a local bank. This loan was guaranteed by Hubbard and Shaler, both American citizens. With the departure of the Colombians, the revolution was complete. The following day, to express their gratitude to the United States, an American Army officer, Major William Murray Black, was asked to raise the new Panama flag over the prefecture of Colón. Soon after, an official cable arrived from Hay at the State Department. As the people of Panama had 'resumed their independence', it read, the US consuls should 'enter into relations with it as the responsible government of the territory'. '*Viva La Republica de Panama!*' exclaimed the *Star and Herald*.

The revolution had succeeded with American connivance, but it still relied on the United States to make it irreversible. The news of the uprising caused a sensation in Colombia, where the initial fury was aimed at Marroquín. His residence was pelted with stones, the police were called in, leading to the wounding of several protestors, and martial law was declared. But soon, as detailed accounts of the events became known, the anger was redirected towards the United States. A heavy guard was thrown around the American embassy and Beupré was told he should leave the country for his own safety. As ambassador Herrán delivered a formal protest to Secretary of State Hay, thousands of Colombians volunteered to take part in an expedition to recapture Panama. Reyes threatened that unless recognition was withdrawn from the breakaway republic, the United States would have 'a second Boer War' on its hands.

But only hours after the declaration of independence American troops had been landed and there were half a dozen US gunboats on either side of the Isthmus. Roosevelt was wielding his 'big stick' – naval power – for the first time. The Colombians were forbidden to land soldiers anywhere in Panama. On 19 November Reyes arrived off Colón as head of a commission charged with offering Panama anything short of independence. But he was not even allowed to go ashore, and proceeded to Washington to try his luck there. Meanwhile, the Colombians equipped a force to try to make it overland to Colón through the Darién jungle. The men started off the following month, exhorted by their general that 'It is preferable to see the Colombian race exterminated than to submit to the United States.' But Darién proved impenetrable and, ravaged by disease, the troops soon turned back.

Claude Mallet's take on the extraordinary events of earlier in the month is pretty much spot-on: 'I have come to the conclusion,' he wrote on 20 November, 'that the scheme for a Republic was planned here, supported financially by persons interested in canal affairs in Paris, and encouraged by the Washington officials.' Nor was he unaware of the implications: 'The Americans, by their action here, have cast international customs to the winds, and henceforth, a new example has been set how to acquire the territory of your neighbour or friend.'

As Cromwell and Bunau-Varilla had anticipated, the US newspapers on the day after the 'revolution' were dominated by domestic election news. On 5 November, however, Panama was on every front page, and many papers would concur with Mallet's reading of the events. It was, said one, 'revolution of the canal, by the canal, for the canal'. 'It is another step in the imperial policy,' said the Pittsburgh *Post*. '"Might makes right" – steal from the weak.' There were many echoes of five years before, when the war with Spain had led to the formation of a widely supported Anti-Imperialism League, and complaints that the United States had abandoned its founding principles. For the Baltimore *News*, the 'Panama Affair' had, like the US actions in the Philippines and Hawaii, brought the US down to the sordid level of the land-grabbing European powers. To blame, said another paper, was the 'hot-headed and immature' Theodore Roosevelt. 'It begins to look as if nobody can touch that Panama ditch without being defiled,' concluded the Salt Lake *Herald*.

The criticisms of American aggression, connivance in the revolution and over-hasty recognition of the new republic would be led by the *New York Times*, then a fiercely partisan Democratic paper. To the *Times*, the canal was 'stolen property', and it soon focused its guns on the shady role of Cromwell. One of his partners, Edward B. Hill, when approached replied in classic style, 'You can quote me to the extent of saying that I have nothing to say.'

Others took a more pragmatic line. Even if the policy was wrong, said the Houston *Post*, 'The thing is done, there is no way of undoing it, and the least said about it the better.' For the San Francisco *Chronicle*, it was a sign of the times, but not therefore a cause for regret: 'The world must move,' it wrote. 'It is an age of power. The weak will be protected, but they will not be permitted to obstruct, whether upon the continent of America, the isthmus of Panama, the isles of the Pacific, the plains of Manchuria, or the valleys of the Ganges and the Indus. It is manifest destiny.'

In all, about two-thirds of the United States' newspapers supported Roosevelt's actions, buying into his theory of 'eminent domain' and his portrayal of the Colombians as blackmailers and extortionists. Those opposed tended to be Southern and Democratic-leaning. Certainly, public opinion never quite reached the level of opposition to the action in the Philippines. 'The disheartening fact is that the connivance of our administration in the dismemberment of a sister republic is accepted so phlegmatically,' wrote a correspondent to the New England Anti-Imperialist League. 'The country ought to be ringing with the protests of citizens in mass-meetings assembled.' But the man in the street's verdict, as reported by a Yale professor of law, was that 'it served Colombia right'. With the general acceptance of the US action over Panama, one of the founding principles of the United States passed away forever and the stage was set for US aggression and expansion throughout the region and, indeed, the world.

Leading the country away from its historical anti-colonialism was, of course, Theodore Roosevelt. While many hoped that there had not been direct involvement in the revolution, they also admired the President's 'virile' and 'strenuous' response to events. One congressman was quoted as saying to Roosevelt, 'Mr President, I am glad you did not start the rabbit to running, but as long as the rabbit was going to run anyhow, it's a good thing we did not have a bow-legged man in the White House who couldn't catch it.'

Roosevelt, of course, defended his actions and denied any role in the uprising. 'I did not lift a finger to incite the revolutionists,' he declared. 'I simply ceased to stamp out the different revolutionary fuses that were already burning.' His first task after the *fait accompli* was to bring his cabinet on board, and he gave a long, detailed statement of his position. When he had finished he turned to his

Secretary of War, Elihu Root. 'Well,' he asked, 'have I answered the charges? Have I defended myself?'

'You certainly have, Mr. President,' replied Root in a jokey tone. 'You have shown that you were accused of seduction and you have conclusively proved that you were guilty of rape.'

But in the changed political climate, it did not matter. On 10 November Roosevelt and his wife went to the opera to see *Barbette* at the National Theatre. One of the lines was 'What, a diplomat steal? A diplomat never steals. He only annexes!' The entire audience turned towards the President's box, and Roosevelt laughed as heartily as anyone and waved his hand in glee at the admiring crowd.

The next day the French ambassador Jules Jusserand had lunch with the President. When the talk inevitably turned to Panama, Roosevelt declared, 'It is reported that we have made the revolution; it is not so, but for months such an occurrence was probable and I was ready for it. It is all for the best . . . Everything goes on there as we would wish; I am about to receive Mr. Bunau-Varilla.'

On 4 November, Bunau-Varilla had received a cable from Amador asking for the immediate transfer of the promised $100,000 to pay for the bribery of the Colombian troops. No mention, however, had been made of the agreed appointment of the Frenchman as Panama's Minister Plenipotentiary or Envoy Extraordinaire. Bunau-Varilla reluctantly released $25,000, which was transferred to a Panama bank for the use of the junta. The next day another cable arrived, again pressing for more money and for Bunau-Varilla to expedite the recognition by the US of the new republic. But, to Bunau-Varilla's growing suspicion, there was still no mention of the diplomatic appointment. In fact, the junta was preparing to send its own commission to Washington to negotiate a new canal treaty, just as Bunau-Varilla feared. The Frenchman was determined that only he should have the honour of seeing his name on the canal treaty, and was not about to let anyone else 'mess up' the negotiations.

Bunau-Varilla knew very well that the US could not recognise Panama until the Colombian troops had left Colón, but his reply implied that both the advance of the rest of the money and the recognition from the United States, so crucial to Panama in its first days, depended on his appointment as Panama's minister in Washington. When *de facto* recognition arrived just after midday on 6 November, the Panamanians were under the impression that this had been arranged by Bunau-Varilla and later that day, wanting to keep him on side and secure formal recognition from the US (which required a reception by the President), the junta at last gave him the appointment he wanted.

But three days later, just as Amador and Federico Boyd were preparing to sail for the United States, Bunau-Varilla was cabled detailed instructions about the sort of treaty Panama wanted. The terms included joint tribunals in the Zone, the reversion to Panama of land leased to the New Company, and powers of raising duties at the terminal ports. The clear implication was that Bunau-Varilla was to start negotiations, but to discuss all matters with Amador and Boyd when they reached Washington.

It is not known whether this cable was ever seen by Bunau-Varilla. By 9 November he was already in Washington, 'to begin there,' as he puts it, 'the last and supreme battle'. The same day he lunched with Hay, having informed the Secretary of State of his appointment as Envoy Extraordinaire as soon as he had received it. At the meeting Bunau-Varilla urged Hay to organise quickly his official reception by the President. Hay agreed to this, but then asked the Frenchman about reports that a Commission was setting off from Panama to come to negotiate a canal treaty. Bunau-Varilla had seen the same newspaper story that morning and had his answer ready: 'Mr. Secretary of State, the situation harbours the same fatal germs – perhaps even more virulent ones – as those which caused at Bogota the rejection of the Hay–Herran Treaty.' The same 'intrigues' of 'politicians' were active in Panama, as in Colombia. The situation could only be saved, Bunau-Varilla exclaimed, by 'firmness of decision, and lightning rapidity of action. It is necessary to leave the enemy no time to perfect his plans.'

The 'enemies' – the 'fatal germs' – were, it should be stressed, the Panamanians themselves, the leaders of the country he was supposed to be representing. But to Bunau-Varilla, the Commission was a 'manoeuvre', and 'intrigue ... Amador was a party to it. I knew his childish desire to sign the Treaty.' Bunau-Varilla was determined that such 'childish' politicians should not stand in the way of 'the last and supreme battle' being fought, and won 'for the triumph of the Panama Canal' by Bunau-Varilla himself.

Hay did not miss the urgency, producing for circulation a draft treaty the very next day. He also took on board the Frenchman's tone, and realised that as long as he was dealing with Bunau-Varilla rather than the incoming Commission of Amador and Boyd, Panamanian interests could be largely discounted. Both men were also aware that the treaty faced its sternest test at home in Washington, in a Senate that had only narrowly approved the choice of Panama. In addition, the rumours of improper US involvement in the 'revolution' had provided ammunition to enemies of the administration and/or the canal, what Bunau-Varilla called 'the passions of parties and of contradictory elements'. But with Panama prostrate – through its dependence on the US military for its survival, as well as because of its *extraordinaire* representation in Washington – a deal could be rushed through whose terms would be irresistible to the Senate.

This is reflected in the articles of Hay's first draft treaty, produced on 10 November. Its basis was the Hay–Herrán Treaty, including the one-off $10 million payment and the annuity, but substantially modified in favour of the US. Nowhere was Panamanian sovereignty acknowledged and the proposed Canal Zone was increased in area by 60 per cent and included the 'terminal' cities of Colón and Panama. Within this Zone, now to be American 'in perpetuity', the US would have total military and civic control. Every possible objection that the Senate could raise was dealt with head on. In fact many of the measures echo those amendments proposed to the Hay–Herrán Treaty by Morgan with the explicit purpose of making the deal unacceptable to Colombia.

On the same day that Hay composed this draft, the Commission of Amador and Boyd set sail from Panama. They were due to arrive in New York seven days later. With them they carried orders for Bunau-Varilla that he should 'adjust' a treaty, but that 'all clauses of this Treaty will be discussed previously with the delegates of the Junta, M. Amador and Boyd'. That Bunau-Varilla had not been explicitly told this by cable shows how, overestimating his importance, the junta feared antagonising its 'friend' in Washington; and also that they never suspected that he would move with, as he put it, such 'lightning rapidity of action'.

On Friday 13 November, in a hastily assembled uniform of the official representative of Panama, Bunau-Varilla was presented to Roosevelt. To witness history in the making, and the *de jure* recognition of the new Republic, the Frenchman's son went along too. After formal statements, Roosevelt took Bunau-Varilla's arm and asked him, 'What do you think, Mr. Minister, of those people who print that we have made the Revolution of Panama together?' Bunau-Varilla replied with a rush of satisfactory rhetoric about 'calumny' and 'the mist of mendacity'.

As he left the reception, Bunau-Varilla, aware that Amador was now only four days away, gave Hay another nudge. 'For two years you have had difficulties in negotiating with the Colombians,' he said. 'Remember that ten days ago the Panamanians were still Colombians . . . You have now before you a Frenchman. If you wish to take advantage of a period of clearness in Panaman diplomacy, do it now! When I leave the spirit of Bogota will return.'

In fact Hay was operating at breakneck speed. A week later he would write to his daughter, 'As for your poor old dad, they are working him nights and Sundays. I have never, I think, been so constantly and actively employed as during the last fortnight.' He rushed his treaty round the departments and had a revised draft with Bunau-Varilla by late on 15 November.

Bunau-Varilla was at one with Hay on the need to placate the Morgan party in the Senate – he vainly tried again to convert the Alabama Senator – but even he had to object to the inclusion of the terminal cities in the proposed US zone. Panama City was, after all, the seat of government of the new Republic. But he offered instead the right to expropriate property in Panama City or Colón on public health grounds and to enforce sanitary arrangements therein. And to see off any possible objection about the lack of US control, he went even further than Hay had dared, adding this amendment: 'The Republic of Panama grants to the United States all the rights, power and authority within the zone mentioned . . . which the United States would possess and exercise if it were the sovereign of the territory . . . to the entire exclusion of the exercise by the Republic of Panama of any such sovereign rights, power or authority.' The 'inflammatory, unnecessary and offensive' clause goes to show how little Bunau-Varilla weighed Panamanian dignity against pleasing the US Senate.

Within twenty-fours hours, helped by his hired lawyer Frank Pavey, Bunau-Varilla had completed his new draft and was on his way round to Hay's house. But finding it in darkness he returned early the next morning and delivered the treaty. That same morning, 17 November, he learnt that Amador and Boyd had landed at New York.

Then, yet another happy accident: Boyd and Amador were met off the boat by Cromwell's agent Roger Farnham. Cromwell himself was due back from Paris later that day. Could they wait, as he wanted to speak to them? Aware of the lawyer's power and influence, the Panamanians delayed going straight to Washington, and met Cromwell later that day, and were persuaded to appoint him Panama's financial agent.

By coincidence or not, it gave Bunau-Varilla a precious further twenty-four hours to close the deal. But there was no word from Hay as the Frenchman waited nervously in his hotel suite for the entire day. At last, at 10.00 p.m. he sent a note to the Secretary of State's house. He would tell the Panamanians to stay in New York, he wrote, but had to sign the treaty the next day. Hay replied immediately, inviting Bunau-Varilla to come that night.

When they met, Bunau-Varilla again urged speed. He was utterly open about the reason: 'So long as the delegation has not arrived in Washington, I shall be free to deal with you alone, provided with complete confidence and absolute powers. When they arrive, I shall no longer be alone. In fact, I may perhaps no longer be here at all.'

Hay was happy with Bunau-Varilla's draft, but knew that what looked like a great deal for his country might not look so good to a Panamanian. As he would write to Senator Spooner, the new treaty was 'very satisfactory, vastly advantageous to the United States, and, we must confess, with what face we can muster, not so advantageous to Panama . . . You and I know too well how many points there are in this treaty to which a Panamanian patriot could object.' If the Hay–Herrán deal had been unfair on Colombia, the new treaty was many times worse for Panama, as Hay later admitted.

At lunchtime the next day, Hay consulted with the Attorney General and the Secretary of War, Elihu Root, and in a frantic afternoon the final drafts were drawn up in the State Department. At 4.30 the two Panamanians, blissfully unaware to what was going on, boarded a train for Washington, but at six o'clock Bunau-Varilla arrived at Hay's office to sign the treaty. To Bunau-Varilla's delight, waiting reporters addressed him as 'Your Excellency'. At 6.40 p.m., the treaty was signed, with a pen owned by Cromwell and ink from Abraham Lincoln's inkwell. 'We separated not without emotion,' Bunau-Varilla later wrote, 'having fixed the destiny, so long in the balance, of the great French conception.'

At 11.00 p.m., Bunau-Varilla was at Union Station in Washington to meet Amador and Boyd. As he recounts, 'I greeted the travellers with the happy news! "The Republic of Panama is henceforth under the protection of the United States. I have just signed the Canal Treaty."'

The Panamanians were stunned. According to Bunau-Varilla, 'Amador was positively overcome by the ordeal' and nearly fainted. Neither did Boyd respond as he should have done to 'a happy event which ought to have filled their hearts with joy'. In fact, having been at first disbelieving, the Panamanians were soon furious, all the more so when they learnt the terms of the treaty. Reportedly Bunau-Varilla was spat at by Boyd. They realised they had been betrayed. 'Cherish no illusion, Mr Boyd,' Bunau-Varilla said when the Panamanian suggested that fresh talks could be had on various points. 'The negotiations are closed.'

Amador and Boyd did try to reopen talks two days later, but without success. In the meantime, Bunau-Varilla attempted to bully them into ratifying the treaty there and then, without further recourse to Panama, at one point pressing a pen into the hand of Amador, who reacted by angrily hurling it across the room. Bunau-Varilla then cabled Panama offering immediate credits of up to $100,000 from the bank of the House of JP Morgan if they ordered Amador and Boyd to ratify. Everyone knew that General Rafael Reyes would soon be in Washington and would offer pretty much anything to get Panama and the canal back for Colombia. Although Reyes' mission would prove fruitless, in spite of the high-profile support of ex-President Cleveland, it provided Bunau-Varilla with the leverage to force super-quick ratification of his treaty by Panama. In fact, the junta agreed to sign on 26 November, before they had even seen the treaty, which was on its way by boat, wrapped in a Panama flag and sealed with the family crest of John Bigelow.

At 11.30 on the morning of 2 December, less than twenty-four hours after being brought to Panama City, the convention was ratified. There cannot have been time to make a Spanish translation of the English text or to make copies for distribution to the nine men due to confirm the agreement. The likelihood is that the Hay–Bunau-Varilla Treaty was not even read by the signatories of the ratification decree, though it was to reduce their new country to little more than vassalage.

For the moment, however, the signing was welcomed in Panama. Then, as the rush and adrenalin of the last month subsided, a new view emerged. 'What do you think of the canal treaty?' Mallet wrote to his wife soon afterwards. 'Here the people are disgusted, and one of the prime movers in the independence movement was heard to say "*nos han vendido*" ["We've been sold out"]. Well, the Yankees have got them at last, and they have been foolish enough here to think those hardened and practical people were governed more by sentiment than by their interests.' Soon a view solidified that national rights had been signed away by a foreigner, and that perhaps Panamanians had merely changed an impotent overlord for a powerful and determined one. The brief honeymoon period was over, even before the first spade load of the American canal had been dug.

In spite of all the efforts made to contrive a treaty to the liking of the US Senate, the debate and division there were fierce, ironically in part as an embarrassed reaction to the meanness of the deal. The treaty, one senator pointed out, gave the 'United States more than anybody in this Chamber ever dreamed of having . . . we have never had a concession so extraordinary in character as this. In fact, it sounds very much like we wrote it ourselves.' Most of the opposition, however, was directed at the way Roosevelt had behaved towards Colombia. Some argued that the President had effectively declared war, something only Congress was authorised to do. Democratic Senator Thomas Patterson of Colorado declared that the Canal Zone was 'stolen in the most bare-faced manner from Colombia'. 'The president has denied with some heat that he had any complicity in this business,' said Senator Edward Carmack. 'He does not conceal the fact that he desired this insurrection. He does not conceal the fact that he intended to aid it if it occurred, and he can not conceal the fact that he did aid it.' There had been a lot of talk, Carmack continued, about the people of the Isthmus 'rising as one man', 'but the one man was in the White House'.

On the Senate floor Carmack went on to warn that the action against Colombia was 'but the beginning of systematic policy of aggression toward the Central and South American states'. 'I fear,' declared another senator, 'that we have got too large to be just.' An amendment ordering a payment of compensation to Colombia was defeated, but almost succeeded in uniting the Democrats against the measure.

The Democrats were undecided how to vote on the treaty. They had 33 of the 90 Senate seats, and, if united, could have blocked the deal and given Roosevelt a severe setback less than twelve months before the presidential elections. But many were in favour of a canal, which was also popular in the country. One Texas senator explained the dilemma by telling the story of a dog catching a rabbit in violation of its previous teaching and rules laid down for its conduct: 'You might whip the dog, but would you throw away the rabbit?'

In the event less than half the Democrats voted against the Hay–Bunau-Varilla Treaty, which passed by 66 votes to 14 on 23 February. Two days later the treaties were officially exchanged.

It was one of the most important deals in the history of American foreign relations, as it gave to the United States absolute control over the future Panama Canal, and thus over the strategic and economic crossroads of the Americas. A contemporary historian, Wolf von Schierbrand, stated that the treaty's importance 'to our future political, commercial, and naval expansion, in the Pacific as well as the Caribbean Sea, can scarcely be overestimated. It will be the main pillar of our future strength in those all-important regions.' 'From the point of view of world politics,' said another distinguished commentator, 'the construction and operation of the canal as a government undertaking means the extension of the political control of the United States over the Spanish-American nations.'

On 2 May 1904 the assets of the Compagnie Nouvelle were signed over to the United States. The sale was handled by JP Morgan (thanks to an intervention by Cromwell), who arranged for $18 million in gold to be shipped to France, and bought exchange on Paris for the balance in several European markets, paying the money into the Bank of France. Together with the $10 million paid to Panama, it dwarfed the purchases of Louisiana ($15 million), Alaska ($7.2 million) and the Philippines ($20 million). The actual physical handover on the Isthmus occurred early on 4 May, when a US Army engineer, Second Lieutenant Mark Brooke, met with a representative of the New Company at the old Grand Hotel. After a few perfunctory words, the Stars and Stripes was hoisted. After all the ceremony of the French years, the amazing razzle-dazzle of Ferdinand de Lesseps, it was something of a disappointment to the Panamanians.

The $40 million paid to the French company converted to 206 million francs, of which 128 million went to the credit of the Old Company and 77.4 million to the New Company. None of the shareholders of the Old Company got anything. Bondholders, of whom 226,296 claimed, got on average 650 francs, or $156, approximately ten cents on the dollar of their investment. The New Company shareholders received 129.78 francs per 100 franc share, which worked out at an interest of less than 3 per cent per annum, but must have been much more than they expected. Thus Bunau-Varilla had got not only his name on the treaty, but also got back the money forcibly invested more than ten years before.

The Frenchman had resigned as Panama's minister on 2 March 1904, his job, 'The Resurrection of the Panama Canal', complete. Cabling the decision to Panama City, he asked for his remuneration (which at $1000 per month was in total less than $5000) to be put towards the cost of erecting a statue of Ferdinand de Lesseps, 'the great Frenchman, whose genius has consecrated the Isthmus to the progress of the world'. As he crossed the hotel lobby to take the message to the telegraph office, he reports, 'somebody unexpectedly seized my hands to express to me his congratulations. It was the lawyer Cromwell.'

Roosevelt, never one for self-doubt, conceded in a private letter that there was 'great uneasiness caused among my friends by my action', but in reality he had few qualms about the path taken. 'The one thing for which I deserved most credit in my entire administration,' he would write, 'was my action in seizing the psychological moment to get complete control of Panama.' 'It was a good thing for Egypt and the Sudan, and for the world, when England took Egypt and the Sudan,' he wrote to his old friend Cecil Spring Rice at the British Foreign Office. 'It is a good thing for India that England should control it. And so it is a good thing, a very good thing, for Cuba and for Panama and for the world that the United States has acted as it has actually done during the last six years. The people of the United States and the people of the Isthmus and the rest of mankind will all be the better because we dig the Panama canal and keep order in its neighbourhood. And the politicians and revolutionists at Bogota are entitled to precisely the amount of sympathy we extend to other inefficient bandits.'

In the November 1904 election Roosevelt saw the canal as a benefit, rather than a hindrance to his campaign, even though Henry Davis, the Democratic vice-presidential candidate, criticised Roosevelt's actions over Panama as belonging 'more to an empire than a Republic'. 'Tell our speakers to dwell more on the Panama Canal,' Roosevelt told an aide during the campaign. 'We have not a stronger card.' It had become a symbol for his active, vigorous leadership.

On 8 November 1904, Roosevelt got seven and a half million votes to Parker's five million. The victory was attributed to Roosevelt's personal appeal, but also to the popularity of his activist Panama policies. A dismayed member of the New England Anti-Imperialist League commented, 'We stand today, apparently in the shadow of a great defeat. Theodore Roosevelt represents today the temper and point of view of the American people, as to armies, navies, world power, Panama republics and American police duty on the Western Hemisphere.'

But in spite of this victory, some of the Panama mud had stuck. More dirt would be dug up in the years to come, leading to continued press and Congressional investigations. Most importantly, the events leading up to the start of the US construction effort would put the canal on the defensive in terms of domestic politics. After all the intrigue and politicking, huge pressure would now be bearing down on the canal effort to 'make the dirt fly', with disastrous consequences.

Internationally, the secession and subsequent treaty locked the United States into a cycle of expansion in the region, and its long-range cost in bad feeling and ill will was immense. Had more attention been paid to the legitimacy of many of the Colombian concerns and to the reality of the political situation in Bogotá, rather than to the interests of a private, foreign-owned corporation, a deal could have been hammered out. Once this failed, it was poor diplomacy by Hay to sign a treaty with the new Republic of Panama so patently unfair that it was bound to store up trouble for the future. But as Roosevelt would later point out, while the arguments went on, at least now the canal was being built.

18

'MAKE THE DIRT FLY'

'In America, anything is possible,' Jan van Hardeveld would proclaim to his wife, Rose, and their two small daughters whenever he learnt of some modern miracle of enterprise in his new country. The family lived on a homestead in a remote part of western Wyoming, where Jan worked as the foreman of a gang of largely Japanese workers on the Union Pacific Railroad. A recently naturalised Hollander, he had particular admiration for President Roosevelt's Dutch blood. When he heard about the start of the American canal, he was determined to be part of 'the mighty march of progress'. 'The French gave up . . . but we will finish!' Jan proclaimed. 'With Teddy Roosevelt, *anything* is possible.'

George Martin, a carpenter's apprentice living in Barbados, was eighteen when he heard, he writes, 'A voice from a great people' inviting him to help build the Panama Canal. 'With the others I accepted . . . so I leave father and mother, brothers and relatives, away in the land of the Indies, in the west, and came to this strange land . . .'

As early as January 1904, while the Senate was still debating the Hay–Bunau-Varilla Treaty, journalists from the 'great people' were on the Isthmus reporting back that there 'is nothing in the nature of the work . . . to daunt an American. The building of the canal will be a comparatively easy task for knowing, enterprising and energetic Americans.' Many were confident that it would be a splendid showcase of the ever-growing industrial and technological might of the United States, and the country's new superiority over the old powers of Europe.

With hindsight, the American project might seem to have a 'solid inevitability' compared with the tragically doomed de Lesseps adventure. In fact, the construction was beset by very serious difficulties throughout, but particularly in the first three years, and on several occasions came close to disaster. When the Americans started work they replicated almost all the mistakes made by the de Lesseps

Company: they favoured a sea-level canal; they split authority for the job, as the French had done up until the arrival of Dingler in 1883; their initial site investigation was patchy, leading to unpleasant surprises later on; more than anything, they underestimated or misunderstood the dangers of disease and the simply vast scale of the construction challenge.

On 3 March 1904, a week after the formal exchange of treaties with Panama, Roosevelt appointed a seven-man Isthmian Canal Commission (ICC). Their order from the President was simply that 'the results be achieved'. The chairman was the veteran Admiral Walker, known to many as the 'Old Man of the Sea', who had led previous canal bodies. Although in this respect experienced, he was an old-fashioned figure and had not overseen any really large construction projects. The next most senior figure, and the only member of the Commission who would actually reside on the Isthmus, was another military man, Major General George W. Davis, who was to be Governor of the Canal Zone. The emphasis for the other five appointments was on engineering experience, rather than familiarity with heading up such an immense logistical project. Davis, for his part, had been involved with one of the Nicaragua private canal companies, but was first and foremost a colonial administrator – he had been involved with the organisation of the US military governments of Cuba, Puerto Rico and the Philippines. As in the other newly acquired territories, the government of the Canal Zone, indeed, the entire Commission and canal effort, would report to the new Secretary of War, William Howard Taft.

Before the Commission was appointed, Roosevelt had been lobbied by delegations of prominent US doctors urging him to give the medical challenge in Panama the top priority. They seem to have been preaching to the converted. Roosevelt had been shocked by the death rate from yellow fever among US soldiers in Cuba – five times more men had been killed by illness than by enemy action – and had himself, before he became President, publicised these terrible statistics in Washington. As early as February 1904 he wrote to Admiral Walker, 'I feel that the sanitary and hygiene problems . . . on the Isthmus are those which are literally of the first importance, coming even before the engineering . . .'

Nevertheless, there was no medical representation on the first Commission, effectively the board of directors of the canal effort. But the American Society of Doctors did get their recommended man, Colonel William Crawford Gorgas, appointed chief medical officer. Gorgas was well respected for his work attacking the yellow fever epidemic in Cuba and was the country's leading expert on tropical diseases.

In Panama, those who were able busied themselves with preparing to take advantage of the forthcoming influx of men and money, particularly in respect to rental prices. On 9 March, Claude Mallet wrote to his wife, 'Building is going on everywhere, it is impossible to find a disengaged carpenter or mason . . . owners of properties wish to take advantage of cheaper wages before the canal starts.'

In fact, workers on the railroad were ready to test their strength and ability to influence the labour conditions for the boom time to come. In early April men on the railroad and docks launched a massive strike, demanding better pay and conditions. For a week the transit was brought to a stop, and, amid fears of rioting, US marines, 450 of whom had been stationed at Bas Obispo since December the previous year, were brought in to guard the Railroad's property. At first the police were used to try to force the men, mainly West Indians, back to work. When this failed, the Railroad bosses shipped in more than a hundred workmen from the tiny Fortune Islands in the Caribbean. This broke the strike and kept wages at the same level.

The same month the grandees of the Commission descended on Panama, accompanied by Gorgas and another sanitary officer and fellow old Cuba hand, Louis La Garde. The doctors wasted no time in diagnosing malaria as an even greater threat to the canal builders than the dreaded yellow fever. Gorgas visited the marine barracks at Bas Obispo, a seemingly healthy, breezy spot, and was told that 170 of the 450 had caught malaria since the beginning of the year. The source of the infection was not difficult to find. When Gorgas and La Garde examined the inhabitants of the nearby 'native' village, they discovered that some 70 per cent had the enlarged spleen of the malaria carrier.

The engineers of the ICC were accompanied by Roger Farnham, the press agent of the ubiquitous William Nelson Cromwell. As well as getting himself appointed Panama's US counsel, Cromwell had become an 'all-purpose trouble-shooter for the Republican Party'. He did not need to be told what failure in Panama would do to the party's fortunes in the forthcoming presidential elections and was determined to keep an eye on the canal effort. Mallet did his best during the Commission's two-week stay to discover what he could about the Americans' plans. One of the commissioners told him that 'he was sure every member of the Commission hoped that a sea-level canal would be built if it be practicable'.

In fact, nothing concrete was decided by the trip. The question of the design of the canal hinged on the suitability of a variety of sites for the construction of dams and/or locks, namely Gamboa, Bohío and Gatún. Until proper, deep borings were made, all the engineers could propose in the meantime in the way of 'making the dirt fly' were harbour improvements at Colón and designs for waterworks for the two terminal cities.

On 6 May, two days after the official handing over of the French properties to the United States and the raising of the Stars and Stripes on Ancón Hill, John Findlay Wallace was appointed to the job of Chief Engineer, in charge of all canal construction work, although without a seat on the Commission. Wallace, a Midwestern-railroad veteran and first-rate engineer, had been tempted to Panama by $25,000 a year, a larger salary than any other government employee except the President.

The new Governor of the Zone, Major General Davis, arrived to stay on 17 May. A man of few words or courtesies, he soon found that what could be got away with in Cuba or the Philippines would not do in Panama. Davis threw himself into the work, issuing orders and rushing about, but failed to call on President Amador for several days and even refused to attend the many ceremonies inevitably arranged in his honour by the Panamanians. Complaints reached Washington and eventually Davis was ordered to change his attitude. There was further bad feeling as the marking out of the new US Canal Zone boundaries was begun and the reality of the Hay–Bunau-Varilla Treaty sunk in. 'They have taken all the meat and left the bone,' one 'disgusted' Panamanian complained to the British consul.

'The canal employees are coming,' Mallet wrote to his wife on 2 June. 'The Isthmus is swarming with Yankees already,' he reported a week later. 'From ocean to ocean you see them everywhere and American flags hoisted on all sides.' Most excitement was attached to the arrival – at the end of the month – of Wallace and Gorgas to take up permanent residence. 'The medical board declares that not one mosquito shall survive,' wrote Mallet. 'Panama without mosquitoes? What a blessing . . .'

On 21 June the steamer the *Allianca* sailed from New York. As well as Wallace and Gorgas, on board was William Karner, a colleague of Wallace from Chicago who had been appointed Assistant Engineer, two other senior sanitary officers, a new Head Nurse, Eugenie Hibbard, with two other nurses, and about a dozen clerks and sanitary inspectors. Karner, who would later play a crucial role as the chief recruiter of labour for the canal, had taken the job out of loyalty to his old boss, in spite of the fact, he writes, that the 'proposition [of the job] and a residence in Panama was not very alluring to me'. Eugenie Hibbard, a Canadian, had made her name in hospital and training-school administration. Forty-eight in 1904, she had served in Cuba – and survived an attack of yellow fever. In spite of her great experience of difficult postings, Hibbard reports that she was quite daunted by the prospect of Panama. Her friends had asked her why she would want to go to 'such a God forsaken place where the French have so finally failed!'. 'We felt,' she wrote of herself and two fellow nurses, Miss Markham and Miss McGowan, 'that we were going to a country of swamp and jungle, filled with drawling and flying death, where any white woman was sure of destruction.'

After a rough and uncomfortable voyage, the *Allianca* arrived at Colón on the morning of 28 June. 'The rainy season had just commenced,' William Karner remembered, 'and a shower that morning had left the streets of Colon and Cristobal in thick, impassable mud. It was not a pleasant introduction to a strange country and city, both of which we knew had a bad reputation as to health and sanitation.' The party was met by Governor Davis and piled on to a train to Panama City.

The outskirts of Colón made their customary shocking impression on the new arrivals. The conditions, Hibbard remembered, 'beggar description, the houses being huts of wood built on piles 2 or 3 feet above the most filthy water, foul smelling and covered with green slime filled with most objectionable refuse . . . Leaving Colón behind,' she continues, 'we saw for quite a distance alongside the Canal long neglected and broken and abandoned machinery, with here and there many graves surrounded by small railings (wooden) and marked with a rough white wooden cross impressed one forcibly of what had taken place. Soon we began to climb the mountains and the remarkable beauty of the country, the hills and foliage thrust itself upon us. I had never seen such luxuriant growth of trees and shrubs . . . flowering plants and many beautiful orchids growing at random.'

The transit completed, the nurses were conveyed to the French-built hospital at Ancón. Once the pride of the tropics, the grounds had reverted to jungle and the buildings were in a woeful state of decay and dilapidation. The nurses were shown to their quarters, 'a strange and unattractive abode,' says Hibbard, 'the first night passed in these quarters was sufficient to have broken down the courage that has brought us here. The following day, I asked the Commanding Officer for something to guard ourselves with and he gave me a pistol (a Colt revolver) too heavy for me to handle in one hand. I placed it on a chair at my bedside at night and looked lovingly upon it as a possible protector.'

The wards, in which about thirty patients, mostly incurables, were being cared for by the French Sisters of Charity, were filthy by Eugenie Hibbard's standards. The task of cleaning them up, and clearing out the ancient, bug-infested horse-hair mattresses, fell to the nurses. 'There was, I realised, a stupendous piece of work before us,' Hibbard writes, 'and so it proved to be.'

The day after the arrival of the *Allianca*, Joseph Le Prince, one of Gorgas's sanitary inspectors, carried out a check on potential mosquito breeding sites near the Ancón wards. The bottom of the hill, he discovered, was continually soggy, and adjacent pasture had hoofprints full of water. Nearby drainage ditches were choked with weeds, which retarded the water flow and provided the environment mosquitoes needed to lay their eggs. In fact, he concluded, 'A more prolific source would be hard to imagine.' The result was that the hospital was swarming. 'The *Anopheles* [malaria-carrying mosquitoes] were so numerous,' writes Le Prince, 'that night work had to be done in relays; one set of men using fans to protect those working.' No fewer than fifty-four *Anopheles* were noted on the upper panel of a single screen door. Furthermore, the patients in the wards were located according to nationality rather than to the nature of their illness. 'Had it been intended,' says Le Prince, 'to spread yellow fever and malaria with the greatest rapidity among the patients as soon as they arrived, no better plan could have been adopted.' Within just weeks, all but a couple of the small hospital staff had come down with malaria, Gorgas included.

The engineers among the first arrivals would find a similarly melancholy scene. According to Wallace, there was 'only jungle and chaos from one end of the Isthmus to the other'. Panama was a gigantic scrapheap. All along the line of the old French canal, abandoned excavators and dredges, some of huge size, slumped lopsidedly, half-submerged in swamp or stream. Over everything the voracious returning jungle had draped a thick web of vines. Discarded locomotives and spoil cars were piled in huge mounds of rust and twisted metal. Materials lay scattered everywhere, as if abandoned by a hastily retreating army. The buildings, which at one time had housed more than twenty thousand canal workers, had been reclaimed by termites, rot or vegetation. Inside one building, where the rafters had decayed and the roof had collapsed, Joseph Le Prince found several trees growing with trunks of more than 10 inches diameter. In one place, an entire village had been completely buried by the returning jungle.

Evidence of failure, mistakes and waste lay all around. When the newcomers took over the old French administration building, they were surprised to come across twelve unopened packing cases filled with steel pen nibs. Looking at the books William Karner discovered 'records to show that they bought at one time sixty thousand dollars' worth of steel pens'. It was not clear whether this was inefficiency or corruption, but Karner reports hearing many stories of the 'graft' of the French period from old-timers on the Isthmus.

Although the wreckage of the French effort scattered everywhere had an unnerving effect on the first American canal builders, once they started to go systematically through their inheritance, the picture brightened considerably. Many of the two thousand buildings would be repairable. Six working machine shops provided a nucleus for later expansion. The French had left their successors maps and surveys, 'excellently recorded [which] proved to be of great use'. According to Wallace, there was a significant amount of materials and supplies safely stored in warehouses and unissued, which were in 'a fairly good condition, and were systematically stored, arranged, tabulated and properly cared for'. Vital spare parts and machine tools had been liberally coated in grease against the corroding warmth and damp of the climate. 'Splendid workmanship was shown on these machines,' one American engineer conceded, 'and good material was used in their construction.' Although, 'obsolete,' he went on, 'they were good appliances of their date.'

On second glance, the actual digging achieved was impressive also. As well as an eleven-mile passage from Colón to Bohío, there were vast excavations where Eiffel's locks were to have gone. On the Pacific side a passage had been dredged from La Boca to deep water, and 'considerable work had been done on the channel from La Boca to Miraflores'. In addition, over thirty miles of diversion channels had been created for the Chagres River. There had clearly been a lot of very hard work. After everything they had heard, most new arrivals were surprised by the 'magnificence of the French failure'. The Europeans had achieved, it was apparent, 'vastly more that the popular impression'. How much of this immense excavation – nearly 50 million cubic metres (73 million cubic yards) – would be useful would, of course, only be determined when a definite plan for the American canal emerged.

Wallace himself wrote that his approach to deciding the type of canal that should be built was determined by the 'great amount of work already performed by the old and new Panama Canal companies' as well as 'the tentative plans developed by the former Isthmian Canal Commission'. The Walker Commission of 1899–1901 had been heavily influenced by the French New Company plan put together in 1898 to increase the saleability of the canal concern. The French proposal had allowed for a dam and locks at Bohío, some fifteen miles upriver from where the Chagres meets the Atlantic. This, it was planned, would create an artificial lake at 68 feet above sea level. With a surface area of just over thirteen and a half thousand acres (5500 hectares), this would stretch for thirteen miles through the Culebra Cut at the Pacific end of which would be built at Pedro Miguel further dams and locks to return shipping to sea level. The lake would provide water for the locks and also, it was hoped, absorb the seasonal floods of the Chagres. Alternatively, a further raised section could be created through the Continental Divide at 96 feet above sea level between Obispo and Paraíso. Clearly the thinking of the French was in part shaped by the sacrifices made in the 1880s. Both of these options had the merit of avoiding 'the loss of any work already performed'. The idea of a plan that would render much of the French digging superfluous, as eventually adopted, was, for now, too ghastly to contemplate.

Although the Walker Commission had dispensed with the high-level option – feeding the top level of the canal from the upper Chagres would have been too difficult – they had remained wedded to the idea of the main dam at Bohío, albeit with a lake of the higher elevation of 85 feet above sea level. This is what the influential engineer Morison had argued for and defended at the Senate Hearings before the Spooner debate. But the Spooner Act had not specified the type of canal except that it should 'afford convenient passage for vessels of the largest tonnage and greatest draft now in use, and such as may be reasonably anticipated' and should use 'as far as practicable the work heretofore done by the New Panama Canal Company, of France, and its predecessor company.' Indeed, during the 'Battle of the Routes' in the Senate, one of Mark Hanna's arguments in favour of Panama had been that only there was a sea-level canal possible. In fact, de Lesseps' dream of an 'Ocean Bosporus' still held a great appeal, even after the disasters of the 1880s. 'A sea-level canal alone satisfies the requirements of the case,' pronounced an engineer during a debate on the question at the London Institute of Civil Engineers.

There were more realistic voices. A sea-level canal 'would be an ideal solution of the question,' mused another engineer, 'only if the Chagres River did not exist.' One correspondent to the Institute during this debate was J. C. Hawkshaw, son of Sir John, the engineer of the Severn Tunnel who had been an invited expert at de Lesseps' Congress of 1879. Like father like son, Hawkshaw junior warned that the river presented an insurmountable obstacle to a sea-level canal. He did, however, 'wish success to those who were proposing to complete the canal, on which so much has been spent, and which could not fail, if carried out, to be a benefit to the commerce of the world in general, and more especially to that of this country'.

The upshot was that Wallace and the ICC members went to Panama with the most basic specifications of the canal – lock or sea-level – undecided. Nothing could be ruled out, and therefore there was much work to be done investigating all the possible options.

When the New Company was handed over to the Americans on 4 May, they took over a skeleton workforce of about five hundred men, most of whom were employed in the Culebra Cut, where two French ladder excavators were carrying out intermittent work. Others were maintaining such machinery as had been stored away. At the beginning of June, five different US-led parties were established and set to work: one to survey Colón Harbour; another to start planning waterworks for Panama City. The three other parties were instructed to carry out the deeper borings demanded by the Commission to test for the suitability of various sites for dams and started work at Gatún, Bohío and Gamboa. Although Gatún had not been mentioned by Walker as a possible dam site, there had been several recent papers published in US engineering journals which had suggested it. The Gamboa group was also charged with mapping the routes of spillways to carry the flood waters of the Chagres away from the line of the canal, as would be required for a sea-level canal. In July a base camp was established at Bas Obispo and the twelve Americans, accompanied by two dozen locals recruited to do the machete work, started to search for a route to link the upper Chagres with a small river which flowed from the heights of the Continental Divide into the Pacific.

Effectively, they were recrossing the ground covered by Henri Cermoise back in 1881–2, and the same conditions prevailed. But the laboriously cut *tranches* had long disappeared. 'It was not possible to advance a foot without hacking one's way through a tangle of creepers,' an engineer wrote of the expedition. 'Lizards and gaudy snakes crawled and scuttled everywhere . . . insect pests were superabundant.' At the end of each day a space was cleared in the jungle and a makeshift shelter was improvised using poles with canvas or palm fronds for a roof. Night-time was a torment of itching and scratching at the festering sores caused by ticks, 'red-bugs', 'jiggers' and other parasitic insects which specialised in laying eggs under the skin of their victims. Supplies were carried by canoe upriver to the surveyors, but once in a while a monkey was cooked and eaten. However, 'the appearance of the skinned animal was so suggestively human that few of the college-bred men from the North fancied the diet'. And where Cermoise had discovered exhilaration and comradeship, the Americans, as one of them wrote, experienced only 'demoralization'.

As field work continued, three of the old French excavators were overhauled and set to work in the Culebra Cut, both to provide visible proof that they were 'making the dirt fly' and also to provide data on the effectiveness and unit costs of different types of machine. At the ocean termini of the line, a new Divisional Head Engineer, Frank Maltby, had started work on dredging the harbours. Maltby was originally from Pittsburgh, but had worked for three years as head of dredging operations on the Mississippi. He got the job through a friend of his on the Commission. He had never been to sea before and was sick for the entire seven-day voyage to Colón. He found the town 'indescribably filthy' but was given lodging in the old 'De Lesseps Palace', an imposing residence built at Cristóbal for *Le Grand Français*'s 1886 visit. Within two days he was hospitalised for a week with diarrhoea, but once at work he quickly demonstrated that whatever its outward appearance, much of the supposedly 'obsolete' French plant was eminently usable, even machines that had spent the last fifteen years semi- or entirely submerged under water. 'As the use of cheap steel had not become the practice at the time of their creation,' explains one US engineer, 'they were built of a superior class of iron – a much better metal to withstand the ravages of time and sea water'. Within a couple of months Maltby had six of the old Scottish-built ladder dredges back working, crewed in the main, as during the French years, by Greeks. The locomotives that lay abandoned all around were also found to be 'built like a watch in workmanship, of splendid material', and were similarly pressed back into service.

But there were also numerous frustrations. Soon after his arrival, Maltby tried to organise the building of a short length of railway track. For some reason there seemed to be plenty of ties lying around. Rails could be picked up from abandoned track, and spikes could be pulled out of rotten ties. But as there was no spike maul to be had, his men had to bang in the spikes with axes.

In fact, it was just as well that so much equipment was suitable to put back to work, as precious little was arriving from the United States. In Washington the Isthmian Canal Commission was from the outset in a state of paranoid ineptitude. The sensational events of the 'Panama Affair' in France in 1890–92 had been watched around the world. So, however unfairly, the French had left behind in Panama, along with everything else, the taint of waste, extravagance and corruption. The instinct of the US Commission, therefore, was to query and triple check every requisition. 'When this whole thing is finished,' the Commission chairman Walker announced, 'I intend that those fellows on the hill shall not find that a single dollar has been misspent.' As well as critics in Congress, Walker was acutely aware that the anti-Canal or anti-Roosevelt press was waiting to pounce on any example of 'waste' in the French style. Therefore, everything had to be cleared by all seven members of the Commission, each of whom felt personally responsible that no 'graft' would be tolerated on their watch.

The result was chaos and deadlock. A system of purchases was created that required a nightmare of forms in triplicate. The work on the Panama sewers was hampered by the fact that it took the filling in of six vouchers for the hire of a single horse and cart. That meant that the engineer in charge had to spend his Saturday night filling in no fewer than 1200 such forms. There was a further, more serious setback to this work when it emerged that the pipes for the job had been sent in the wrong order. Then it was discovered that some vital equipment had been sent, for economy's sake, by sailing schooner and would not arrive for months. When Wallace cabled Washington to protest at the lack of equipment coming through he was sharply reprimanded by one of the Commissioners that sending cables cost money.

Among an ever-thickening blizzard of paper, orders became duplicated or lost. Alternatively they were pared down or simply filed away. When Walker left his job the following year, over 160 requisitions were found stuffed into drawers in his desk, some many months old. Those on the Isthmus responded by attempting to predict their needs far into the future or simply bumped up their orders expecting them to be adjusted downwards. On one occasion, Wallace's chief architect, his twenty-nine-year-old nephew O. M. Johnson, calculated that he would eventually require 15,000 doors, for which he needed 15,000 pairs of hinges. He might have expected to receive a fraction of that, but somewhere in the paper storm in the Washington office the order took on monstrous proportions and, soon after, 240,000 perfectly made hinges turned up at Colón. It was just like the French and their pen nibs.

The architect's office was one of the many bottlenecks in the initial organisation. There was a great deal of work involved in repairing the French quarters, let alone designing and building new accommodation. But it was a chicken-and-egg situation. As Wallace complained, 'Suitable quarters and accommodations could not be provided without organization, supervision, plans and material, which of course, rendered a large force necessary almost at the commencement of the work, which had to be provided with suitable quarters and accommodation.'

Demand for labour was acute while the need for quarters massively outstripped the available supply. Cities of tents were created on the slopes of Ancón Hill and elsewhere, but these were soon full as the workforce expanded to 3500 by November 1904. To 'make the dirt fly', the Washington office was sending hundreds of men to the Isthmus every week, 'before there was any way to care for them properly, or any tools or material to work with,' as Frank Maltby complained.

Soon after her husband's departure for the canal project, Rose van Hardeveld received her first letter from Jan, who had given up his post on the Union Pacific Railroad to accept a job with the Commission and thus become part of Teddy Roosevelt's 'great march of progress'. Having sailed from San Diego, he arrived at Panama City and made his way to Culebra. 'A heavy suitcase in each hand, no light anywhere, the sweat rolling down my face, I stumbled along the wet slippery track, which I had been told to follow until I found a place to turn off,' he wrote. 'I could sense that the water was on both sides. If my foot slipped from the ties, it landed in soft mud. In the deep darkness I seemed to have walked miles, and I never dreamed there could be such unearthly noises as came to my ears from all around. Thick croaking, hoarse bellowing, and strange squeaks and whines leaped at me from the blackness. I have learned since that these swamp noises are made by lizards, frogs and alligators, but to me they sounded like the howling of demons. Well, I decided that turning back looked almost as hard as going on, so here I am.' As she read the letter, remembered Rose, 'tears stood in my eyes . . . My Jan was not a man to contemplate turning back from any goal he had elected to pursue – unless obstacles loomed virtually insurmountable.'

His accommodation turned out to be 'a big bare lumber barn, not quite so well constructed as the horse stables on the ranches at home' divided into cubicles just big enough for two men to share. A week later another letter arrived. 'The food is awful,' Jan wrote, 'and cooked in such a way that no civilized white man can stand it for more than a week or two . . . Almost all the food is fried. They feed us fried green bananas, boiled rice, and foul-smelling salt fish. It rains so much that honest to goodness my hat is getting mouldy on my head . . . I haven't had on a pair of dry shoes in weeks.' In the next letter, he reported that disease was rife. Rose's parents started raising objections to her plan for the rest of the family to join Jan in Panama.

There are numerous such examples of the shock and instant demoralisation experienced by new arrivals during this early period. Some, however, greeted these challenges with a cheerful determination resonant of the early French period. Jessie Murdoch landed at Colón along with a party of other young nurses in mid-1904 feeling, she admits, a mixture of 'apprehension', 'homesickness' and 'dread of what the future might hold'. Colón was alarming with its 'narrow, dirty, half deserted streets, with the native element running about half clothed', and at Ancón hospital, in spite of the warm welcome from Eugenie Hibbard, she was dismayed by the 'old rusted iron French beds, with mildewed mattresses'. On her first night she ventured outside, only to be 'eaten alive' by mosquitoes. Retreating to bed, 'Each had a candle, but it was soon found that it was not wise to keep these burning, as they attracted moths and all sorts of flying insects.' 'Yet in spite of these many difficulties,' she would write later, 'we were not disheartened, but thoroughly enjoyed the novel experiences.' 'We found upon our arrival here,' wrote young engineer James Williams, 'the wreck of the French companies, a foreign language, strange people, poor food, no ice, no lights, no drinking water, no amusements, or decent living quarters . . .' But more important for Williams was the 'thrill and the knowledge that we were working for Uncle Sam, accomplishing something that the eyes of the world were focused upon and something that every citizen of the United States were interested in.'

Others employed in the US by the Washington office were less impressed with the patently yawning chasm between what they found and what they had been led to expect. 'We were supposed to have furniture issued to us, my allotment being nominally six chairs, a bed, three tables, washstand and tin pitcher, and a clothes rack,' explained John Meehan, who arrived in early December 1904. 'What we really got was a cot, and a dynamite box.' Meehan, like van Hardeveld, was living in Culebra, which consisted of several labourers' bunkhouses and a smattering of cantinas and 'chino shops', which sold canned food at high prices. Everyone lived on tins of sardines and soggy crackers. The only two-storey building was a 'hotel' run by 'Cuban Mary', a 'disorderly place, very dirty, crude in every way'. There was one muddy main street; 'chickens walked about inside the stores and native shacks, a few pigs and a million goats wandered about the streets'. Reading in the evening was impossible, Meehan complained, because of the 'army of bugs'. The only thing to do was to go to bed or to one of the bars.

Meehan would remain on the project for many years. Others took one look around and simply headed back home again. Charles L. Carroll, a graduate from Pittsburgh, arrived in Panama in August 1904. A month later he wrote to his mother: 'I am thoroughly sick of this country and everything to do with the canal

. . . Everyone is afflicted with running sores. We are compelled to sleep in an old shed, six to a room . . . The meals would sicken a dog . . . Tell the boys at home to stay there, even if they get no more than a dollar a day.' Weeks later Carroll left the Isthmus for good.

In October 1904, the Italian minister in Panama reported back to his government on the dismal start to the US canal effort: 'The managers are said to be dishonest and incompetent,' he wrote. 'There have been many errors and much wastage and pilfering of money. The workers of all nationalities are treated inhumanely. As a consequence of all this, most people look back on the French administration, with all its defects, as more capable, more honest and more just towards the workers.'

Wallace would complain that the delays in sorting out the problem in accommodation were down to 'supplies taking for ever to arrive' and the labour 'immediately available and who could be secured from the surrounding countries [being] incompetent, shiftless and lazy'. But much of the intake from the United States was also seriously below standard. A request was posted for twenty-five track foremen and when they arrived there were only two that could drive a railroad spike. In further echoes of Blanchet's experience with his new arrivals in 1881, William Karner complained that recruits employed by the Washington office 'were not examined at all'. One young man presented himself to Karner as a rodman. It was quickly realised that he was nothing of the kind having no training or experience at all. It was then established that he had received his appointment through the efforts of his member of Congress.

In spite of all the worries about corruption or graft, this was the origin of many of the Panama appointments not just in the early years but throughout the American period. The Swinehart family were typical. At the end of 1904 Swinehart senior, the chair of a local Republican group in Steamboat Springs, Colorado, wrote to his congressman: 'I have two sons who wish to go to Panama to work on the Canal . . . I will consider it a great favor if you will see some member of Canal Com.' Less than a month later, the congressman delivered two plum appointments. 'Please place my name on your list of working Republicans and command me at any time,' responded the delighted father.

Along with idealists, professionals and the beneficiaries of political favours, the Isthmus was also drawing in 'railroad men who were blacklisted on the American railroads, drunks, and what we called tropical tramps, American drifters in Latin America'. US diplomat William Franklin Sands sailed for Panama in early October 1904. On his ship, a British Royal Mail Packet Company steamer, he was taken aback to read a notice outside the dining room which ordered: 'Americans will put their coats on for meals'. Why pick on Americans?

Then he discovered that the captain had frequently had to arrest US mechanics on the way to the Isthmus for drunkenness, gambling and even leading mutinies against the officers.

As in the days of the Gold Rush, such new arrivals inevitably caused friction and difficulties with the Panamanians. One British journalist reported that 'the people of Panama look upon Americans as noisy, grabbing bullies'. In return, admitted a senior American administrator, 'The average American has the utmost contempt for a Panaman and never loses an opportunity, especially when drunk, to show it.' In fact by the autumn of 1904 relations between Americans and Panamanians were strained at every level. This had led to dangerous, potentially violent fractures within the new Panamanian political establishment, as well as dissent between senior US Zone officials. It was to report on the origins of this mess, and to suggest solutions, that diplomatic 'troubleshooter' William Sands – only twenty-nine but a veteran of diplomatic posts in the Far East – was sent by Taft to the Isthmus in October 1904. His mission was to ensure that nothing in the Panamanian political firmament got in the way of the building of the canal.

After independence at the beginning of November 1903, the leaders of the plot who were Panamanian nationals, rather than Americans, had formed themselves into a temporary ruling junta. Led by Arango and Amador, the junta contained several token Liberals but was otherwise firmly Conservative.

In early January 1904, William Buchanan had arrived on the Isthmus as a US special envoy. At a dinner in his honour he gave the assembled Panamanian politicians a stern warning: the civilised world had determined to enforce order and peace, he said. 'Panama must conduct itself as a civilized nation or it will cease to exist as an independent country.'

This threat did much to keep tension between the Conservative administration and their Liberal enemies in check, and a National Assembly was elected which, although Conservative-led, included an almost equal number of Liberals. Nevertheless, the Conservatives set about removing potential enemies. Several senior Liberals were offered plum diplomatic postings to get them out of the country; General Huertas, a Hero of the Republic for his part in the events of November 1903, was now seen, because of his popularity and Liberal sympathies, as a threat and was sent on a lengthy fact-finding mission to the United States and Europe.

But as soon as one potential enemy was removed, another emerged. The undisputed leader of the Liberals was Dr Belisario Porras, an arch-enemy of Amador, described by the American consul in Panama as a 'revolutionary firebrand' and 'notorious hater of foreigners'. Porras, who had worked as a lawyer for the French Company, had opposed the Hay–Herrán Treaty as giving too much control to the United States, and was appalled by the terms of the subsequent Hay–Bunau-Varilla Treaty. Panama, he said, had been 'swallowed up' by the US; national sovereignty had been sacrificed for the benefit of a few wealthy Conservative Panama merchants. In June 1904, he returned to Panama City from exile abroad to be greeted by a huge crowd in Santa Ana Square. Although he admired, he said, the 'greatness and harmony of North American institutions' he believed that 'any Latin American nation who fused her destiny with that of the United States would suffer greatly and rue the day of their alliance'.

In the meantime, the Americans themselves had been providing plenty of fuel for anti-US sentiment on the Isthmus. Even before the revolution, United States interests – in particular the United Fruit Company, with its huge banana plantations in Chiriquí – had seemed to dominate the Panamanian economy. Once the Canal Zone was established, there were further indignities. In May 1904, the Zone authorities successfully demanded that an American doctor be allowed to inspect all ships arriving at Colón and Panama. The man appointed did not even speak Spanish. The following month it was announced that the domestic tariff laws of the United States would be applied to the Zone. This meant that goods from the US arrived free of duty, while imports from other countries, including Panama, were forced to pay very high rates. As it was simple to smuggle merchandise from the Zone to the Republic, the measure would slash the Panama government's vital customs revenue at the same time as infuriating the country's merchants. Then ports were opened at La Boca and Cristóbal, both adjacent to the terminal cities but within the Zone. As well as threatening to ruin Panama City and Colón, this seemed to be contrary to the terms of the Hay–Bunau-Varilla Treaty, which had specifically excluded the 'terminal cities and the harbours adjacent to said cities' from the US-controlled area.

These measures provoked furious clashes between Amador and his Liberal first *designado*, or vice-president Pablo Arosemena, and gave many the impression that the annexation of the terminal cities, and, indeed, the entire Republic, was imminent. 'I look upon the Republic of Panama as doomed,' Mallet wrote to his wife at the end of June. 'The Yankees are playing the same tricks here as they did with Colombia in regard to the canal question . . . The U.S. Government are behaving here like highway robbers; they neither respect treaties or persons.' A month later he reported 'Opinion here amongst the natives is spreading, and they now think the bargain with the Americans has been a bad one for them and their country; they would prefer, I think, to return to Colombia than continue this way.'

William Buchanan had left Panama and had been replaced by John Barrett, who was attempting to negotiate a way out of the impasse. Then Governor Davis became involved, which, as he was not an accredited diplomat, the Panamanians saw as a further slight. The negotiations did not go well. For one thing, Davis was intensely unpopular, and Barrett was not much better, variously described as 'very loud spoken', 'vulgar' and 'full of self-assurance'. Furthermore, the two Americans both felt that they should be the leading US voice on the matter and relations between them broke down completely.

In August, the US published a letter, reportedly written by Bunau-Varilla to Hay back in January, that clarified the 'ports question' in his treaty in favour of the Americans. The Frenchman was no longer in the employ of Panama, so he seemed an ideal candidate for taking the blame. But then other correspondence emerged, showing that the provisional government, specifically the arch-Conservative Tomás Arias, had authorised Bunau-Varilla's concession. Soon after, an anonymous fly sheet was distributed in the streets, accusing Arias of having sold the country's interests, and ordering him to resign or be assassinated.

The following month Hero of the Republic General Huertas returned to Panama, having cut short his trip after hearing that Amador planned to replace him. On 28 October, Huertas wrote to Amador demanding the removal of Arias, and of another ultra-right minister. Arias resigned, but Amador refused to release the second man. Then the President learnt that Huertas, with the backing of Belisario Porras, planned to arrest him at a forthcoming military function. A severely rattled Amador appealed directly to Roosevelt for help in avoiding a military coup, and Huertas and other opposition leaders were sent a firm message from the US legation saying that revolutionary changes would not be tolerated. At the same time, a detachment of marines was moved from their barracks to Ancón Hill. The military function went ahead without incident; Amador stayed at home.

Fortified by US support, the President then made his move against Huertas, demanding his resignation. The general held out for a while, but threw in the towel on 18 November. After consulting with Barrett, Amador then decided that a standing army, albeit of only 250 men, was not needed by the tiny republic, and the Panamanian Army was disbanded.

The US was subsequently blamed for the loss of the army and the national prestige that went with it. But while no doubt demonstrating the ruling elite's dependence on US support, the move was orchestrated by Amador and his Conservative allies. They had worried about the power of the army even in the heady days of November 1903. Now, with the assistance of the US, the only force in the land that could eject them from power was no more.

William Sands arrived in the aftermath of the failed coup, and found Panama 'festering with intrigue', the American canal officials and diplomats at loggerheads, and growing anti-Americanism in many quarters. The existence of the Canal Zone, Sands would write, 'made for an ambiguous and most delicate diplomatic situation. Canal affairs and interests were constantly overlapping or overshadowing the Republic's affairs.' To many Panamanians, the Americans seemed determined to tell them what to do, and there was deep concern about 'whither Theodore Roosevelt and his "Yankee imperialism" might be tending'. Sands likened the position of the US minister in Panama with that of a Resident Minister from Great Britain in India, where the threat of being 'incorporated into the Raj' was ever-present. Nevertheless, he was shocked at how much the Americans were disliked on the Isthmus.

The fault, Sands wrote, was with the 'confused and chaotic foreign policies of the United States . . . our diplomacy had been operating in a political jungle of its own creation'. Furthermore, Sands argues, Roosevelt had opened a 'Pandora's Box' in Panama, which would have serious political consequences. Sands had served in both Japan and Korea before Panama. Writing during the Second World War, he suggests that the US action in 'the creation of Panama, as in the annexation of Hawaii' had offered the Japanese 'a ready-made politico-diplomatic pattern for expansion'. In 1905, Korea became a protectorate in much the same way as had Panama in November 1903. Five years later, Korea was formally annexed as Japanese power was extended over much of Manchuria. The end result, Sands suggests, was the fearsome attack on Pearl Harbor. 'The large stone which Theodore Roosevelt dropped into the coastal waters of the Caribbean jungle sent its unperceived ripples out into the Pacific,' he writes, 'and up against the far islands of the Pacific and Asiatic mainland, whence they turned back in a tidal wave upon our own shores.'

One of Sands' first actions was to meet the new Secretary for Foreign Affairs, Santiago de la Guardia, to request that the position of the Governor and the US minister to Panama might be combined in one person. In marked contrast to Davis, Sands was careful to adhere to proper formalities, donning full diplomatic garb, top hat included, and hiring the best two-horse carriage he could find for the one-block journey between the legation and the Secretary's office. The approach worked, with the Panamanian happy to allow Sands' request, although de la Guardia did confide to Sands his fears about future relations between their two countries. 'Don Santiago was aware,' wrote Sands, 'that a new North America had come into being since the Spanish War, one that was not very well understood as yet even by the North Americans themselves.'

Meanwhile, other Panamanians had taken the arguments, particularly about customs revenues, out into the open. A delegation was sent to Washington, and articles by prominent Isthmians started appearing in New York newspapers. Then, in October 1904 a leading Panamanian Liberal, Dr Eusabio Morales, secured a commission from the influential *North American Review* for an article critical of the Hay–Bunau-Varilla Treaty. When word of this got to the Republican Party,

Morales was approached by men representing the Treasurer of the Party and offered a bribe to spike the piece. It was just before the presidential election and there was concern that revelations about Roosevelt's connection with the independence of Panama might come out. Morales declined the money but withdrew the article when Roosevelt publicly instructed Taft to go to Panama and settle the tolls question.

The Secretary of War, accompanied by Cromwell, arrived in Colón at the end of November. The highest-ranking American to set foot on the Isthmus so far, Taft was a good choice to mollify the Panamanians, in spite of the fact that, privately, he referred to their country as 'a kind of Opera Bouffe republic and nation'. At over 300 pounds, Taft was hugely, disarmingly fat and jovial-looking, and could deploy 'the most infectious chuckle in the history of politics'. Sands would call Taft 'a cold man, despite the legend which grew up about him . . . his geniality was wholly of the surface', but most people found him charming. On 1 December he used a speech at a welcoming banquet to assure his audience that the United States had no imperialistic designs on the republic, and then over the next ten days he combined touring the works with negotiating with the Panamanians over the points in dispute. Every evening there were banquets or balls, where Taft entered heartily into the spirit, amazing the Panamanians with his enthusiastic dancing. 'Though the heaviest man, in weight, in the room,' remembered William Karner, 'he was as buoyant and light on his feet as a feather or a rubber ball.'

Over the negotiating table, similarly light-footed, Taft managed to placate the Panamanians without actually conceding too much. A controversy over postage rates was settled in Panama's favour; imports from Panama to the Zone were to be duty free; there was a pledge of money for a stretch of road and a new hospital in Panama City. Taft promised that only essentials for the canal would be imported tariff-free from the US. Also included in what became known as the 'Taft Agreement' was an assurance that employees of the ICC from tropical countries would not be allowed to purchase food from Commission shops. Loopholes in the deal ensured that canal business would not suffer from any of these concessions, but it is the spirit of the Taft Agreement that is important. In the Philippines, despite the defeat of the main pro-independence forces outside Manila in 1902, an insurgency had arisen in the Muslim south that was daily claiming American lives. For the moment at least, the US felt that their best policy in Panama was to save the face of the incumbent, pro-American leadership and project a stance of compromise and respect.

Taft's next stop was Jamaica. Taking consul Mallet with him as a go-between as well as Chief Engineer Wallace, the Secretary met Governor Sir James Swettenham to ask permission to recruit workers for the canal.

The labour problem was among the most serious facing the first American canal builders. In spite of the stream of new arrivals, many soon left or turned out to be unsuitable, and by the autumn of 1904 departments had taken to offering inducements for men to leave one sector of the work to join another. One shipment of labourers was met by agents of the Municipal Engineering Division and others from the Building Department 'and so keen a competition developed to obtain the men that there ensued a street fight and the subsequent arrest and jailing over night of the principals'. There was also a longer-term question: who was going to build the American Panama Canal?

The leaders of the project, of course, like the new machinery to be deployed, would be American. It was also 'recognized' that 'most of the superintendents, foremen, and the higher grades of skilled labor would have to be brought from the United States'. According to the canal's quartermaster, Major R. E. Wood, 'there was no surplus throughout Central or South America' for this sort of work, and 'in many classes there were no men at all available' locally.

It was initially hoped that the American canal effort would be characterised by machines rather than men. Early plans estimated that some eight to ten thousand workers would be required, and Wallace told Taft that he wanted to restrict the number to ten to fifteen thousand. A gross underestimate, as it turned out – at one point there would be more than fifty thousand on the payroll – but still a sizable force to be found.

White workers from the United States were ruled out from the unskilled jobs early on as too expensive, too unionised and vulnerable to tropical diseases. It would be 'useless to discuss the question of utilizing the white race for heavy out-of-door work with pick and shovel in the mud and rain,' wrote Governor Davis in November 1904. 'American working men have no call to Panama,' suggested a US commentator, 'any more than English working men have to the plains of India.'

Characteristically for the time, the question was seen in terms of the fitness of different races for the job ahead. As chief recruiter William Karner would write, 'It has been an interesting job – experimenting in racial types.' Brigadier Peter C. Hains, who served on the Walker Commission of 1899 and would become a Canal Commissioner in 1905, laid out official thinking on labour in a 1904 article for the *North American Review*. 'Where will the labour come from? . . . The native Isthmian will not work. He is naturally indolent; not over strong; has no ambition; his wants are few in number and easily satisfied. He can live for a few cents a day, and he prefers to take it easy, swinging in a hammock and smoking cigarettes. The native population is wholly unavailable.' The 'Chinese coolie', who had built railroads all over the United States, was considered able to cope with the climate, 'industrious' and easy to manage, but could rarely speak English, and 'as soon as he gets a few dollars,' wrote Hains, 'he wants to keep a store'.

Hains' preference was for black workers from the British West Indies, whom he characterised as 'fairly industrious; not addicted to drink; can speak English . . . he is willing to work, [and] not deficient in intelligence'. There were other advantages: the islands were reasonably nearby and well served by steamer services; education levels were relatively high; the Antilleans had some immunity to some tropical diseases. But more than anything else they were cheap – wages and conditions on the islands were such that virtually anything the Americans offered would be an improvement. To Taft, there was another great advantage to the West Indian worker. Despite his being 'lazy', he had been taught by the British respect for discipline and authority. 'He does loaf about a good deal,' the Secretary of State wrote, 'but he is amenable to law, and it does not take a large police force to keep him in order.'

Thus Taft chose Jamaica as the nearest and largest of these 'natural markets for unskilled labor' to visit in person. But the meeting with the Governor did not go to plan for the Americans. Swettenham remained immune to Taft's charm; in fact, he seemed rather anti-American. The United States party got the distinct impression that he would quite happily see the US canal effort fail. It probably did not help that Admiral Walker, chairman of the ICC, had been musing, according to Mallet, that Jamaica should be taken over by the US as part of the canal's outer defences.

To the Americans Swettenham stressed the negatives of the Jamaican experience in Panama during the French period: the 'able-bodied emigrants returned enfeebled, sick, infirm, or maimed [who] had to be kept alive at the expense of the parish'; the huge cost to the Jamaican government of repatriating workers after the de Lesseps Company failed. He may have also had in mind the destabilising effects of mass migration – serious worker unrest had been bloodily suppressed only two years before. Then there were the powerful planter interests on the island, ever reluctant to see their pool of cheap labour reduced. The upshot was that the Governor announced he would only give permission for recruiting if the US government deposited in the Jamaican Treasury £5 for each labourer shipped to Colón, against the possible costs of repatriating him. Taft, as expected, was appalled at these terms. He called the meetings to a close and returned to the United States.

Wallace arrived back on the Isthmus and, after his customary two days of prostration from sea-sickness, summoned William Karner and ordered him to get on the next boat to Barbados and to set up a recruiting office there as quickly as possible.

Barbados, although the most distant of the 'natural markets for unskilled labor', had several advantages over the other British West Indian islands. The

Windwards and Leewards had relatively low populations and were considered to be ruled directly from the Colonial Office. Barbados, though, seemed to have more sympathetic and independent officials and was massively overpopulated, with 200,000 people living on just two hundred square miles. Karner arrived at Carlisle Bay, near Bridgetown, on 31 December.

The original jewel in the Imperial crown, Barbados had been a British sugar plantation colony for nearly three centuries. The novelist George Lamming called it 'the oldest and purest of England's children'. In the eighteenth century its sugar crop, tilled by slave labour, had made huge fortunes for planter families, but by the time of Karner's visit the industry, which provided 97 per cent of the island's exports, was in ruins. Because of the abundance of cheap labour, the planters had long managed to avoid modernising, and the business was hopelessly backward compared to its competitors, with most of the sugar mills still wind-powered.

By 1902 the price of sugar was a third of the already low level of 1882. Planters responded by reducing wages further, which fell to as low as a shilling (twenty-five cents) a day. Because of the oversupply of labour, the workers were powerless. In addition, with the majority of land held by the planters, many workers were tied to the estates – only if they laboured for the plantation owner on his terms were they allowed to tend small plots on his acres. It was essentially a pre-Emancipation arrangement. Harsh anti-vagrancy laws dealt with anyone who tried to step outside the system.

The shortage of space for the population, together with the exceptionally high dependency on the nearly worthless sugar crop, combined to make Barbados miserably poor in 1904, even in comparison to the other islands. 'Chronic pauperism . . . like a chronic disease is . . . undermining the population of this island,' reported parish inspectors. People were becoming ill, a Poor Law report wrote in 1903, because of 'the difficulty in obtaining an ample supply of nourishing food'. Diseases that might affect the whites, or bring quarantine down upon the island and thus hurt the important steamer business, were dealt with effectively, as in the case of the smallpox epidemic of 1902. But ailments that only really affected the poor, undernourished black population, such as typhoid and dysentery, were allowed to flourish.

In the period 1901–4, average mortality for infants under one year was 282 per thousand live births, three times that of the UK at the time and by some considerable distance the highest in the British Caribbean. This rate would get worse. In 1906 in one parish, St Michael, it was up to 499 per thousand. Prompted by shocking statistics such as these, Joseph Chamberlain, British Secretary of the Colonies, labelled the West Indian islands, formerly so prosperous, as the 'Empire's darkest slum'.

When William Karner arrived he checked into the best hotel on the island, and his first impressions of the place were that it was 'quaint', a 'Little England', 'quiet and restful'. The white Barbadians he met were 'intensely and seriously English', 'cordial and courteous'. During the following years he would become familiar with almost all of the Caribbean islands, but professed Barbados the place where 'the color line was drawn closer . . . than in any of the others'. Intermarriage 'was equivalent to ostracism for the white contracting party' and although officially the blacks were entitled to the same privileges as white people, the reality was radically different. 'The race domination is frankly acknowledged,' an American journalist visiting Barbados wrote. 'The island has always been and still is run for the whites . . . it is a heavenly place to live for the white man who can ignore the frightful misery of the negroes.'

For the black working class, living in squalid poverty in dreary wooden shacks amid the monotonous landscape of endless cane fields, the great hope was to save enough money to buy some land of their own and take some control over their lives. Barbadians were and are famously thrifty and long before the arrival of Karner it was taken as understood that the best way to save up some money was to work abroad. Throughout the second half of the nineteenth century some thousand men a year migrated to work in Trinidad or British Guiana (now Guyana). With wages so low, desperate poverty everywhere, and a tradition of emigration already established, it looked like fertile recruiting ground for the Americans.

Soon after his arrival, Karner was introduced by the US consul to the Colonial Secretary, who told him that the government had been looking at setting up an agency themselves to aid work abroad, as there was a 'large surplus' of labourers on the island. In turn, the Secretary took Karner to meet the Governor, Sir Gilbert Carter. Carter was keen that there should not be a repetition of the situation two years earlier when Barbadians working on a railroad project in Brazil had become stranded and had to be brought home at the government's expense, but otherwise was far more open to the ICC than had been Swettenham. Karner reckoned that the fact that Lady Carter was an American was helpful.

Karner than arranged for transportation through the Royal Mail Steam Packet Company which had a large office on the island, employed a local agent, S. E. Brewster, and had medical and contact forms printed up. The contract, agreed with the government, was that each labourer would be employed for five hundred days at a rate of ten cents US an hour. This was about half the minimum that would have been acceptable to a North American labourer but generous by Barbadian standards. Work was ten hours a day, six days a week with time and a half paid for overtime and Sundays. Passage to Colón was paid for – with food for the voyage provided – and 'medical attendance, medicine, and quarters without furniture, [were] to be furnished free to the laborer, while in the employ of the Commission'. At the end of the contract, or if the worker was incapacitated while employed by the Commission, repatriation would be free.

Eventually some twenty thousand Barbadians – the engine room of the canal effort – would be employed under this same contract, with only small modifications. The initial uptake was disappointing, however. It seems that the memories of the French period, when over a thousand Barbadians had travelled to Panama, the bad political reputation of Central America generally and the recent experience in Brazil combined to create suspicion of the new canal project. These fears were fanned by stories of unemployment and hardship on the Isthmus spread by planters and managers not keen to see their all-important labour surplus disappearing.

'There was no rush and on the steamer sailing for Colón, January 26th 1905,' wrote Karner, 'I shipped only sixteen laborers.' It would have been seventeen, but at the last moment 'one man got stage fright, shouted that I was sending him into slavery and that he would rather kill himself there than to go to the Isthmus and die in slavery'. The next shipment had a few more after Brewster 'did some hustling in the adjoining parish'. But Karner was hindered in his efforts by a loss of contact with Wallace in Panama due to a broken cable. When he did receive a message from his boss his orders were vague. Then there was a problem cashing the ICC's cheques and further administrative hurdles. Karner estimated that he needed a full-time employee just to fill out his myriad requisition and expenses forms. It was a similar frustration as was being suffered on the Isthmus.

Those first Karner recruits arriving right at the beginning of 1905 joined a canal project mired in confusion and disillusionment. In Panama City, the job of installing running water and sewers was supposed to have been completed by January 1905, but was still months away. Much important equipment had not arrived. The Americans tried to make do with reconditioned French tools and 'scrap'. The building department was having the same difficulties. When materials were reluctantly supplied, they were often held up by the inadequate facilities at Colón for unloading, storage and distribution. Building and dock workers were in short supply as the majority were taken up with 'actual canal work'.

What this meant was 'making the dirt fly' in the Culebra Cut. This is what journalists from the US came to see, and they wanted to report back that the canal was being dug. But the effort from an engineering point of view was an almost total waste of time. In August, by which time three old French excavators had been chugging away for about a month, Gold Hill began to slide into the trench below it, and work was suspended for four weeks. In November the first American-built excavator arrived. This was a 95-ton Bucyrus steam shovel,

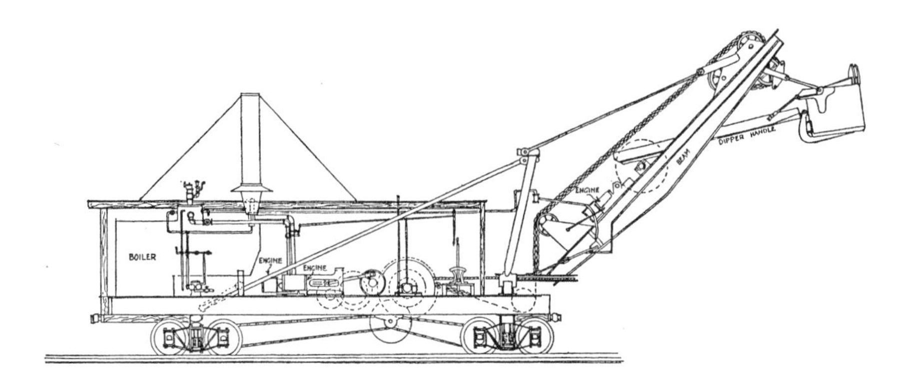

Bucyrus steam-shovel

which could scoop up nearly 5 cubic metres of spoil at a time. Wallace was off the Isthmus, so it was William Karner as Acting Chief Engineer who ordered the new shovel into action in the Cut. 'Ambitious to "make the dirt fly",' writes Karner, 'I came near to a mortifying and embarrassing failure, for the night after starting the shovel at work there was a slide in the cut which nearly buried the shovel from sight.' The Bucyrus shovel, three times as powerful as any equivalent used by the French, would become the workhorse of the canal, but it was an inauspicious start.

The following month there were six of the old French ladder excavators at work together with an increasing number of resurrected Belgian locomotives and French dump cars. But for all the sound and fury, an engineer reported, 'the impression made on the soil in comparison with the entire mass of earth to be handled' was mere 'hen scratches'. On wet days, the locomotives could only pull four cars at a time, and there were frequent derailments, blamed on the 'long and rigid wheel base of the French and Belgian engines', as well as the 'poorly ballasted', largely improvised, tracks. As there were also inadequate traction and dumping facilities, the excavators were often idle for lack of spoil cars.

The results of the test borings along the line had also provided disappointment. At Bohío, where the Walker Commission had planned a dam for the lock-canal option, drilling teams had discovered a deep geological gorge, the original bed of the Chagres River. Thus the bedrock on which the dam would have to be anchored was a seemingly unworkable 168 feet below sea level. Above the bedrock was a porous mixture of gravel and other alluvial detritus. Wallace concluded that 'there is little probability of finding a satisfactory location for a high dam in this vicinity'. There were borings also at Gatún, which had been suggested as a possible site for a dam in spite of the great width of the valley at this point. But here there were two underground gorges even deeper than at Bohío. The Americans drilled to 200 feet below sea level and still did not reach bedrock. 'As 200 feet was considered in excess of the practicable depth at which it was advisable to construct a foundation for a dam,' Wallace reported, 'it was not considered necessary to go deeper.' For him, the test results precluded 'the economical or safe construction of a dam in [the] general vicinity' of Gatún. Ominously, other scattered borings along the canal line offered only geological confusion: 'Practically no regular stratification exists,' Wallace reported.

In the absence of an obvious site for a dam along the line of the canal on the Atlantic side, Wallace reported that he favoured the adoption of a sea-level canal plan. This was, he wrote, 'self-evidently' 'the most desirable in economy of maintenance, operation, time of passage through it, and simplicity of design, plan and execution . . . the deterrent factors being time and cost'. But more than anything he wanted a decision, to 'remove the principal elements of uncertainty now existing in regard to the project as a whole'.

Hampered by more than just uncertainty, the effort was on the brink of petering out, strangled by red tape, bickering and incompetence. Wallace had underestimated the Isthmus. Under pressure to 'make the dirt fly', he neglected proper preparatory work. He also failed to provide leadership. His idea of management was to delegate. When he employed Frank Maltby, he told him: 'I want you to build up an organization so complete and efficient that you won't have to do anything but sit on the veranda and smoke good cigars.' Wallace himself was hardly ever seen out on the line; he rarely left his office, and every small suggestion from one of his subordinates would be demanded in writing. Soon his desk was as covered in chaotic mounds of paper as that of Admiral Walker in Washington.

With no improvement in sight in the grim living conditions, discontent was spreading all along the line of the canal. Then at the end of 1904 it got dramatically worse: yellow fever broke out on the Isthmus.

19

YELLOW JACK

The American doctors and sanitary inspectors had arrived in Panama full of confidence. They might have felt underappreciated by the engineers among the first arrivals, but they were sure they had the knowledge and experience to rid the Isthmus of disease-carrying mosquitoes. Before departing for the tropics Gorgas and Le Prince had met a senior US entomologist, who had asked them to send back samples of Panamanian mosquitoes. 'I will have to do it soon, Doctor,' Le Prince had exclaimed, 'for in a year or so there will be no mosquitoes there!'

Consul Mallet had met Gorgas the day after the doctor's arrival. 'He says you and the children will soon be able to live here,' Mallet wrote to his exiled wife, 'and that not only will the sanitary conditions be improved but there will be regular American schools established. He will bring his family, and many others will do the same.'

Gorgas had been to Paris to study the medical records of the French companies. He knew the story of the Dingler family, of the massive losses and demoralisation from disease during the 1880s. But he was not downhearted, as he was aware that the Americans had a crucial advantage over de Lesseps' men. For between the time of the French on the Isthmus and the start of the American canal effort, a massive advance had occurred in the control of malaria, and, even more so, yellow fever: the miracle of Havana.

As in the fields of technology and engineering, the twenty years after the beginning of the French canal saw astounding advances in medical science. In 1880 the germ theory of disease – pioneered by Pasteur, Koch and Lister – was still the subject of debate in the medical profession and derision by the public. But by 1900 a revolution had occurred. The new world discovered under the microscope had ushered in fresh understanding of diseases that had baffled man for centuries. For the US effort in Panama no advance was more important that the understanding of the mosquito transmission of malaria and yellow fever.

The entire idea of an insect vector was quite new. In 1878 a Scottish doctor, Patrick Manson, working in southern China, had discovered that mosquitoes carried the developmental stage of a parasitic worm that caused elephantiasis. It was the first proof that a bloodsucking insect could harbour, and presumably transmit, organisms of human disease. Then in 1881, in Havana, the western hemisphere's 'yellow jack' capital, an article was published in a medical journal that not only identified the mosquito as yellow fever's carrier, but also the particular species, *Aëdes aegypti*. Its author was French-Scottish doctor Carlos Finlay, who had worked in Cuba for twenty years. He had studied the literature of yellow fever epidemics and had noted that they all mentioned the unusually high prevalence of mosquitoes. He then identified the particular species by analysing the factors the sites of the epidemics had in common – temperature, elevation above sea level – and matching them to a mosquito that thrived only in these conditions.

But the theory – with hindsight a stroke of genius – was either ignored or ridiculed. The problem was not just resistance to a new idea; Finlay himself carried out numerous experiments the results of which seemed to disprove his own theory. Countless times he attempted to infect a patient using a mosquito that had bitten someone with yellow fever, but no one ever got ill as a result. Finlay, soon disparagingly known as 'the Mosquito Man', was written off as a crank.

More than fifteen years later, by which time the Isthmian mosquito had seen off de Lesseps and his French canal dream, an obscure English army physician, Ronald Ross, stationed at a remote field hospital in India, worked out the life cycle of the malaria parasite. In the summer of 1897 he dissected an *Anopheles* mosquito that was known to have bitten a malaria sufferer. Under a microscope he found in the insect's stomach the same circular cells identified as the malaria parasite *Plasmodium falciparum* by a French doctor Alphonse Laveran, in Algeria in 1880. The following year Ross located the mosquito's salivary gland and found the parasite there, confirming that the insect vector passed the disease on with a subsequent bite. Thus to control the disease, which at the time took a million lives a year in India, it was necessary to prevent the mosquito biting an infected patient as well as keeping them away from healthy humans.

In the same year, 1898, Henry Rose Carter, a maritime doctor with experience of yellow fever quarantines, made an observation that offered to come to the rescue of old Dr Finlay's theory. While working in a small Mississippi town during an outbreak of yellow fever, he noticed a strange pattern: there was usually a period of twelve days to three weeks between the appearance of the first case of yellow fever in a community, and subsequent cases apparently derived from it. This led him to carry out case studies on dwellings where a patient had become infected. By carefully noting visitors to the house, he established that those who came in the first two weeks were fine, but thereafter, and even when the original patient was removed, there was a risk of infection. Not only did this point to an insect vector, but it also established that the virus needed a period of 'extrinsic incubation' inside the mosquito before it became dangerous. This neatly explained the failure of Finlay's experiments.

After the end of fighting in Cuba in August 1898, some fifty thousand US military remained on the island for 'pacification'. Their continued losses from disease highlighted the fact that yellow fever probably presented the single greatest threat to US expansion in the tropics. For almost everyone there was little doubt as to what was to blame – the filthy streets of Havana. In January 1899, William Crawford Gorgas, who had worked in a yellow fever camp during the war, was charged with cleaning up the city.

Gorgas, the son of a Confederate general, had become a doctor as the only way to secure an army commission after his rejection by West Point. Posted all around the frontier areas of the United States, Gorgas was on the Río Grande near the border of Mexico in August 1882. It was, wrote his wife whom he met that year, 'yellow fever's first encounter with one who became its implacable foe, and whose life was to be concentrated on its extermination'. Gorgas caught the disease but survived and was therefore subsequently immune. Thereafter yellow fever epidemics became his speciality.

But for now, Gorgas was firmly with the majority about the cause of the disease: rotting rubbish and filth, together with 'fomites' – everything the patient had touched while infected. Gorgas met and befriended Finlay in Havana but, he later wrote, 'we were rather inclined to make light of his ideas, and none more so than I'.

Backed to the hilt by the city's military governor, he went to work. Streets were swept clear of dead animals and rubbish. Every house was cleaned and disinfected with 'chlorinated lime', essentially bleach. The exhaustive sanitation campaign was closely followed by the US press. With plenty of 'before and after' pictures the work made good copy and showed the American intervention in a good light. The cleanup also slashed the previously high rates of illnesses such as typhoid and dysentery. Furthermore, the yellow fever epidemic of the previous year showed no signs of returning.

But in August 1899 some 12,000 non-immune Spanish workers arrived on the island. Within weeks yellow fever was back. It seemed that all Gorgas's hard work had not made the slightest difference. Worst of all, the disease seemed to hit hardest in the smarter 'expatriate' areas of Havana (where, of course, the highest proportion of non-immunes was living). The deaths of high-profile members of the US community caused widespread panic.

Governor Leonard Wood ordered Gorgas to redouble his sanitation effort and made $50,000 available for the purpose. But the renewed round of sweeping and scrubbing had no effect. It was time for a new approach.

In June 1900, a special Yellow Fever Commission was appointed by Washington. Dr Walter Reed, an expert bacteriologist as well as an experienced 'frontier doctor' like Gorgas, was in charge, assisted by Drs James Carroll, Aristides Agramonte and Jesse W. Lazear.

Reed seems to have landed in Havana with an open mind, amenable to persuasion by observed fact. The questions facing the Commission were simple: what agent causes yellow fever? How is it spread? How can it be stopped? On the first of these, the doctors drew a blank early on. They looked through the microscope at the blood of yellow fever patients but could not find the 'germ' or 'bacteria'. Viruses, of course, were unseen and unknown in 1900. On the second question, however, there was an early breakthrough. Soon after his arrival Reed was presented with a curious case. A soldier at the nearby military base had been locked in a cell following a disciplinary infringement. A month after having been imprisoned, he had come down with yellow fever and died six days later. But none of his eight cell mates had contracted the disease, even the one who had subsequently slept in the dead man's bunk. The men had remained sealed in the cell, but it did have a window. Reed concluded that something must have carried the infection in through the window, passed it to a single subject, and then departed: it had to be an insect.

The Commission turned to Doctor Finlay, whose theories were suddenly in favour at last. Finlay was happy to provide *Aëdes aegypti* eggs so that specimens could be bred for experiments. While several of the doctors tried to find out all they could about the insect's habits and life cycle, others attempted to achieve the demonstration of the theory that Finlay's experiments had failed to provide. For this volunteers were needed who were prepared to be bitten by infected insects. Although those selected would, of course, be the strongest and healthiest specimens, and the best care available would be ready for them, they would still run a heavy risk of death. Five hundred dollars was offered as an incentive, although the first US Army volunteers nobly refused the reward, setting a precedent that was followed throughout the experiments. Reed's team of doctors also subjected themselves to the same tests.

Early results were disappointing, but after belatedly taking into account Carter's findings on 'extrinsic incubation', yellow fever was successfully transferred from one patient to another. Along the way both Carroll and Lazear became infected. Carroll narrowly survived, but Lazear, with what Gorgas called the worst case of yellow fever he had ever seen, died an agonising death on 25 September.

Reed was much chastened by the gruesome sacrifice of Lazear, a hugely popular member of the team, but was convinced that his experiments had yielded up yellow fever's secret. However, when Reed presented the mosquito theory to a Public Health Association meeting in Minneapolis in November 1900 he was greeted by a stony silence, followed by scathing criticism. The Washington *Post* was condemning in its report of the new theory: 'Of all the silly and nonsensical rigmarole about yellow fever that has yet found its way into print – and there has been enough of it to load a fleet – the silliest beyond compare is to be found in the mosquito hypothesis.' The centuries-old, 'common-sense' theory that yellow fever was caused by dirt was not to be moved.

Reed returned to Havana and set up a new base, named Camp Lazear, in an isolated spot outside Havana. Here it was possible to achieve a far higher level of control and scientific rigour over experiments. Reed was determined to provide data that even the most die-hard believer in the filth theory would have to accept. At the end of November an experiment was started wherein a doctor and three volunteer soldiers were confined for twenty days to a mosquito-proof wooden shack. Inside, they slept on the soiled and *vomito negro*-plastered bedding of previous yellow fever victims. No one got ill, and the 'fomites' theory, which had governed quarantine law for centuries, was demolished forever.

Other meticulous tests demonstrated again that *Aëdes aegypti* was the vector of the disease, but there remained doubters. One was Gorgas himself. Even if the insect did spread the disease, what was to say that this was the sole method of transmission, or even the most common one? The only way to demonstrate the theory for sure was to take it out into the field: to get Gorgas's sanitation squads to destroy the *Aëdes aegypti* of Havana and see what happened.

This Herculean task was something of a last resort. But tests with inoculations had ended when a young American nurse died from her injection, and to screen all yellow fever patients – to keep the mosquitoes from becoming infected – was impossible as many cases went unreported. Therefore the only option left was to take on the mosquitoes themselves.

Called the 'aristocrat of mosquitoes' *Aëdes aegypti* is distinctly marked. Its body has a series of silvery half-moons, its legs are alternately black and white, and it has four brilliant stripes on its thorax. Only the female of the species is a bloodsucker – she needs it to mature her eggs – and she has a marked preference for human blood to the extent that she knows where to attack a person: under the wrist or on the ankles, where the skin is thinnest, and never on the face or top of the hand as these places are easily slapped. Also, she is only very rarely found away from human habitation and always lays her eggs near a ready supply of human blood.

The researchers in Havana also discovered that she is very meticulous, preferring to lay her eggs in clean water in man-made containers, such as earthenware jugs or water butts. And this, they concluded, was her great weakness. So when Gorgas went to work in February 1901 on his new campaign directed entirely against *Aëdes aegypti*, he started by directing that all such containers should be removed or screened. Where this was impossible, a layer of oil was poured on the surface of the water to suffocate any larvae, or 'wrigglers' as they were commonly known, that might be lurking inside. At the same time, the entire city was divided into districts and meticulously fumigated by burning sulphur or pyrethrum, a dried flower used as an insecticide. The results were truly spectacular – a dramatic reduction in yellow fever from 1400 known cases in 1900, to only 37 in 1901, none of them after October. The eradication procedures didn't just kill of *Aëdes aegypti*, but reduced the *Anopheles* population as well, thus decreasing malaria cases by more than half.

Havana had suffered some five hundred deaths a year from yellow fever for as long as anyone could remember. Now, quite suddenly, it was over. Furthermore, a controversial and important new theory had been proved correct, and the methods selected to exterminate the disease vector had worked. When the scale of the triumph began to sink in, Gorgas wrote to Reed, 'When I think of the absence of yellow fever from Havana for a period of fifty days, I begin to feel like rejoicing that I was ever born!'

All of this hard-won knowledge and expertise was brought by Gorgas to his task in Panama. Indeed most of the doctors who arrived with him in June 1904 were also veterans of the miracle in Havana. Yet by September, Gorgas would be utterly frustrated, despondent and full of dread. 'The Commission have their own ideas about sanitation, and do not seem much impressed with mine,' he wrote. Gorgas accurately predicted what was to come: 'I fear an epidemic is inevitable. If only we could convince them! If only they knew!'

Gorgas had been confronted by many medical challenges at the start of the American effort in Panama. On his reconnaissance mission in April 1904 he had noted the filthy streets and the extraordinary prevalence of malaria. But he had selected yellow fever as the first of his enemies to be attacked. For a start, the defeat of the disease was an achievable goal; Gorgas had demonstrated in Havana that a spectacular and very press-friendly victory could be won. He had also seen enough yellow fever epidemics to know the disproportionate panic they could cause and how they were ignited by an influx of non-immune personnel, just as was about to happen in Panama. Furthermore, yellow fever attacked the bosses and spared the workers, affecting in the main white outsiders, perceived to be the most valuable and least expendable of the canal workers. The health of this group would remain Gorgas's top priority. Writing in an American medical journal in 1909, and sounding more than a little like Alfred Mahan, he boasted, 'our work in Cuba and Panama will be looked upon as the earliest demonstration that the white man could flourish in the tropics and as the starting point of the effective settlement of these regions by the Caucasian'.

Gorgas got to work within two weeks of his arrival, dividing Panama City up into districts and beginning house-to-house inspections, as he had done in Havana. But from the very start his efforts were severely hampered. The problem was that the gentlemen of the Commission simply did not believe the mosquito theory. The year before, a scientific congress in Paris had reviewed Reed's yellow fever work and proclaimed it 'scientifically determined fact', and Ronald Ross had received a Nobel Prize for his work on the mosquito transmission of malaria. Ross even visited Panama in mid-1904 and pronounced Gorgas's plans sound. Nevertheless, to Walker, the theory was 'the veriest balderdash'. Gorgas should concentrate on cleaning up the filth and smells, the chairman of the Commission pronounced. The Zone Governor, General Davis, concurred, saying to the doctor in a fatherly tone, 'I'm your friend, Gorgas, and I'm trying to set you right. On the mosquito you are simply wild. All who agree with you are wild. Get the idea out of your head. Yellow fever, as we all know, is caused by filth.'

Ross himself wrote, 'The world requires at least ten years to understand a new idea, however important or simple it may be.' Nevertheless, it is not hard to sympathise with Gorgas's frustration. Wallace was no better, sounding like old Jules Dingler when he announced that 'clean, healthy, moral Americans' would not contract the disease.

So the Commission believed that Gorgas's efforts were a waste of time and money, and his requisitions for supplies suffered even worse than anyone else's, despite Roosevelt's express order that the medical effort be given top priority. Vital copper screening and medical supplies ordered in October 1904 had still not arrived the following April. Gorgas's department was also given the workers that no one else wanted and ordered to pay the lowest wages. 'Consequently,' Joseph Le Prince complains, 'only poor and unintelligent labour could be obtained . . . some of them could not even climb a ladder. This was the only class of labor allowed the sanitary department. We had to use it, and succeed or fail.'

The fumigation squad visited Claude Mallet's house on 27 July 1904. His rooms were swept clean and his yard cleared of anything that might hold water. His tanks were covered with wire netting and taps installed, and he was shown how to pour oil on the surface of the water. He was pleased with the result, writing to his wife that he was now able to read 'without having to wipe the mosquitoes off every second', and pronounced himself 'a convert to the mosquito theory'. But he also reports the more than slightly haphazard way in which the house was fumigated – windows and doors were left open – and that *Aëdes aegypti* larvae were still found in his water tanks on another inspection a week later.

Gorgas had taken on a huge task. Unlike Havana, in Panama mosquitoes bred all year round. The first inspection showed *Aëdes aegypti* larvae 'existed at practically every house in town'. There was no running water, and, with ice expensive, locals cooled water by keeping a supply indoors in earthenware vessels, called *tinajas*. 'In these,' reports Joseph Le Prince, 'larvae thrived in great numbers.' Gorgas simply did not have sufficient men and materials to carry out the job as thoroughly as required.

Some Panama City residents like Mallet welcomed the chance to have their homes cleaned out at the expense of the US government. The British consul even tried to get the gang to paint his house while they were about it. But the measure was not compulsory. At the beginning of July a fine of five dollars was introduced for anyone who bred larvae in their house, but there were hardly any prosecutions, and usually tenants simply emptied water vessels at the back door while the inspector was entering at the front. Alternatively, the offending containers were hidden away. As most Panamanians were immune to yellow fever, they felt little compulsion to assist in the eradication programme, and many did not believe it was possible anyway. 'To attempt it is a dream, an illusion, perhaps simply a case of American boasting', wrote the local Liberal newspaper, the *Diaro de Panama*. It was very different from Havana, where Gorgas's squads had had a sympathetic governor and martial law to back them up.

Not that the Americans, who were most at risk, were much more helpful. Visits to Panama City were strongly discouraged, but this was widely ignored, particularly after the banning of gambling in the Zone in August 1904 and the subsequent steep reduction in liquor licences. Most of the United States personnel on the ground shared the Commission's scepticism about the mosquito theory. A particular concern of the Sanitary Department was the Administration Building, where some three hundred non-immunes, mostly young Americans, were working. Screen doors were left propped open, and many rooms were unscreened altogether. But when Joseph Le Prince pointed this out to Wallace's nephew O. M. Johnson, the young supervising architect, and alerted him to the fact that the bowls of water used to dampen the architects' copying brushes were ideal places for mosquitoes to breed in, he was laughed at. 'Le Prince,' Johnson said, 'you're off on the upper storey!'

'But suppose,' the inspector argued, 'we have twenty deaths here. Who'll be responsible?'

The architect laughed. 'I'll stand the responsibility,' he said.

The yellow fever onslaught, as predicted by Gorgas, began at the end of the year. In November an unemployed Italian labourer was brought into the San

Tomás hospital in Panama City with a serious case of the disease. He survived, but then news came that two other Italians, members of an opera troupe that had been touring the line of the works, had died from yellow fever on the boat home. The following month Gorgas was forced to suspend his fumigation programme in Panama City due to lack of supplies and six new cases emerged, several from the areas left untreated. 'Some yellow fever cases exist in the San Tomas hospital and the Yankees are much scared,' wrote Mallet to his wife. Indeed, with the first yellow fever death in December, the Americans were getting jittery. In August US minister John Barrett had been making plans to bring his mother to live with him in Panama. But on 20 December he wrote to her that he had changed his mind: 'If you should be unwell here or if anything should happen to you I could never forgive myself for bringing you to Panama.' Wallace tried to calm the workforce by ostentatiously riding around town in the company of his wife, newly arrived from the United States. But his efforts were undermined when it became known that the couple, fearing the worst, had imported two smart metal coffins.

The sense of impending doom deepened with the appearance in mid-January of six cases on the US cruiser the *Boston*, anchored in Panama Bay. An inspection 'revealed a dishpan of water standing outside the cook's headquarters. The thing was so thick with mosquito larvae that it was practically a purée.' On 16 January the recently arrived young wife of Wallace's secretary John Seager died after a short and very sharp attack of the disease. She had only been married for two months and her death caused deep shock to the expatriate community. Governor Davis called it 'the saddest incident in the history of the American colony . . . I attended the funeral this afternoon in the chapel of the Hospital, and as I looked at young Seager overcome with grief at the loss of his wife, I had great difficulty in restraining sympathetic tears. Nearly all the eyes in the room were moist . . . Naturally this death among those that are so well known has almost caused a panic in the ranks of the American employees of the Canal Commission.' There were further high-profile deaths, and by the end of January, after nearly twenty cases had been identified that month, it was clear that an epidemic was under way. 'How glad I am that I did not bring you down to Panama,' Barrett wrote to his mother on 30 January. 'With malaria and yellow fever rife . . . not knowing who will be the next victim. I am glad that you are safe in "God's Country" . . . About the only subject of conversation here in Panama is yellow fever.'

Everyone had seen the cemetery on the hill outside Colón, 'one of the saddest graveyards in the world, acres of little white crosses falling over and rotting under the jungle of tropical growth'. Now the shadow of Monkey Hill, where the dead from the French period lay buried, 'darkened the whole Isthmus'. 'The rush to get away,' wrote Gorgas's wife Marie, now with him in Panama, 'quickly assumed the proportions of a panic. The canal force – labourers, engineers and office men alike – seemed possessed of one single view: "Let's get out of this hell hole," . . . and men arriving one day would take their departure the next, frequently on the same boat.' Inevitably the news of the exodus reached the United States, where a newspaper dubbed Panama 'the place where the "ghost walks" . . . it seems that almost everybody is at a standstill down at Panama save the paymaster'. The *Panama Star and Herald* concurred: 'Unless something is done and done quickly,' the paper wrote in late February, 'all hopes of building a canal across the Isthmus of Panama may be set aside.'

At Ancón hospital cleaning and painting of the wards had, according to Eugenie Hibbard, 'advanced quite rapidly'. The water guards around the bed legs were removed, and the grounds cleared, revealing the statues of saints scattered around the garden, previously hidden by dense undergrowth. The French nurses were shipped out as well. For all their compassion they were untrained and hindered in their effectiveness by the vows they had taken. Even if instructed not to feed men awaiting surgery, they would ignore the order as it went against their vow to give food to the sick. The nuns also had little time for the mosquito theory. American nurses would put mosquito nets in place around the beds at night, only to find in the morning that they had been tied back by the nuns with bows of brightly coloured ribbon.

The hospital staff was suffering from the same frustrations as the rest of the sanitary department. The nurses complained about the poor salaries and lack of quarters. Jessie Murdoch was still having to sleep wrapped in citronella-soaked blankets to deter the mosquitoes. According to Hibbard, 'In the early days the securing of absolutely necessary articles was difficult and tedious.' As everywhere on the Isthmus, she was forced to 'make the best use of the materials left by the French'.

The beginning of the year saw a 'steadily increasing number of patients'. The epidemic had spread from Panama City to Colón and along the line of the works. Soon two wards and several private rooms were given over to yellow fever patients. Each had a wire cage built around their bed to prevent them infecting mosquitoes and the wards themselves were protected by three screened doors, with pyrethrum powder burning continually between them and a guard outside to see that they were kept closed. With a tripling of yellow fever cases from December to January, 'death seemed to dominate the situation,' Jessie Murdoch remembered, 'almost causing panic . . . [yet] no one showed the white feather but all stood faithfully to their tasks'.

Then the epidemic stuttered: there were fewer cases in March than in February and for the first two weeks of April there were none at all. On 18 April Gorgas wrote, for a piece to be published by *Harper's Magazine*, 'I personally believe that we have seen the last case of yellow fever in Panama.'

The following day, Gorgas was summoned to the bedside of O. M. Johnson, the architect who had laughed at the warnings of sanitary inspector Joseph Le Prince. The symptoms – the headaches, back pain, terrible thirst and then the *vomito negro* – were unmistakable. Gorgas personally supervised his care, but there was little he could do apart from make the patient as comfortable as possible. In the meantime, there was a stream of new yellow fever cases in the Administration Building. As Gorgas had feared, the disease had struck at the heart of the American project. In ten days, twenty-one cases were carried out of the US canal headquarters. Johnson died on 25 April and was buried in one of the Wallaces' metal coffins. It was, wrote Governor Davis, like 'the ending of many a bright young man I have seen on the battlefield'.

But few of the Americans shared the martial, patriotic determination of the young French engineers twenty years earlier. 'Everybody here seems to be sitting on a tack,' one engineer wrote home. When yet another case was announced in the Administration Building, a young stenographer 'rose from his chair and shrieked, "I want you to understand now that if Tabor dies, I'm going home."' Many were already on their way as 'yellow fever . . . completely filled the atmosphere'. Panama City was abandoned as white Commission staff were moved out to Ancón Hill, but in less than two weeks two hundred resigned. One returning nurse told the New York *Tribune* that, bafflingly, yellow fever was even killing 'well set-up, clean boys with good principles'. 'We were taxed to the utmost in the effort to care for the sick and keep hope and encouragement alive,' writes Frank Maltby, who was still labouring away dredging the ends of the canal. It didn't help that Chief Engineer Wallace was off the Isthmus in Washington, pressing his case to be given more control over the project.

'Many resigned,' read that year's ICC annual report, 'while those who remained became possessed with a feeling of lethargy or fatalism resulting from a conviction that no remedy existed . . . There was a disposition to partly ignore or openly condemn and abandon all preventive measures.' The progress of the epidemic can be traced through the pages of the *Star and Herald*, which itself lost a senior contributor to yellow fever. 'The latest victim of this dread scourge of the south is Mr J. J. Slattery,' reported the paper on 15 May. 'He was a man who made friends rapidly and knew how to keep them . . . the news of his untimely decease came with astounding force.' Slattery was the first to be buried in the newly established American cemetery. Stories were rife of people arriving and leaving on the next steamer. 'Things are looking pretty sour at Culebra, where cold feet has the boys on the jump,' wrote the paper in June. 'According to the latest reports, as each steamer leaves for the States quite a few bid a tearful goodbye to their comrades, many of whom are only waiting for pay-day to take their clothes and go.'

In all, some five hundred US employees resigned during April, May and June 1905, about three-quarters of the white workforce after what was, in historical terms, a fairly mild flare up of yellow fever. Amid the chaos came a reluctant reappraisal of the French effort, as, in the early summer of 1905, the work came to a virtual halt.

In Washington the wheels had been turning, albeit slowly. After his visit to Panama in November, Taft had reported to Roosevelt that the Commission under Walker and Davis seemed unwieldy and obstructive. As an engineer on the ground put it: 'The military regime in Panama, in so far as the furtherance of engineering efficiency is concerned, is a failure; in so far as the maintenance of official orthodoxy is concerned, it is a great success.' This was backed up by Wallace, who requested that the Commission be reduced to three men: Governor and Chief Engineer in Panama, with a Chairman in Washington. In January the President requested that Congress amend the Spooner Act to make this possible, and the following month sent an eminent US surgeon to investigate what was seen as Gorgas's failure to control disease on the Isthmus.

The surgeon, Dr Charles A. L. Reed, toured the works during February, and his report was leaked to the US press. One particular example he gave of the nightmare of red tape afflicting the medical department caught the eye and was reproduced across the country: a newborn baby needed a nursing bottle; the nurse applied to her superior, Major La Garde, who, finding the requisition of the previous September still unfulfilled, then made out another order, which had to be endorsed by Gorgas himself as well as the chief of the bureau for materials and supply, Mr Tobey. Then the order was copied and at last a messenger was permitted to go to a chemist and buy the nursing bottle, which finally reached the infant two days after the necessity for its use had arisen. The bottle should have cost no more than thirty cents, but counting the money value of the time of the nurse, of Major La Garde, of his clerical help, of Colonel Gorgas, of Mr Tobey, of Mr Tobey's clerks and of the messenger, the cost to the government of the United States, Reed calculated, was around $6.75. For all Walker's parsimony, it seemed the waste of the French canal period had never gone away.

Roosevelt was furious at the leak, but even angrier about the blow to the prestige of his canal project. When the Senate refused to agree to the change in the Commission, the President, acting by executive decree, demanded the resignation of Walker and his six colleagues. On 1 April he announced a new governing body for the canal, composed, as Wallace had suggested, of an executive committee of Chief Engineer, Chairman and Governor, with four essentially sleeping partners added in to pretend compliance with the terms of the Spooner Act. Wallace, still highly thought of in the press, remained as Chief Engineer, and now had a seat on the Commission. Theodore Shonts, a 'gruff, domineering' Pennsylvanian who had built, run and owned a number of Midwestern railroads, was made chairman, and Charles Magoon the new Governor. Magoon had made his name as a lawyer specialising in colonial administration and had served as legal counsel to the first Commission. He had helped Davis set up the Zone government, and had visited the Isthmus with Taft the previous November. While in Panama he had made a very favourable impression on the American chargé d'affaires William Sands, who had recommended him for the new dual role of Governor and American minister to Panama.

Wallace was summoned to Washington for the first meeting of the new executive committee on 10 April. Inevitably the recent yellow fever outbreak was discussed. It seems that there was little confidence in Gorgas. Shonts suggested that he be replaced with a friend of his, an osteopath with no experience of tropical medicine. Magoon agreed that a more 'practical doctor' was required, one who would deal with the 'smells and filth'. This suggestion was passed on to Taft, who approved it and sent it to the President.

But Roosevelt consulted medical authorities in the US, all of whom backed Gorgas as the best man for the job. Finally he sought the advice of a close friend and hunting companion, Dr Alexander Lambert. The Commission and the Secretary of War, Roosevelt told Lambert, were complaining that Gorgas spent all his time trying to kill mosquitoes while Colón and Panama were as dirty and stinking as ever. 'Smells and filth, Mr President,' Lambert replied, 'have nothing to do with either malaria or yellow fever. You are facing one of the greatest decisions of your career. You must choose between Shonts and Gorgas. If you fall back on the old methods of sanitation, you will fail, just as the French failed. If you back up Gorgas and his ideas and let him pursue his campaign against the mosquitoes, you will get your canal.'

It was a bold step to go against the advice of his own Commission and Secretary of War, but Roosevelt overruled their recommendation and ordered Magoon, who was about to leave for Panama, to give Gorgas all the backing he could.

Accompanied by Wallace, the new Governor arrived at Colón on 24 May, replacing Davis and Barrett, both of whom had already left the Isthmus, the former in the throes of severe malaria. Magoon was an immediate improvement on his predecessors. Sands describes him as 'huge in all three dimensions . . . and he had the gentle nature which so often accompanies vast bulk'. At 230 pounds, he was about as far from Roosevelt's vision of the new 'strenuous' American as you could imagine. But he was immensely clubbable, from the very first ceremony in honour of his arrival, Magoon 'displayed great interest in the people', and, according to the *Star and Herald*, won the confidence of the Panamanians more than any other American had done so far. This helped immensely in his diplomatic task, defined by Sands as achieving 'a truce among the personally jealous political leaders and between the racially hostile political parties long enough to get the Panamanian Republic in working order'.

In his role as Governor of the Canal Zone, he had an even harder job ahead. Publicly he oozed confidence and optimism, but he reported in private that conditions were deplorable. Earlier that month the *Star and Herald* had written, 'It would perhaps be difficult to find any spot on earth where discontent reigns so supreme as on the Isthmus of Panama.' Magoon wrote to Shonts back in Washington that he found the men working on the canal 'ill-paid, over-worked, ill-housed, ill-fed, and subjected to the hazards of yellow fever, malarial fever', and other diseases. Whatever exuberance had fired the first Americans in Panama, it was now long gone, replaced by bickering and demoralisation.

One of Magoon's first acts was to listen to Gorgas about the problems he had been having with requisitions. The Governor then cabled through to Washington the doctor's demands, and within forty-eight hours long-denied supplies were on their way. It looked as if Gorgas would at last have the tools he needed to complete the job. Magoon also made speeches promising that schools and churches would be built, and families encouraged to move out to the Isthmus.

Then, after only two weeks back in Panama, Wallace applied for urgent leave to attend to personal affairs in the US. He also requested a private interview with Taft. Although Wallace had professed himself pleased with the recent changes to the Commission, it appears that it irked him that he had to answer to Shonts. He believed that the canal effort would be best served by concentrating all power in one man, namely himself. He also confided to Magoon that he had been offered a job with a salary of some $50–60,000 back in the US. Wallace told Magoon that as he considered himself essential to the canal effort, he was going to try to squeeze Taft for a higher salary. In fact he wanted to be both chairman and Chief Engineer with the right to come and go between Washington and Panama as he wished. Failing this, he would be happy to leave the 'god-forsaken' yellow fever ridden Isthmus for good and take the money offered at home. All of this Magoon promptly cabled to Taft.

So the usually genial Secretary of War was in a black temper when Wallace arrived to meet him at the Manhattan Hotel in New York on 25 June. At the request of Taft, William Nelson Cromwell was present as a witness, much to Wallace's annoyance – one of the complaints he wanted to make was against the lawyer's disproportionate influence on canal affairs. Taft bluntly asked Wallace what could be so important to cause him to leave the Isthmus again at such a crucial time. Wallace replied that he wanted to resign as he had been offered another job with none of the risks of living in Panama. Perhaps Wallace expected Taft to offer him the top job to keep him in the organisation, but if so he misjudged the Secretary of War. Taft exploded: 'For mere lucre you change your position overnight without thought of the embarrassing position in which you place your government by this action. By every principle of honour and duty you were bound to treat this subject differently . . . Great fame attached to your office, but also equal responsibility, and now you desert them in an hour . . . I am exceedingly sorry that you cannot see what a dreadful, dreadful mistake you are making. It pains me more than I can tell.' Wallace offered to stay on in some capacity to minimise the upset to the construction effort, but Taft ordered him to resign and have nothing more to do with the canal.

Wallace's resignation sparked fresh panic in Panama. 'We felt like an army deserted by its general,' Frank Maltby would write. The rainy season had started, and yellow fever cases in June had nearly doubled on the month before. Men started frantically checking themselves every morning for signs of the illness. According to Marie Gorgas, 'the effect [of Wallace's resignation] upon the workers at the Isthmus was deplorable. It seemed to inspire the labouring and the executive forces with one ambition: a determination to scuttle. There was only one reason why they did not get away en masse, and that was the lack of shipping space to carry them.'

20

RESTART

The arrival, in July 1905, of Wallace's replacement, the rugged and ingenious John Stevens, marks a turn in fortunes for the beleaguered canal. Stevens had built the Great Northern Railroad across the Pacific Northwest. The foremost railway man of his day, he had proven his tenacity in rough territory from Canada to Mexico, surviving attacks by wolves and hostile Indians along the way. Although he would leave under a cloud two years later, his new plan of action would ultimately save the canal.

Initially Stevens, who had been about to leave for the Philippines for a railway construction job, turned down the offer to become Chief Engineer in Panama. 'Then I was asked to meet . . . Cromwell,' Stevens writes. The lawyer 'seemed to have a deep and heartfelt interest in the success of the proposed work . . . after listening an hour or two to his silver-tongued arguments I consented . . . with conditions.' Stevens laid these out when he met President Roosevelt. There could be no interference from above or below, and he could not promise to stay on the job beyond the time 'I had made its success certain, or had proved it to be a failure.'

Stevens sailed from New York accompanied by Theodore Shonts, the new Commission chairman. Shonts remembers being shocked at discovering that there were more canal employees booked on the steamer's return journey than were on the outward trip. Little fanfare greeted their arrival on 24 July, and Stevens straightaway started assessing the situation. For all his experience, what he discovered was profoundly shocking. 'The condition of affairs on the Isthmus,' he would later write, 'can truly be described as desperate; even by many well-wishers it was regarded as hopeless.' Hungry men were foraging the swamps for sugar cane and the jungle for bananas, and in the faces of office workers and labourers alike he saw fear and disillusion. According to Stevens they were 'scared out of their boots, afraid of yellow fever and afraid of everything'. Many believed the 'history of the Americans on the Isthmus would be a repetition of the De Lesseps failure'. It was widely rumoured that Shonts had been sent down to tell them to pack up and go home.

Stevens found 'no organization worthy of the name,' he said, and 'no co-operation existing between what might charitably be called the departments'. In the Culebra Cut, Wallace had been pushing the work ahead in order to satisfy what Stevens would call 'the idiotic howl about "making the dirt fly"'. But when Stevens surveyed the work from a hill above la *grande tranchée*, he saw all seven steam shovels idle for want of spoil trains. There were seven locomotives, but they were all derailed. 'I believe I faced about as discouraging a proposition as was ever presented to a construction engineer,' Stevens would write.

The first meeting between Stevens, Shonts, Magoon and Gorgas was a chance for a fresh look at the problems facing the enterprise. Magoon explained that just about everyone who could get off the Isthmus had left. The Governor reckoned the main difficulty was with food supply – shortages caused by crop failures and an exodus of men from the countryside to work on the canal had driven prices to levels twice that of New York. On two recent occasions wages had been raised for the American workers, and both times the Panamanian merchants had increased their prices to match them. Magoon reckoned some of them were making profits of up to 100 per cent. Shonts' reaction was to order the establishment of 'commissary', ICC-run shops along the line, free to sell to anyone, even though, as Magoon quickly pointed out, this was against the agreement made with the Panamanian merchants by Taft. Shonts, who clearly thought that Magoon had become a little too friendly with the locals, brushed this aside, telling the Governor to 'keep his eye on the ball . . . Our sole purpose on this Isthmus is to build the canal.'

The other priorities identified during this meeting were to improve the accommodation available to the workforce and to confront the spectre of fever. 'There are three diseases in Panama,' Stevens proclaimed to the assembled white staff the next day. 'They are yellow fever, malaria and cold feet; and the greatest of these is cold feet.' From the start the new Chief Engineer projected a hardy image. 'I have had as much or more actual personal experience in manual labor than any one here – surveys, hardships, railroad construction in all its details and operation,' he announced, calling on his subordinates to display 'dogged determination and steady, persistent, intelligent work'. According to Frank Maltby, he was 'enthusiastically cheered; the men looked at each other appreciatively as if to say, "That's the man to follow"'.

Stevens cancelled the plan to build him a large official residence at Ancón, and ordered the removal of the canal headquarters to Culebra, right on top of the work and away from the temptations of Panama City. In his battered hat and rubber boots and overalls, and with an ever-present cigar, Stevens trudged up and down the works, assessing equipment and personnel, trying all the time to spread calm and determination among the workforce. The evenings were spent dealing with the administrative mess left by Wallace and hammering some sort of organisation out of the chaos.

Frank Maltby fully expected to be replaced by a new man, but heard nothing from Stevens for a week after his arrival. Then he had a brisk summons by telegram: 'Come to Panama on the first train. Stevens.' 'We sat out on the veranda under a full tropical moon and among the magnificent Royal Palms of Ancon Hill,' Maltby writes of their meeting. 'Everyone else disappeared. Mr. Stevens did not talk much but asked questions – and could that man ask questions! He found out everything I knew. He turned me inside out and shook out the last drop of information I had.' Maltby learnt subsequently that Stevens had had someone waiting in Washington to take his place, but, uniquely among Wallace's department heads, he kept his job. Maltby, in turn, was impressed by Stevens – the fact that he had gone through 'the hardships of a pioneer in the rugged West' and his quick grasp of the issues at stake. 'His desk was always clear,' Maltby writes; 'one could get a quick decision.' It was a great improvement on Wallace.

Steven's first major decision came after only a week on the Isthmus. In the full knowledge that it would not play well to the press and his political bosses back in the States, on 1 August he ordered a stop to the excavation work in the Cut. No more dirt would fly until proper preparations had been made. In the Cut, half of the steam shovels were shut down, and the free workforce transferred to sanitation and accommodation.

Stevens thought little of the French machinery, however well restored. 'I cannot conceive how they did the work they did with the plant they had,' he wrote of the de Lesseps effort. In its place he put in huge orders for new American machines. Wallace had been planning to try out various different designs of excavators, spoil cars and locomotives, but Stevens just trusted to his experience and went ahead and ordered what he wanted. Stevens respected Wallace as an engineer and put his failure down not just to the 'severe case of fright' from yellow fever that had precipitated his departure, but also to his unwillingness to take on his superiors. Stevens, unlike Wallace, had no seat (for now) on the Commission, but expected, and got, whatever he demanded. 'I determined from the start,' he would write, 'that the only line of policy that promised success was one of going ahead and doing things on my own initiative, without waiting for orders or approval.'

To be fair to Wallace his experimentation with excavation had provided useful experience. The Bucyrus shovels had proved to be strong and reliable, and now

Stevens ordered dozens more, including several state-of-the-art 105-ton monsters. Where early models had gone wrong, changes were made, with speed and strength the primary requisites. Steel replaced iron, and parts liable to break were enlarged and strengthened. The result, according to Stevens, was 'a machine in every way superior to any in existence'.

Stevens also ordered 120 locomotives and 800 cars to carry away the spoil, as well as new drills, vastly superior to those used by Wallace and the French. A compressed air pipeline was planned for the length of the Cut to power them. Stevens had by now seen enough to identify the bottleneck in the excavation effort. As he put it, 'The problem was simply one of transportation.' To maximise the effectiveness of any excavator, it had to be serviced efficiently by spoil removal trains. Thus, for Stevens, the railway, the Panama Railroad, was the key to success on the Isthmus, something never fully appreciated by the French.

Stevens' first impression of the Railroad had not been favourable. He described it in July 1905 as 'two streaks of rust and a right of way'. Its management was 'thirty years behind the times'. Most of it was still single-track, there were practically no sidings 'worthy of the name' and the rolling stock had been obsolete for years. In the summer of 1905 traffic was almost at a standstill with thousands of tons of freight piled in cars, on docks and in warehouses, some of which had not moved for eighteen months. Even shipping papers and other records had been lost. Stevens was told the good news that there had been no collisions on the line for a year, but replied dismissively, 'A collision has its good points as well as its bad ones – it indicates there is something moving on the railroad.'

Stevens brought in a new manager, W. G. Bierd, and laid plans for the double-tracking of the whole line, the installation of a new telegraph line and for the rebuilding of culverts and bridges. At the termini he ordered that new sidings and yards be started, along with extra warehouses, docks and new coaling plants.

All this would need a massive influx of labour. While he was still in New York, Stevens had met William Karner and instructed him to return to Barbados and to keep the recruits flowing to the Isthmus (although Stevens did warn that he was a 'crank' for Chinese labour). In addition, recruiting agents in Martinique and Cartagena were told to step up their work.

Since the disappointing first shipment in January, the attitude to working in Panama had been transformed in Barbados. Karner puts this down to an incident in early May 1905. On the return trip of a Royal Mail steamer from Colón to Bridgetown, 'two colored men stepped from a rowboat to the landing,' he writes, 'almost in front of the window in my office . . . They attracted considerable attention from the men working on and around the landing. It was quite unusual to see a negro laborer riding in a cab, and when these two men, who were smoking, got into a cab and started uptown, the crowed of colored men and women stood aghast and wondered. They soon learned that the two men were in the first shipment I made, January 26th. They had saved money, paid their return passage and still had money for a good time at their home-coming . . . From that time on there was a steady increase in numbers in our shipments.' 'Colón Man' had arrived in Barbados.

The 'Panama Craze' that suddenly took hold of the island led to a repeat of the scenes witnessed at Kingston docks a generation earlier. On one recruiting day, described by American journalist Arthur Bullard, more than two thousand people crowded into Bridgetown's Trafalgar Square, where the police could barely control the crush. 'I wanted to go and do you know the reason why?' asks Barbadian Benjamin Jordan, who was nineteen in 1905. 'In Barbados they wasn't paying you nothing. Even getting ten cents an hour to come to Panama was better than staying in Barbados.' Another canal old-timer, interviewed for a television documentary in 1984, simply wanted to be part of the great project: 'one of the greatest engineering feats the world has ever undertaken.' Soon groups of young men, wearing their best suits and heading on foot for Bridgetown docks, became a familiar sight all over the island. Panama offered excitement, adventure, an escape from the tiny, claustrophobic country with its poverty and rigid social stratification. It was almost an act of rebellion. One group of young men on the way to Bridgetown, it was reported in the Legislative Council, passed a field where a gang of sugar estate workers were being supervised by the plantation manager. One of the young men in the Panama-bound crowd shouted: 'Why you don't hit de manager in de head, and come along wid we!' It may have been meant in jest, but it sent a shiver of anxiety through the island's white elite.

According to Arthur Bullard the men waiting in Trafalgar Square were ushered a hundred at a time into a large warehouse where they were given a medical examination. Weeded out first were the very old and very young, as well as the obviously sickly. No one who had worked for the ICC before (and had therefore returned to the island without fulfilling their five-hundred-day term) was to be allowed a new contract. The doctor checked the remaining men in turn for trichina, tuberculosis, and hernias, rejecting a number each time. 'All during the examination I had never seen a more serious-looking crowd of negroes,' Bullard continues, 'but when at last the doctor told them that they had passed, the change was immediate. All their teeth showed at once and they started to shout and caper about wildly . . . as they expressed their relief in guffaws of laughter and strange antics.'

Benjamin Jordan had passed all the medical checks, but had a final question to answer: 'The doctor was examining us ten in a row,' he remembered. 'He said he wasn't sending anyone to Panama under twenty. I was the smallest one in the bunch. When he came to this fella, he said, "How old are you boy?" "Twenty."

"How old are you boy?" "Twenty." Everyone was twenty. When he come to me, I was the youngest of the group, you know. All of the other nine were twenty. I was the smallest. I was only nineteen and had got to put on a couple more years. "I am twenty-three, doctor." "No, boy, you're not twenty-three years old, but you'd like to go to Panama, wouldn't you?" "Yes, doc." "Well I'll send you. But you nine, go home." Well I had to get out of the city fast,' says Jordan. 'They were going to break my this that and the other to prevent the doctor sending me. I had to hurry up and get out of the town.'

Loading the steamer to Colón started at nine in the morning and took most of the day, as the men's contracts were checked and possessions carted on to the deck. Huge crowds gathered to see the men off and wish them good luck. 'I never saw so many negro women in all my life,' wrote Bullard of the sailing day he witnessed. 'All of them in their gayest Sunday clothes, and all wailing at the top of their voices.' Bullard was taking the same steamer to Colón. On board were seven ship's officers and more than seven hundred blacks. 'Every square inch of deck space was utilized. Some had trunks, but most only bags like that which Dick Whittington had carried into London. There was a fair sprinkling of guitars and accordions.' As the boat got underway the singing of hymns started, with one side of the deck Church of England and the other nonconformist. 'There was only one song, a secular one, on which they united,' writes Bullard: 'Fever and ague all day long/At Panama, at Panama,/Wish you were dead before very long/At Panama, at Panama.'

The journey to Colón took about twelve days, and all the time the passengers remained on deck, in rain and sun. 'There was no room to stretch themselves,' wrote another journalist, Englishman John Foster Fraser, who also took a Karner steamer from Bridgetown to the Isthmus. 'They slept higgledy-piggledy in the most cramped of postures.' On Bullard's ship, he writes, the decks were often hot enough to fry an egg, and the passengers had to be hosed down to keep them cool. Even so, in the heat tempers flared and on one occasion the captain had to threaten a man with his revolver to break up a fight. Seven men were clapped in irons and put in the baking-hot brig.

There was a more serious outbreak of disorder when the food delivered to the men turned out to be inedible. The situation was only defused when land was suddenly sighted. On Fraser's boat, 'The shipowners did not provide plates or cups . . . it was an amusing sight when feeding time came along. The ship's servants brought up cauldrons of soup or tea. The passengers, presenting their meal tickets, had food poured into their fists, and then, sheering off, they huddled together and gulped the meal as though they were ravenous.'

Young Benjamin Jordan arrived in Panama in October 1905. 'They brought me here, put me down in the Cut, put a pick and shovel in my hand. I had never seen a pick and shovel before. I started to cry.' His boss took away his tools and gave him a job as a waterboy, fetching drinks for the other workers. Jordan was not the only new arrival to find the confusion, noise and mass of unfamiliar machinery overwhelming and frightening. 'On the appearance of the place, I thought I'd go straight home,' remembered one digger who arrived at the end of 1906. 'Everything looked so strange, so different to home. I felt that I would go back home. But it wasn't so easy to do that, you know, so I continued.'

Martinique, devastated by the volcanic eruption of May 1902, proved to be another fertile recruiting ground. In August 1905 alone, there were 628 arrivals from the French island. But the following month a group of 140 Martiniquans refused to leave the steamer at Colón. They were frightened of being vaccinated and the longer they stayed in the port the more stories they heard to put them off going ashore. After persuasion by the French consul failed, the men were deprived of food for twelve hours, and then forcibly removed by club-wielding police. There was a further problem in November when the arrival of some five hundred female Martiniquans, imported to work (officially) as laundresses and domestic servants, caused the local correspondent of the New York *Herald* to write that, scandalously, the ICC was importing prostitutes at the US taxpayers' expense. (William Sands says, 'these women had been imported for a very definite social purpose'.)

Recruitment from Martinique, initially almost as high as that from Barbados, fell away following these incidents. Karner in Bridgetown, however, would continue to increase his shipments dramatically. During the first nine months of 1905, fifteen steamerloads carried some three thousand Barbadians to the Isthmus under contract. Recruitment was halted in September due to a lack of accommodation for the workers, but was restarted at the beginning of 1906, during which twenty-one steamer voyages carried over six and half thousand new recruits. In March 1906 Karner even chartered a steamer, the *Solent*, to be engaged in nothing but carrying workers from Bridgetown to Colón.

On 15 September 1905 the Kingston *Daily Telegraph* printed a letter supposedly from a Canal Commissioner that stated, 'we have no immediate use for the Jamaican', but soon after, the Panama Railroad, a separate body from the Commission, even though its stock was wholly owned by the US government, recruited nearly six hundred Jamaicans, paying their departure tax of £1 5s. This neatly side-stepped the impasse created by the falling out of Taft and Swettenham. The Railroad was often useful in this respect. As Frank Maltby explains: 'As a corporation it could perform many functions impossible or difficult for the Government to do directly under existing law, rules and regulations. It became quite the habit in discussing projects of doubtful legality to say "You can't do that. Let the Panama Railroad do it."'

Many more Jamaicans travelled to Barbados to be recruited there, and the Commission also assigned an agent to Kingston to encourage voluntary migration to Panama. Thousands took up this option. As they had to be wealthy enough to pay the departure tax and the steamer fare of around £2, the Jamaicans on the Isthmus tended to be the more skilled and better-educated workers. There were Barbadians travelling to the Isthmus under their own steam as well, in roughly equal numbers to those going from the island under contract. This included those who did not like the idea of signing themselves up for nearly two years or had been rejected on medical grounds, as well as men who hoped to get better terms than the basic unskilled labour deal Karner was offering. Harrigan Austin was an experienced carpenter, and arrived in Panama on Sunday 9 October 1905 having paid his own steamer fare. Austin was one of about a hundred men and women – nearly all West Indians – who responded to a competition staged in 1963 for 'The best true stories of life and work on the Isthmus of Panama during the construction of the Panama canal'. The initiator was the Isthmian Historical Society, run by American Zonians. The competition was publicised in newspapers in Panama, Central American and the Caribbean islands, and offered a first prize of $50. Although some respondents might have tailored their accounts to suit the purpose of winning the money, the large collection contains many diverse attitudes and opinions as well as a wealth of detail.

Harrigan Austin's Panama adventure did not start well. To raise the fare of £2 10s. ($14) for the passage, he had to pool the savings of his extended family, and took virtually nothing with him apart from his carpentry tools. He landed at Colón, 'having had a hazardous trip, of thirteen days in bad weather, poor accommodation in general with sparing meals on a crowded ship, we were all more or less hungry. We saw after landing on the dock, a pile of bags of brown sugar. And the whole crowd of us like ants fed ourselves on that sugar.'

Austin was taken on straightaway and loaded with the other men into freight cars, then 'hurried off, and distributed at various Stations. My lot happened to be Las Cascadas.' The men were led to a camp of tents, and each was assigned an army-style cot. The next morning he was put to work repairing old French quarters at Bas Obispo. His foreman was a white man, but he appointed one of the experienced West Indians as sub-foreman 'as really the only thing he knew to do was to watch us but really very little about handling or directing a carpenters' gang'.

Other West Indians were less cynical about their American bosses. John Butcher arrived from Barbados on 12 January 1906, and, like Austin and some two thousand more workers, was employed on assembling and renovating buildings, a Stevens priority. His first assignment was as a plumber's helper. He instantly got on with his immediate boss, a young American, Edward T. Nolan (they were still in touch with each other by letter in 1963). Nolan's superior, the assistant quartermaster, was 'a real pusher', Butcher remembered. 'He always promised permanent work to the better workmen. Hearing this, I tried my best to work harder and more than anyone else. Carpenters, plumbers, electricians, painters – all accepted this challenge but of course, as far as plumbers were concerned there were none better than the Nolan–Butcher team. Through hard work, we excelled in whatever jobs we were assigned . . . As a husky, strong, active young man who was never afraid of work, I was always in demand. I still have the joy of just knowing that there are so few of those early houses on which I did not work.'

Many of the West Indian accounts show similar pride in their work and participation in the Panama project. Others take pleasure in the new skills they were rapidly forced to acquire – operating tools such as hammers and drills that they had never even seen before. But Harrigan Austin, already an artisan when he arrived, was deeply disappointed when he was put on the same wage level as the unskilled labour: 'We were often forced to work in the rain,' he writes, because if they stopped, their wages were cut. 'Indeed to some degree life was some sort of semi-slavery, and there was none to appeal to, for we were strangers and actually compelled to accept what we got.' In the case of an argument, says Austin, 'the bosses and policemen right or wrong would always win the game, and those men who had the chances filling such position were generally of the dominating type who tried to bring others into subjection for their fame.'

The police of the Isthmus had given the massive influx of West Indians, or 'Chombos' as they derogatorily called them, the same sort of welcome afforded to the arrivals of a generation before. There was particular dislike of their presence in Panama City, where large gangs were at work for the whole of 1905 building sewers and waterworks, and paving the main streets. On one occasion at the end of April a group of about 150 men staged a sit-in when their American foreman ordered them back to work before most had eaten their lunch. The demonstration grew more heated, with loud complaints about low wages and their late payment. The foreman summoned the local police, who charged the strikers and then chased them across the square in front of the old Administration Building. In the ensuing mêlée twenty-one workers were seriously enough injured by bayonet stabs and rifle butt blows to require hospital treatment.

'The disgraceful scenes of yesterday will live long in the memory of those who witnessed them,' wrote the *Star and Herald* on 29 April. 'The senseless attacks upon inoffensive persons was enough to make anyone's blood boil . . . Carelessness and utter indifference to the wants of the men who are brought here to do the actual work of digging the Canal is silently, but nevertheless, surely breeding trouble. Who is responsible? There is no denying the fact that the men have been brought here under false pretences and misrepresentations and have been badly paid and underfed . . . and in many cases put to work under incompetent men possessing an inherent hatred and contempt of the colored races.' The Colón *Independent*, a tri-weekly founded in 1899 by a Barbadian, Clifford Bynoe, would write a year later of this time, 'One could scarcely breathe God's free air without being clubbed and kicked . . . The American occupation then was a terror and a disgrace.'

Consul Mallet, whose job it was to try to protect British citizens from such attacks and to seek redress from those responsible, blamed both the Americans and the Panamanians for the incident. Writing to the Foreign Secretary about the 'serious disturbance' he explained: 'the majority of Americans here – I refer particularly to those who hold subordinate positions – exhibit an extraordinary contempt for the Jamaican negro, and most of the trouble so far has been due to the blustering behaviour of the foremen in charge of the gangs.' But he wrote too about the 'intense dislike and jealousy felt by the police towards the coloured natives of Jamaica on the Isthmus'. There had been increasing complaints, he reported, about the violent methods resorted to by the native police when making an arrest for some trivial offence. 'In fact,' he ended, 'their arbitrary conduct is no better to-day than it was during the worst period of the Colombian regime on the Isthmus of Panama.'

After much prevarication Mallet at last extracted an official apology from the Panamanian Secretary for Foreign Affairs. Magoon, of whom Mallet had a high opinion, made conciliatory noises, but the Canal authorities asked the Panamanian police to arrest Jamaican Charles Schuar, who was seen as the leader of the workers' action. He was promptly imprisoned and then deported.

There did follow, however, a reorganisation of the police by an American instructor, and, on Mallet's suggestion, some West Indian constables were taken on to help police their own neighbourhoods. But the West Indians remained wary. 'In Jamaica a constable is a peacemaker,' one told an American journalist. 'Here he just hits a man with a stick.'

From the very start, and for much of the construction period, the question of food would remain a leading catalyst for dissent and dissatisfaction among the West Indians. While prices rose fast to 'Klondike' levels, wages had stayed the same for the workers. In November 1904, the PRR, flush with its success in breaking the mass strike six months earlier, had actually lowered its minimum wage to bring it into line with the Canal's pay rate. The money saved was spent on rises for the white skilled railroad employees. Further pay hikes for the Americans on the Isthmus had not been matched for the West Indians, many of whom now found themselves in a desperate situation. 'Instead of the canal bringing with it those good old times it is bringing hard work and starvation pay for the majority and fortunes for the few,' complained the Colón *Independent* at the end of 1904.

'The poor laborer has had his life blood gradually crushed out, heavier and heavier his expenses and eventually his very hopes and air castles dashed completely to the ground and broken forever! To day he can hardly live . . .'

On a wage of seldom more than a dollar a day, the West Indians were being asked to pay seventy-five cents for a dozen eggs and sixty cents for a chicken. Coffee and bread brought to the works by West Indian women cost an hour's wages. In their desperation, many were surviving on a diet of sugar cane and were becoming seriously malnourished. At the beginning of 1906, two American journalists from the New York magazine the *Independent* interviewed a man they described as of 'unusual intelligence'. He declined to be named in the article, and is referred to only as a 'Jamaican carpenter'. But he is of particular interest as he had been on the Isthmus since 1894, and had worked for the French at the height of the activity of the New Company. 'Things were very different in those days,' he told the interviewers. 'The workmen are more afraid of the Americans than of the French . . . there are no loafing jobs now, such as there used to be. It is like running a race all the time. You don't mind it for a day, but you can't keep it up.' Although the basic pay, he said, was better from the Americans, there were no longer chances to increase this through piecework. 'Besides,' he went on, 'the blacks had more chance of promotion under the French. They could get to be timekeepers or checkers then, but they can't now.'

The biggest difference for him was the cost of living. Yams used to be sixty for a dollar, but now you only got sixteen. 'If they starve themselves,' he said of the West Indian labourers, 'they can save a good deal. If they are well fed they don't save. Out of 80 cents a day it takes 50 to buy food, and then there are washing, clothing etc., besides. Some of the men try hard to save; buy 2 cents bread, 2 cents sugar, and go to work all trembly and can't lift a thing.'

William Karner backs this up. 'In their anxiety to save money to send back home,' he wrote of the Barbadians, 'they were literally starving themselves.' An American journalist, John Foster Carr, who visited the Isthmus in mid-1905, investigated this for himself: 'I have looked into hundreds of their pots boiling over bonfires, as they crouched beside them,' he wrote. 'A very large number contained nothing but rice, or a piece of yam, or some plantains. Others had added a small piece of salt pork, beef, or codfish. In the rainy season, when with damp wood their primitive fires fail them, they tell me that they have often to choose between half-cooked yams and rice – which the doctors say is not digestible – and biscuits . . . weak and anaemic are these poorly fed laborers. They fall easy victim to malaria, and on this account alone, the Chief Sanitary Officer maintains, pneumonia with seventy to ninety per cent of fatal cases is prevalent among them, when the better-fed white man is nearly immune.'

After a steady average of four penumonia cases a month, suddenly in October

1905 there were twenty-six, all West Indians. On 13 December the headline in the Colón *Independent* read: 'Pneumonia Rampant On the Isthmus. Has taken an epidemic form. Fatal Among Colored People'. A doctor was asked by the newspaper for the attribution of the disease. 'A severe cold,' he replied. 'The laborers are generally wet with perspiration and will sit in the wind to cool off. Very often they go to bed in their wet clothes and when the chilly part of the night comes on the body becomes cold.' With the return of the wet season in spring 1906, the rate would jump again. Because viral pneumonia was almost unknown on the islands, the new arrivals were particularly vulnerable.

In November 1905, journalist Poultney Bigelow made his famous trip to Colón. An experienced reporter (and former law-school classmate of Theodore Roosevelt), Bigelow had made his name with stories on labour conditions in South Africa and the Far East. His father was John Bigelow, who had visited the Isthmus in 1886 together with de Lesseps and remained a life-long supporter of the canal. Poultney Bigelow, however, seems to have been drawn to the story by the increasing rumours back in the United States that the project was in trouble from labour strife, confusion and corruption.

In Colón, whose condition he described as worse than anything he had ever seen, even in the 'slums of Canton', Bigelow wandered the streets, unescorted by any official, interviewing those he came across. One Barbadian he met he described as wearing 'a clean collar, a black derby hat and a good suit of clothes – an educated and prosperous example of his race'. From him he heard complaints about the high cost of living and the late payment of wages. 'But his main grievance was that as a man of color he received no encouragement for his work; no one seemed to care whether he got good work out of his men or not – all the white men about him were trying to see how little they could do, each for himself.' Others reported that they received unequal justice and were treated with rudeness by the American bosses.

Worst of all, however, was the sickness. 'Throughout my pestiferous excursion up and down this filthy city,' Bigelow wrote, 'I could not find a single man or woman who had not suffered or was not suffering from fever of some kind.' One worker he met was 'a splendid specimen of manhood, a negro such as would have been recruited with pride into the Tenth United States Cavalry'. But the man was sick and could only walk with difficulty. He felt he had been deceived, Bigelow reports. The place was 'unfit to live in'. He was trying to get back to Jamaica but the next steamer was already full, taking away, '400 negroes, all returning to Jamaica in disgust'.

The journalist's own steamer from Colón carried the same number of returnees, and when Bigelow met Swettenham and the Chief Justice of Jamaica they confirmed to him 'what is denied by official authority in Washington' that 'negroes are returning from the canal in portentous numbers' and that the men 'were not honestly or humanely treated'.

In mid-December Stevens wrote to Shonts: 'Notwithstanding nearly six thousand new laborers were brought in between August 15 and November 15, our force shows little or no increase . . . our forces are being constantly depleted by departure from the Isthmus . . . The Jamaicans are returning almost universally.'

The West Indian workers, hampered by disease and malnutrition, were criticised by the American canal authorities as weak and idle. Official ICC reports wrote of their lack of vitality and frail 'disposition to labor'. 'The West Indian's every movement is slow and bungling,' wrote one American journalist, echoing the prevailing sentiment. 'Every small object a subject for debate; anything at all a sufficient excuse for all hands to stop work.' But, as the writer went on, there was a 'certain and unjustified cruelty' in forcing 'poor half-fed fellows' to work eight to ten hours in such heat. 'Until you have tried to do a good fifteen minutes' work with a pick and shovel during the rainy season . . . you can have no idea of the exhaustion that tropical heat brings even to the laborer who is used to it.'

Stevens declared that the value of the West Indians as labourers was low under any conditions, but admitted, 'They were not getting proper food in sufficient and regular amounts to give them strength for continuous work.' The situation was improved a little with the gradual opening up of ICC-run 'commissary' stores along the line, where, using coupons issued against pay, employees could buy food imported from the US at 'cost price', or at any rate cheaper than prevailing locally. There was much praise in some quarters for the fact that the Commission, through bulk buying, could ship goods two thousand miles and still sell them at prices comparable to those in New York City. This may have been a great boon for the well-paid white workers, but the wages of the West Indians, although generous by Barbados or Jamaica standards, would not have gone far in the US. It did not help either that cooking was forbidden in the workers' barracks.

Construction had been ongoing on several hotels, where the white employees were served meals. Soon after taking over, Stevens decided to offer the West Indians cooked food as well in a bid to improve their productivity. Originally the entire catering operation was farmed out to a private company, but when the contractor calculated that he needed to charge $30 a month per white employee and $12 a month for a labourer, the Commission decided to set up and run the operation themselves.

While more 'hotels' were built for the American workforce, for the blacks there were 'messes', kitchens established near the work sites. No tables or chairs were provided in these, so the diners were forced to squat or stand with their food. By February 1906, there were over fifty in operation, feeding 7–8000 workers a day. Breakfast invariably consisted of coffee, bread and porridge; lunch was usually bread, beans and rice; supper was more bread with potatoes, soup, coffee and perhaps some meat.

But if malnutrition declined, grumbling continued. The food served seems to have been indescribably awful. One sympathetic American called it 'the leavings from the hotels . . . [which] are not fit to eat before they are leavings'. Barbadian John Butcher, who otherwise generally speaks well of his treatment on the Isthmus, described the rice he was given as 'hard enough to shoot deers'. When they had meat, 'many men spent an hour trying to chew or eventually threw away because it was too hard'. Harrigan Austin remembers the food as 'poorly prepared, almost raw'. But anyone who protested was arrested for bad behaviour by the policemen sent to keep order in the kitchens at meal times.

In order to maximise the efficiency of their workers, the Commission sought to take more and more control over the lives of the West Indians. When he arrived at Colón at the beginning of 1906, Jules LeCurrieux was, at sixteen, even younger than Austin Harrigan. Although born in French Guiana, LeCurrieux's family had emigrated to Barbados, so he came under a Karner contract from Bridgetown. Straight off the steamer, the men were piled into railway freight trains for distribution along the line. 'To our surprise,' writes LeCurrieux, 'we were unloaded off the train as animals and not men, and almost under strict guard to camps.'

These 'camps' might have been of tents, like Harrigan Austin's, or simply box cars from the railway, but LeCurrieux was deposited in one of the newly assembled workers' barracks. This consisted of a separate toilet and a hut of about 50 by 30 feet. Into this space were crowded seventy-two men. No furniture, sheets or pillows were provided. Jamaican journalist Henry de Lisser visited one of the barracks later that same year: 'Inside the houses themselves you find groups of men seated round a box and playing cards; you find some listening to one of their number playing softly on a flute; you find others in bed,' he wrote. 'These beds are canvas cots fixed onto iron standees which can open and shut as required. Each standee is about seven feet high, and has three cots hung on either side of it, each cot being six-and-a-half feet long by two-and-a-half feet wide . . . when they were all occupied the room cannot be the most pleasant place in the world to sleep in.' According to the 'Jamaican carpenter', 'There is no privacy or quiet in the bachelor buildings . . . Some of the men are noisy at night and have no sense of decency . . .'

'We were taken to a kitchen,' continues Jules LeCurrieux, 'and each of us were given 1 plate, 1 cup, 1 spoon, and a meal, then those utensils were ours – the price to be taken out of our first pay.' That same afternoon LeCurrieux was put to work with a gang in the Cut. His job was to drill 20-foot-deep holes at the top of Gold Hill to be stuffed with dynamite to 'tear the old hill down'.

When work finished for the day LeCurrieux was given a thirty-cent ticket, the value of which was taken from his pay. This entitled him to three meals and accommodation in the barracks. After two weeks he got his first wages, which he used to buy a pillow and blanket 'and a few cakes of soap to wash our dishes and clothes'. 'The discipline maintained in the labour camps is severe,' reported Mallet back to London. LeCurrieux remembers that at 9.00 p.m. an old piece of rail was knocked with a metal bar, signalling 'go to bed, no sound'. At 5.00 a.m., they were woken by more loud knocking, and after a hurried breakfast they were on a labour train by six. To keep the men at work, they were denied food or shelter if they could not produce the ticket that said they had laboured that day. 'This rule worked well and tended to drive out the undesirable class,' reported Stevens. No one was allowed in the barracks during working hours – you had to be on the job or in a hospital. Those who broke this rule were arrested and fined three days' wages.

Furthermore, about once a week there were spot checks on those sleeping in the barracks, as the Colón *Independent* complained: 'At midnight when everyone is asleep, suddenly the cry of "tickets" is heard. The laborer, frightened out of his sleep, very often cannot remember at the moment what he has done with his ticket, and is hustled off to prison.' 'This system has been adopted to keep loafers out of the camps,' continued the paper, 'but it would be better to allow a few loafers to get in than that so many innocent men should suffer. The system is a rotten one and must be changed.'

'I grew careless last week,' wrote Jan van Hardeveld to his wife Rose in August 1905. 'Before I realised it I was one sick hombre – stomach out of order and my blood full of malaria bugs.' There was better news – he had met a fellow Dutchman, Jan Milliery, who had been in South Africa during the Boer War and had subsequently worked as a track foreman. Milliery was experienced in life in the tropics and was a good cook. The two men were rooming together and had become firm friends.

Against the advice of her family, Rose had decided to take herself and her daughters out to Panama to join her husband. In the meantime, while they waited for quarters to become available, she had been reading up about the history of the canal dream, and the more she learnt the more she began to share her husband's enthusiasm. In his letters, Jan was frank about the difficulties – there had been little headway with the 'real work', the men were 'dissatisfied', the labour problems were 'acute'. The resignation of Wallace had thrown everybody. But Jan had clearly not lost his faith: 'The slowness of the work would be discouraging,' he wrote home in September, 'if I were not certain that our Government can and will accomplish whatever it sets out to do. You know what I always say

– in America, anything is possible . . .' In the same letter he told Rose that the quartermaster had at last assigned him married quarters. Two weeks later the family were on their way to Panama.

The voyage for Rose was accompanied by a mixture of sea-sickness, 'deep foreboding' and great excitement – at seeing Jan again and at the prospect of joining the canal effort. 'This will be our chance to be among those who make history!' she told her girls. There were other wives on board, and the talk was all about whose husband earned what. Everyone saved their best outfits for the landing at Colón, 'and it was as if with these clothes we had donned a more formal manner. Although we were all really going to live in a huge construction camp, a sudden concern over possible difference in social status among us seemed to arise, both ridiculous and pathetic – as if we were trying frantically to carry some vestige of propriety and tradition into the jungle with us,' writes Rose. As the *Advance* neared the shore, the passengers crowded the rail, scanning the dock for sight of their loved ones. For a while Rose could see no sign of her husband, but then, there he was, standing on a pile of logs, 'frantically waving his hat to attract my attention'. Even before the gangplank was in place, Jan had climbed aboard and gathered his family into his arms. 'The children were shy,' remembered Rose. 'They scarcely knew him at first. He was so very thin and burned to a deep brown.'

To the new arrivals Colón was as shocking and overwhelming as ever. The waving, feathery palm trees were as beautiful as Rose had imagined they would be, and the flowers 'larger and brighter' than anything she had seen before. But the horses that drew the carriage to the railway station were 'unbelievably mangy-looking', there was a 'foul smell' cloaking everything, and 'naked brown children' playing on islands in a sea of rubbish, sewage and greenish, filthy water. Once on the train Jan confessed that their intended house was not yet ready and they would be camping out for a few days in another one nearby.

Leaving the train at Las Cascadas, Rose found herself in the 'blackest, darkest place I had ever been in. Not one flicker of light shone anywhere. We stumbled over a number of wet, slippery tracks and walked along a board walk until we reached the steps of the house', obviously long unoccupied. Inside, furniture had been thrust through the door and left, and a large number of bats were living in the rafters. They had clearly been in residence for some years: 'a penetrating stench, so vile it was almost unbearable' hit them as soon as they went inside. Black lizards with bright yellow heads scurried for cover. The first priority was to get the children to sleep. Jan and Rose assembled two beds, and covered them with a mosquito net as the house was unscreened. 'By the time I had undressed Janey and Sister,' writes Rose, 'they were both sobbing forlornly.'

The next morning, as Jan ran for the labour train to take him to his post in the Cut building track for the excavators and spoil trains, Rose set about finding something to give the children to eat. Venturing out of the house, she was confronted with the sprawl of 'dingy, non-descript houses' that constituted Las Cascadas, and was instantly 'hot and uncomfortable . . . as though I was being smothered between wet, evil-smelling sponges'. She identified a shop run by a 'bald-pated Chinese' but he seemed to have little to sell except tinned butter, plantain, yams and soggy, stale English biscuits. At a nearby meat stall, Rose turned away in disgust at the flyblown ribbons of beef. They ended up having fruit for breakfast, bought from a West Indian woman hawking from house to house a tray of oranges and bananas she carried on her head.

At lunchtime, Rose's husband returned accompanied by his Dutch friend Jan, who had been renamed 'Jantje' – 'Little Jan' – to avoid confusion. Rose learnt that as Jantje was not an American citizen he had to find his own living quarters, and was still trying to find something suitable after three months on the Isthmus. His plan was to bring his new wife and their son, born the previous November, out to live with him in Panama, and he had already started the process of getting US citizenship.

Jantje Milliery had arrived at Ellis Island at the end of 1902, having briefly lived in Brussels after leaving South Africa. He married a Dutch woman, Martina Korver, from Rotterdam, the following year and they lived in Jersey City Heights. He had applied to the ICC on 26 June 1905, enclosing a reference from the Dutch Consul General of Orange Free State: 'Applicant is apparently a steady reliable man,' it reads. 'He is conversant with track repair methods, and speaks English very well.' On 5 July he received an appointment, and sailed for Panama on the tenth. The money was not that good – he started on $1000 a year (van Hardeveld, five years older and already a US citizen, was on $1700) – but being part of the canal meant more to Milliery than the money: it was a means to becoming an American.

That evening Jan van Hardeveld returned exhausted from his day's work. Rose and the girls had spent the afternoon trying to work out which was worse – the soggy heat of outside or the eye-watering stench of the interior of the house. Rose begged him to take some time helping her sort out the mess, but he was 'so absorbed in his work that he could not see what overwhelming difficulties confronted me in trying to care for the family'. 'You don't know what you are asking,' Jan replied impatiently. He could not leave his unreliable West Indian labourers for a minute, he said. 'We're here to dig this canal.'

Jantje stepped in to help Rose in the business of rapid acclimatisation, showing her how to hold a damp matchbox at a certain angle to strike and get a light. The local plantains, he explained, could be fried or roasted in their skins. It was all right to buy beef, but it should be cooked for a long time and then only the broth used.

Soon after her arrival Rose was taken to a local party, where she felt very out of place as the only 'white' woman in the room. 'The people of the new Republic of Panama considered us an uncouth race,' she soon learnt. There was at that time only one other white American woman in Las Cascadas, an old tropical hand with an alarmingly pinched and yellow face. It was very much a frontier town. Going to and fro, Rose had noticed 'dingy places bearing across the front the word *Cantina* . . . by the smells around the places, and the conditions of the men and women coming out of them, I knew they must be saloons. It worried me to see the little black children running in and out, freely, to see the black women staggering, laughing, cursing, and to watch our own men going in for drinks.' She had quickly discovered that Jan and Jantje carried a demijohn of rum beside their workplace ice box – 'and that bothered me too'. Soon Rose found herself involuntarily averting her eyes when she passed the cantina near her house. 'A brown woman sat there looking out,' Rose remembered. 'She reminded me of a fat spider waiting for someone to devour. She often smiled and nodded at me in a friendly way, but I hated her.'

Then at last their proper house was ready, or ready enough for them to move in. The first building erected on the site by the French, it was known as House Number One. Despite the grand name, it was small and dingy, with only one bedroom and a tiny kitchen. But from its high position it offered a view all along the line of the canal as far as the Pacific Ocean to the south and Bas Obispo in the north, where, 'the good old Star Spangled Banner, doubly beautiful and precious in this strange country, flew from the pole on top of the hill in Camp Elliott, the United States Marine Corps station'. And just below, indeed virtually under the house, was the Cut, where, Rose writes, 'the French had made a noticeable beginning'. Immediately below the house lay 'pieces of machinery overturned, strings of cars, engines, and twisted rails, all covered with growing vines and brush. Large trees had grown up through the couplings of a string of cars at the foot of the hill.' Further down, however, the Americans were at work, and Jan was able to point out to the children the Bucyrus shovel served by his track-shifting gang.

For the children it was thrilling to be so close to their father at work and to the most spectacular section of the canal. According to Rose they 'quickly assumed something of their father's proprietary feeling' about the project. They were also making the most of being virtually the only American children among many homesick adults. At the new house ice was delivered every day and in November an ICC-operated food store opened at Empire, two miles away. Even so, day after day was the same meal: 'beans, soggy crackers, Danish butter, and fruit. Occasionally a chicken relieved the monotony.' On one occasion Rose took the train to Panama City, where she found that anything could be obtained but at exorbitant prices. She was also unimpressed by the 'crooked gambling houses, filthy saloons and brothels. The well-known "American sucker" had come to the Isthmus and was being properly fleeced by those who knew how.'

Towards the end of the year, a trickle of wives started arriving. One in Las Cascadas had a fright on her first day when she mistook an iguana for a crocodile, and 'remained in a permanent state of terror, . . . every little bug made her hysterical'. Rose did her best to help the newcomers become oriented, and, doing so, she writes, 'I sometimes succeeded in bolstering my own failing morale.' Meanwhile her husband was becoming increasingly tired and grumpy, endlessly complaining about the standard of work of his West Indian labourers. 'There was little I could say or do to make him relax,' writes Rose. 'The Canal about which he was so intensely, feverishly concerned might have been his own personal project – his and Teddy Roosevelt's.'

Other Americans needed more than the belief in Roosevelt's 'great march of progress' to induce them to come to the Isthmus and stay there for any length of time. The first step was to offer exceptional wages. Soon after taking over, Stevens established four recruiting agents, one in New York, one in New Orleans and two roving. These interviewed candidates and promised anything up to double his current salary in the US. They also offered free transportation to Panama, free furnished accommodation and medical care and long, six-week annual paid holidays. According to a senior American administrator on the Isthmus, 'Special inducements were added one after another, until an established system was developed which contained perquisites and gratuities which in number and value far exceeded anything of the kind bestowed upon a working force anywhere on the face of the globe.' But the initial results were disappointing. Jobs were easy to come by in the United States at that time. More than anything else, though, the shadow of fever still darkened Panama's reputation.

21

THE RAILROAD ERA

Crossing the Isthmus by railroad during his first week in Panama, Stevens' fellow passengers had pointed out Gorgas's sanitation squads draining pools, fumigating houses and oiling waterways. Each spotting was cause for new 'ridicule, not only of Colonel Gorgas but also of the mosquito theory, some of these comments reflecting very severely upon the quality of the Colonel's mental equipment. My attention,' Stevens later wrote, 'was repeatedly called to the great waste of money and the utter futility of the whole procedure.'

Since the arrival of Magoon, Gorgas had at last started receiving supplies, and Stevens and Shonts, with Roosevelt's instructions ringing in their ears, had further upped the backing for the Sanitation Department. Shonts, although still sceptical about the mosquito theory, had in July 1905 increased Gorgas's workforce from two hundred to over two thousand. Stevens initially shared Shonts' doubts. But he was so impressed both with the urgent need to combat the fearful disease and by Gorgas's conviction that he could do it, that he decided to provide full support and funding.

So began, in July 1905, one of the most famous sanitary campaigns in history. Mosquito brigades were formed for every district of Panama and Colón. Panamanian doctors were recruited at a local level, responsible for daily house-to-house inspections. A combination of respect for Magoon and Gorgas's charm and tact produced new compliance from the residents of Panama City, who were offered a $50 reward for reporting a case of yellow fever. For each one – and there were forty-two cases in July – their movements during the days before the appearance of the first symptoms were painstakingly traced to find the source of infection.

But nothing was left to chance. Every dwelling, from the grandest *palacio* to the tiniest shack, was now meticulously fumigated. Teams were sent out through the streets, with each man carrying a ladder and a gallon of glue. On their shoulders they had strips of paper 2 metres long and 3 inches wide to stick over doors, windows, holes and openings in the wall to prevent the smoke or the insects escaping when the pans containing sulphur or pyrethrum and alcohol were lit inside. If it was a large building, once it was sealed the sulphur smoke was pumped in from outside, through a tube inserted into the door through the keyhole. In a year, 330 tons of sulphur and 120 tons of pyrethrum – the entire annual output of the US – were used up. 'The fumigation campaign was so intense,' remembers one sanitary squad worker, 'that there was a big, thick white cloud of smoke from the sulphur hanging over the Zone and Panama City, and even the leaves on the trees curled up.'

At the same time other teams checked on the water that people kept in barrels for everyday use, unwittingly providing breeding grounds for the next generation of *Aëdes aegypti*. In early July 1905 running water had at last been connected for Panama City, which was of great assistance to the efforts of the sanitary squads. So with households in the city largely clear of water containers, the squads could concentrate on potential breeding sites among rubbish and elsewhere. 'An empty tin can was the special aversion of the Sanitary Squad,' writes Frank Maltby. 'We became so clean, orderly, and "dried out" that it was painful.' Where water could not be removed – in cesspools, cisterns, puddles or potholes – other teams arrived to spray the surface of the water with kerosene to smother the 'wrigglers'. In a year, over two and a half million gallons of oil was used in this way. Success demanded total thoroughness. Gorgas even had the holy water in the font of the cathedral changed every day once it was found that mosquitoes were breeding there.

Indeed, with most households dispensing with domestic water containers, the yellow fever mosquitoes started laying their eggs wherever they could find pools of water – in hollowed-out stones and even in the leaves of trees, in particular that of the *calocasia*, which grew in weedlike profusion around the houses. As well as clearing all vegetation from the vicinity of living quarters, Gorgas's squads started laying traps. Such is the fastidiousness of the female *Aëdes aegypti* that the bowls of sweet clean water now left out by the sanitarians proved far more tempting for oviposting than a dirty puddle. Every day the water in the traps was simply emptied on to the dry ground and replaced. 'They eagerly accepted the much cleaner tin pans placed out for their particular benefit,' writes Joseph Le Prince, 'and thus involuntarily committed race suicide.'

As well as aiming to exterminate the *Aëdes aegypti* by attacking its larval stage, Gorgas sought to prevent any adult survivors from coming into contact with humans, or more precisely, vulnerable white Americans. The US workforce now found itself thoroughly protected by metal screens. But because of the warm, salty and damp climate of the Isthmus only very pure copper would resist rapid corrosion (even so, the screens on every American home or dwelling place had to be checked by yet another team every week). It was all astronomically expensive. One order for high-grade copper screening, signed off by Stevens in the autumn of 1905 came to $90,000, nearly double the entire Sanitary Department budget for the previous year. Orders also went in for three thousand rubbish bins, four thousand buckets, a thousand brooms and 5000 pounds of common soap. Alongside the specific anti-mosquito measures came a massive expansion of the medical facilities. A second hospital at Colón belonging to the Railroad was taken over and the capacity of Ancón hospital was extended to 1500 beds and the staff to 470. Twenty district hospitals were opened the length of the line, along with forty smaller field hospitals, as work started on refurbishing the old French sanatorium on Taboga. In his first twelve months Gorgas had been allowed to spend $50,000. For the next year, his expenditure topped two million.

If this was throwing money at the problem, it worked. In August cases of yellow fever fell by nearly a half to twenty-seven, with nine deaths. The following month there were only seven cases and four deaths. The last death from yellow fever was reported in Panama City on 11 November 1905. As they gathered in the autopsy room, Gorgas instructed his staff to take a good look at the man: he was, said Gorgas, the last yellow fever corpse they would see.

And so it proved. It may have taken twice the eight months required in Havana, but now Panama was free of yellow fever, for the first time, and forever. It was a massive breakthrough for the American canal.

The following year, 1906, would be one of the most difficult for the project: even if the terrible fear caused by yellow fever had faded, the death rate among the workers and technicians – from malaria, pneumonia, typhoid and accidents – would actually increase to its highest for the whole US construction period. But at the end of 1905 the project's leaders could congratulate themselves on a number of achievements to add to Gorgas's yellow fever victory. Problems with the water supply for Panama City had been corrected, so that Mallet could write to his wife in September that at last 'the strong smell of decomposed fish has gone'. Sewers were in operation and much of the city had been paved. In the six months since Stevens' arrival over six hundred of the old French buildings had been repaired, twice the number Wallace managed in a year. Stevens' other priority, food supply, although not solved, was certainly being addressed. Refrigerated trucks were now carrying ice and perishables along the line to the growing number of commissary stores, hotels and messes. Several of the Panama Railroad's steamers had been fitted out with cold compartments, so that fresh food could arrive in pristine condition from the

United States. At the end of the year Stevens started plans for a cold-st plant and bakery at Cristóbal.

The white, largely American canal community was changing, and the project was beginning to lose its frontier town feel. When American journalist John Foster Carr visited the Isthmus at the beginning of 1906, six months after his first trip there, he found that 'the day of the good-for-nothing tropical tramp had nearly passed'. Certainly Stevens had set about, as he put it, 'weeding out the faint-hearted and incompetent', but for Carr the Isthmus itself had also carried out some sort of selection. 'The men themselves,' he wrote, 'have distinctive virtues as a body that are easily accounted for. Most of them are plucky, for it took pluck to come to the Isthmus and stay when yellow fever was in the land. They are mostly decent, healthy fellows, for the climate is severe on vices of over-indulgence in a northerner. Of those who will not heed the warning many are invalided home; a certain number die. Something like a real moral selection is the result.'

Six months before, Carr had found the young Americans on the project simply surviving. At the end of the day, 'they were too tired to talk. They sat about silently and went to bed at nine o'clock. On Sunday there was nowhere to go, because the jungle hemmed them in as if it were a thousand leagues of ocean.' The one day off would be spent 'on the hotel verandas, smoking, lazily watching the vultures floating high up in the air, "talking shop," and telling tales of Cuba and the Philippines, where scores of them have been'. But now some of the rough edges were disappearing. Food was plentiful and relatively cheap; screens on windows and doors meant that lights could be burned after dark; the shock of the climate and the work ahead wore off; the fear of 'yellow jack' was gone. 'The last six months have seen a great change,' wrote Carr at the beginning of 1906. 'New habits are forming and life is rapidly approaching the normal.'

Carr was something of a drum-beater for the canal project; other journalists give a more nuanced view of the emerging new community. 'Normal family life is becoming established and society is developing peculiar forms,' reported the New York *Independent* magazine in March 1906. 'In some places it resembles official life in India. At the balls married women reign supreme, with abundance of admirers and no debutante rivals . . . After the novelty . . . wears off, life . . . is barren and dull for most of the men . . . It is more from ennui than from viciousness that many of the employees seek for solace in the cocktail and the jackpot.'

Some of the Americans did take things into their own hands. Frank Maltby, still living in the old 'De Lesseps Palace' where every room was now full of young engineers, used trips back to the US to bring to Panama a pool table, card tables and a piano. He also subscribed to every paper he could think of. So was established 'Maltby's Mess', a mini-community remembered fondly by many. Others started a baseball league and bridge clubs.

But for most, particularly the young bachelors, there was little more variety of diversion than had been available for Henri Cermoise twenty years before. 'Most of the young men on the Isthmus have absolutely no places of amusement, recreation, and rendezvous except the saloons and gambling places,' complained a US diplomat. For all Carr's 'moral selection', over six hundred bars were kept busy on the Isthmus, arrests for drunkenness far surpassed any other cause, and a huge number of prostitutes made a good living in the terminal cities.

The corrupting influence on fine young upstanding Americans of life in the tropics, and of Panama in particular, was of constant concern and interest to the public at home. Apart from anything else, it made for good copy. Thus the challenge for Stevens and Magoon in finding respectable diversion for their workers became a question not just of protecting the men from their worst impulses, but also, and perhaps more importantly, of nullifying dangerous domestic criticism. In early 1905, a YMCA representative, touring the Isthmus, had written, in a widely reproduced report, that 'positive forces for evil' were 'wide open in . . . Panama and Colón'. At fault was the legal lottery in the republic, 'saloons and drinking places in large numbers . . . dispensing a most inferior and highly injurious quality of liquor', the popular sports of bullfighting and cockfighting, and prostitution, which, he said, was 'as bad as might be expected in a country of loose marriage relations, lax laws etc'. The solution to the moral crisis was identified as the creation of 'libraries and reading rooms . . . reputable places of amusement, grounds for outdoor games . . . [and] clubs for mental, moral or physical culture'.

This measure had been backed by Wallace and other senior Americans on the Isthmus but the Commission under Walker, although 'heartily approving of the plan', had not felt that the law gave them the authority to 'use money appropriated for the construction of the Canal for the amusement of the Canal employees'. The forceful Stevens had more luck and in November 1905 the ICC decision was overruled by presidential decree, and work started on a clubhouse at Cristóbal, complete with dance hall, card room, bowling alleys, gymnasium, showers and a writing room. Three more were earmarked for Empire, Gorgona and Culebra. At the same time, sports pitches, 'opportunities for wholesome open-air exercise', were planned and started.

But women were considered the key. To break down the all-male, army-camp atmosphere in which 'immoral' behaviour was considered acceptable, Stevens, Magoon and Shonts went out of their way to encourage men to bring their wives to the Isthmus. A heavily subsidised steamer fare was offered, along with superior accommodation and commissary rights. By autumn 1905 work was underway on a number of schools for Zone children, the first of which opened in January 1906.

All this municipal work occupied much of the workforce in Stevens' first six months, but there were important improvements on the engineering side as well. Deeper borings with state-of-the-art drills had at last found bedrock deep under Gatún; at Colón, the docks were overhauled and vast warehouses constructed. The backlog on the all-important railway was cleared by the end of the year. Maltby's coastal division had dug an underwater channel out to sea for large ships to use the port, and the old French canal had been dredged as far as Gatún so that supplies could be brought inland. All this, which involved the removal during 1905 of well over a million cubic yards of underwater material, was achieved by the old, inherited machinery. There would not be an American dredge at work until mid-1907.

In the Cut, a pipeline had been laid carrying compressed air to power the drills and by December 1905 there were nineteen shovels at work in the 'big ditch'. Not that 'dirt' was 'flying' – the floor of the French diggings had not been lowered by an inch. Instead, patience was still the watchword as the shovels carefully widened and prepared the site to Stevens' exact specifications. Terraces were built for further excavators to work on, as teams such as Jan van Hardeveld's laid miles of heavy track for the spoil trains to come.

All the while, the Chief Engineer took pains to remain visible and accessible. Mallet reported to his wife in September 1905 that 'Stevens lives on the line'. The Chief Engineer also put aside three hours every Sunday morning to hear complaints from the workforce and continued to tour the works, dropping in unannounced for lunch with the shovel operators or engineers. According to William Sands, 'Stevens' sturdy, competent presence gradually put new heart' into the workforce.

But whatever public image he projected, Stevens had a number of grave concerns at the end of his first six months in charge, in particular with that selfsame workforce. Stevens had asked his recruiters in the United States to send south some five thousand technicians, mainly railroad men. They had only managed to produce a little more than three thousand, and clearly some barrels had been scraped. On one occasion a consignment of eighteen track foremen reported for work only for it to be discovered that only two had any sort of track experience at all. The rest were sent back on the next steamer. 'I am not running a training school to teach boys engineering and construction,' wrote Stevens angrily to Commission chairman Theodore Shonts. 'What I want is men who can go to work when they get here.' In Washington Shonts had his own problems. For political reasons he was already trying to juggle the lucrative contracts for canal plant around various states. Now he was being steadily lobbied by politicians on behalf of their constituents wanting jobs on the Panama gravy train.

This was immediately apparent to stenographer Mary Chatfield, a formidable looking spinster in her mid-forties, who arrived at Panama City from San Francisco on 30 November 1905. Chatfield would spend sixteen months on the Isthmus, during the course of which she wrote numerous letters to her ladies' literary club in her home city of New York. One of her first letters sets the tone for her reports from Panama: 'I am not running things. If I were there would be some changes, for I never saw such a state of affairs.'

Her most immediate complaint was about the quality of the 'skilled' workforce. Starting her first job in the Hydrography Department, she soon discovered that it was almost impossible to find good recruits to man the gauging stations that measured the velocity and flow of the numerous streams in the canal's path. 'So many men sent down here drink to excess,' she reported back to her literary society. 'I am informed that the Isthmian Canal Commission send numbers of such people down here at the request of senators, congressmen and heads of departments.'

Her own boss, the Chief Clerk of the department, a Scotsman, comes in for particular criticism. 'Like many other people here in positions of authority,' she wrote, he was 'lacking in training and experience for such a position.' A young graduate engineer under the man ended up teaching him his job, 'an everyday state of affairs on the Isthmus,' wrote Chatfield. An American journalist with far less of an axe to grind backs this up, recounting the story of a well-connected clerk starting work on a salary of $2500, more than twice that of his overall supervisor. Good workmen were arriving, Chatfield writes, but 'finding they have not a fair chance against the favourites' they were leaving just as quickly.

Happily for Stevens the beginning of 1906 saw the West Indian workers arriving far faster than they were departing. Much of this was down to Karner's renewed recruitment drive in Barbados from January 1906, and the fact that the West Indians on the Isthmus could now report more regular wages and less police harassment. At the end of 1905, there had been about eighteen thousand on the payroll, compared to three and a half thousand twelve months before. Of the fifteen thousand non-whites, about half were Barbadians.

Such was the exodus of labour from Barbados that by early 1906 it was hardly possible to go ahead with the sugar crop in the St James and St Peter parishes because men 'have returned [from Panama] with money which they are spending in sight of those who did not go', who promptly took off themselves. But if Panama now drew in Barbadians in unprecedented numbers, Stevens was less than happy with the results. Although he applauded the West Indians' 'innate respect for authority', the improved food supply seemed to have had little impact on their productivity or ability to resist diseases. So he started looking elsewhere, seeing the problem, as always, as one of racial characteristics. His first preference was for Japanese or Chinese labour, but a delegation from Tokyo had toured the works in May 1905 and described conditions as 'unsatisfactory'. The Chinese were also unwilling to help, still stunned by the appalling treatment of 'coolies' by the British in the Transvaal. Furthermore, Stevens was aware that the importation of thousands of Chinese would cause political problems at home in the US, where indentured labour was frowned on by public opinion, and in Panama. One of the government's first new laws, introduced at the end of 1904, had been a measure to prevent Chinese immigration.

In Cuba the quality of the large influx of Spanish labour at the end of the war had impressed the Americans, and Stevens saw this as a possible answer to his problems. For him, they had the advantage, unlike the blacks, of 'a capacity to develop into sub-foremen . . . they are white men, tractable, and capable of development and assimilation,' he wrote to Shonts in December 1905. Certainly something had to be done. 'I have about made up my mind,' he went on, 'that it is useless to think of building the Panama Canal with native West Indian labor . . . I do not believe that the average West Indian nigger is more than equivalent to one-third of an ordinary white northern laborer . . . I regard the situation as critical, as the success or failure of our plans rests wholly upon the labor proposition.'

It was all getting on top of John Stevens. He was now working eighteen hours a day and suffering from insomnia, so frustrations about the lack of expertise in his organisation, the high rate of sickness or delays in progress became ever more overwhelming. 'No one will ever know,' he later wrote, 'no one can realize, the call on mind and body which was made upon a few for weary months while all the necessary preliminary work was being planned and carried forward . . . and the only gleams of light and encouragement were the weekly arrivals of newspapers from the States, criticizing and complaining because the dirt was not flying.'

And it was not just the lack of visible, photogenic progress that fuelled newspaper attacks on the canal. Still under scrutiny by the partisan Democratic press was Roosevelt's precise part in the 'revolution' of November 1903. Then, as men returned from Panama sacked or otherwise embittered by the canal leadership, stories starting circulating of out-of-control extravagance and corruption on the Isthmus. The oil supply business, it was alleged, had been given to the Union Oil Company in controversial circumstances. Under the direction of Cromwell, the Panama Railroad had made an undoubtedly illegal bond issue, which afterwards had to be recalled. Some high-ups on the Isthmus were receiving inflated salaries, it was suggested. Then there was the scandalous business of the Martinique prostitutes.

But it was the secrets of 'the lawyer Cromwell', still ever-present in Panama's affairs, that most interested the canal's enemies. Ever since the signing of the

Hay–Bunau-Varilla Treaty, some sections of the press had put forward the theory that the entire revolution had been a ploy by a Wall Street syndicate, a 'Stock Gambler's Plan to Make Millions!' as a New York *World* headline put it. Roosevelt's partners in the 'theft' of the canal, the *New York Times* suggested, were 'a group of canal promoters and speculators and lobbyists who came into their money through the rebellion we encouraged, made safe, and effectuated'. Soon after, the same paper reported that the president of a large French bank had said that roughly half the money paid to the New Company had stayed in the United States. Maurice Hutin, when interviewed, said that the payment had never reached French shareholders, 'as the United States naively thought'. Instead, it was suggested, a syndicate set up by Cromwell had secretly bought up shares in the New Company at rock bottom prices, and then, having persuaded the US government to pay $40 million for them, had pocketed a huge profit.

Alongside the investigation of this story, reports of extravagance on the Isthmus became more and more common in the US press, endlessly conjuring up the ghost of the famously wasteful French effort. At the end of 1905, therefore, a special Senate inquiry was authorised to carry out a full 'investigation of salaries, supplies, contracts, and the general conduct of the commission'. The Senate Committee on Interoceanic Canals, the Panama *Star and Herald* reported on 1 January 1906, was going to 'raise the canal lid'.

Three days later, on 4 January, came the publication of Poultney Bigelow's report from his Colón trip in November. Carried by the prestigious New York *Independent* magazine, and provocatively titled 'Our Mismanagement in Panama', the article caused a sensation. As well as telling of his meetings with ill, disgruntled and departing West Indian workers, Bigelow was scathing about the filthy state of Colón, faulty work on the sewers in Panama, and the shortcomings of the American workforce. 'Our Panama patriots are kept busy,' he wrote, 'finding occupation' for 'flabby young men' with 'political protection . . . who amuse themselves playing the doctor or the engineer.' He dwelt at length on the scandal of the Martinique 'prostitutes' and, in all, found 'jobbery flourishing' and the system in Panama showing 'ominous signs of rottenness'.

The response in Panama was mixed. One engineer exclaimed, 'I do not think there is a place on the face of the globe more lied about than the Isthmus of Panama. But the American people don't want to believe anything good of it, or of those who see fit to undertake the battles down here. However, we are going ahead regardless.' Others, like Mary Chatfield, found that Bigelow had echoed a lot of her own complaints. 'I have heard all those things and many more since I have been on the Isthmus,' she wrote home about his criticisms. It could have been even worse; 'he could not find out much,' she explained. 'People were afraid to tell him.'

The reaction in Washington, however, was swift and ruthless. Bigelow was hauled before the Senate Committee, but not before his report had been viciously rubbished by Taft and several of his sources uncovered and discredited (one, it emerged, had been the veteran American journalist and businessman Tracy Robinson). Bigelow had only been on the Isthmus twenty-eight hours, Taft pointed out, he hadn't left Colón and the West Indians he saw leaving in 'disgust' were simply going home for the Christmas holidays. The Committee followed this line with Bigelow, but in other ways they were less sympathetic to the canal leadership. Magoon and Stevens were summoned from Panama to be interrogated. This particularly irked the Chief Engineer who despised politicians and suffered terrible sea-sickness. Shonts was hauled in as well. Why was Colonel Gorgas, the senators asked, receiving $10,000 a year, far in excess of the salary due to his rank? The careful, diplomatic answers of Magoon alone to the barrage of questions stretch to nearly three hundred pages of published minutes.

Although no longer the Committee chairman, Alabama Senator John Tyler Morgan, now in his eighty-first year, was the driving force behind the questioning. And his main target was not so much the alleged extravagance, but the role of his arch-enemy, the man who now referred to Panama as 'my canal', William Nelson Cromwell. Determined to find out the truth of the lawyer's role in the Panama 'revolution' and the rumoured syndicate, Morgan summoned Cromwell before the Committee. But Cromwell was saying nothing, refusing to answer questions that might affect the privacy of his ex-client, the New Company. Morgan, incensed, brought a resolution before Congress forcing him to testify. The measure passed, but Cromwell was out of the country in France at the time. Soon after, Morgan died, and without his leadership the Senate Committee stuttered and then dropped the investigation. For now, the question of Who Got the Money? remained unanswered. But the story would not go away for long.

For its part, Bigelow's article would cast a long shadow over the next months on the Isthmus, dividing opinion while contributing to an air of uneasiness and crisis. In fact, while the piece contained justifiable criticism, its tone was undoubtedly slanted against the canal project. Although difficult, conditions were simply not as bad as he had made out. A British naval officer, Charles Townley, visited the Isthmus at the end of April 1906, and, considering the press coverage he had seen, was agreeably surprised by what had been achieved. 'Many of the prominent American newspapers have sent representatives to Panama to inquire into the true state of affairs there,' he reported to the British Foreign Office. 'Some of these men have been imbued with an honest desire to tell the truth, but the majority would seem to have realized that criticism of weak spots is more likely to attract readers and increase the demand for their paper than an impartial setting forth of all that has been accomplished. This carping newspaper attitude is beginning to make an impression on public opinion.'

There was one more serious problem identified by Townley, however. In January 1906 Stevens had complained to Morgan about how his efforts were being held back by the lack of a definite plan for the canal. The 'principal elements of uncertainty' in the 'project as a whole', complained of by Wallace over a year earlier, were still painfully unresolved. It was as if, Stevens explained, 'I had been told to build a house without being informed whether it was a tollhouse or a capital.' As Townley reported on 3 May 1906, 'At the present moment the hesitation of Congress to finally decide upon the type of canal to be constructed is hampering the entire labour organization on the isthmus.' Before Ferdinand de Lesseps had even been to Panama, his 1879 Congress had opted for a sea-level canal, with disastrous consequences. Two years into their canal building effort, it was now time for the Americans, in turn, to make their own 'fatal decision'.

22

THE DIGGING MACHINE

For a long while the momentum was with the proponents of an American sea-level canal. In June 1905 Roosevelt had appointed a board of consulting engineers, composed, as in the old French days, of international engineers of undisputed eminence. Of the thirteen members of the board, five were European, including William Henry Hunter, Chief Engineer of the Manchester Ship Canal; two Frenchmen, one a consulting engineer on the Suez canal, the other a senior government civil engineer; the director of all Dutch waterways and a respected Prussian state engineer. The eight American board members included General Davis, erstwhile Governor of the Canal Zone (between attacks of malaria), along with old hands from the various US canal bodies, and one engineer who had helped draft the New Company plans of 1898. The most significant of the three newcomers was Joseph Ripley, who was then working as Chief Engineer of the St Marys Falls Ship Canal, better known as the Soo Canal.

The board did not meet until September 1905, when they were entertained by Roosevelt at Oyster Bay. 'I hope that ultimately it will prove possible to build a sea-level canal,' the President told the assembled engineering grandees. 'Such a canal would undoubtedly be the best in the end if feasible; and I feel,' he added, echoing the late Mark Hanna's arguments during the 'Battle of the Routes', 'that one of the chief advantages of the Panama route is that ultimately a sea-level canal will be a possibility'. But at the same time the President demanded a canal 'in the shortest possible time'.

In the meantime, the engineering leadership on the canal could only speculate about what would be decided. There were plenty of proposals, however, to fill the official vacuum of ideas. According to Stevens, all sorts of plans were 'showered' on him during 1905: 'One genius proposed to wash the entire cut into the oceans by forcing water from a plant on Panama Bay; another to erect a big compressed air plant at Culebra to blow all the material through pipes out to sea [both technologies were seen, at the time, as the coming thing] . . . such schemes provided plenty of amusement to afford relaxation,' Stevens writes.

There were blasts, as it were, from the past as well. Soon after the Oyster Bay meeting Roosevelt received a letter from Philippe Bunau-Varilla, who, like Cromwell, had clearly been unable to let go of his Panama baby. The great Frenchman announced to the President that he had 'discovered an unknown way through this mysterious labyrinth' that was the finding of the best plan for the canal. It was a repetition of his much-cherished 'excavating in the wet' theory, whereby the canal could be operative on a locks basis while being lowered to an open, sea-level channel, the old de Lesseps dream. The sending of the letter coincided with Bunau-Varilla's usual careful attention to publicity.

Getting nothing but polite brush-offs from the President, Bunau-Varilla focused his attention on one of the newcomers among the board's experts, Isham Randolph, who had been Chief Engineer of the Chicago Sanitary and Ship Canal, completed in 1900. But Bunau-Varilla's plan came with estimates of cost and time whose optimism rated with the finest moments of the old de Lesseps propaganda sheet the *Bulletin*. Bunau-Varilla continued to pester away in his own inimitable style, but on 7 November he received back from Randolph his latest missive with the following note attached: 'Mr Randolph . . . advises M. P. Buneau Varilla [sic] that he is not seeking professional advice from him: and further that he deprecates the persistent generosity with which that advice is being urged. He returns herewith unread, the treatise which accompanied M. Varilla's note of the 6th inst.' It is difficult to know whom to feel more sorry for.

Aside from distractions, the sea-level plan seems to have made the early running. Wallace, interviewed by the board, was a firm proponent. Taft, too, had pronounced himself in favour of a sea-level canal while on the Isthmus at the end of 1904. Stevens, also, he says, had taken on the job expecting it to mean digging all the way down. On 4 October, the board of engineers arrived in Panama. It was unusually pleasant weather for the week they spent there, dry with blue skies.

Three months later, the results of the board's deliberations, an enormous report, was handed in to the Isthmian Canal Commission. The *Engineering Record* called it 'The most important document in the engineering history of the Panama Canal to date.' The experts had failed to agree unanimously, but had voted eight to five in favour of the sea-level plan. In the majority group were all the Europeans, along with ex-Governor Davis and two other Americans.

To comply with the instruction of the Spooner Act that the canal had to cope with the largest ships then afloat or being planned would need locks of such size, they argued, as to be 'beyond the limits of prudent design [and] safe and

A steamer carrying labourers from Barbados arrives at Colón. The tiny island provided the bulk of the thousands of workers for the American Panama Canal effort.

An ICC-run mess kitchen for the West Indian workers. No chairs or tables were provided and the food was often inedible. After a short time most made their own arrangements.

A fumigation squad, carrying ladders, paper and paste, assembles in Panama City, 1905

Doctor William Gorgas surveys work on the canal. He declared that his mission was to make the tropics habitable for the Caucasian.

A cemetery on the western slope of Ancón Hill photographed shortly after the completion of the canal

The energetic and resourceful
Chief Engineer John Stevens,
in boater, surveying work along the line

Emptying spoil cars by means of dragging
a metal plough along its surface. Such
ingenious devices saved countless man-hours
for the American canal effort.

Theodore Roosevelt,
in white suit, making
'a Strenuous Exhibition'
on the Isthmus

A West Indian wedding party at Culebra in 1913

A commissary store in Balboa, with its separate sections for 'Silver' and 'Gold'

Claude Mallet, with his Panamanian wife Matilde de Obarrio. Mallet's official reports and private letters give an illuminating view of Panama during the construction period.

Spanish track workers taking a break in the Culebra Cut. Antonio Sanchez is second from the right.

Steam shovels at work on the bottom of the canal

Blasting rock on Contractors' Hill, January 1912

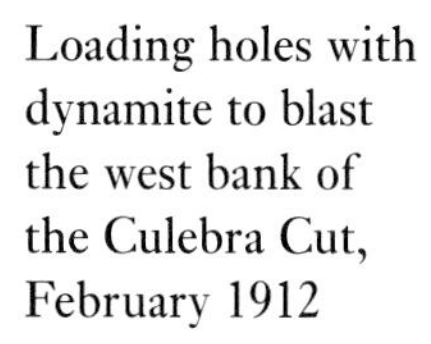

Loading holes with dynamite to blast the west bank of the Culebra Cut, February 1912

At 4.30 pm on 20 May 1913, working at the final depth of the canal, shovels No. 222 and No. 230 meet 'nose to nose' at the centre of the Cut

A slide of 300,000 cubic yards in the Culebra Cut near Empire, 21 August 1912

Gatún Locks under construction, showing the overhead cableway and the huge, rail-mounted structures holding the steel shutters

Workers at the base of the lower Gate of Gatún Locks

The final joining of the oceans being accomplished by pick-and-shovel men digging a channel through the Cucaracha slide

The opening of the canal: S.S. *Ancon* passes the remnants of the Cucaracha slide on 15 August 1914

The U.S.S. *Texas* in Gatún Locks in July 1919, a sight that would have pleased Roosevelt enormously. The military requirements of the United States were instrumental in getting the canal built.

efficient administration'. Even so, however big the locks, they were bound to become obsolete at some time in the future. A sea-level canal, on the other hand, would be 'easily expandable' in the future, and thus would 'endure for all time'.

The plan put forward is close to that of the Old French Company of the early 1880s, with a tidal lock at Ancón and a large dam at Gamboa to regulate the flow of the Chagres, along with some nineteen miles of permanent diversion channels to restrict rivers that would otherwise flow into the canal prism. Even with the numerous levees and embankments envisaged, it was accepted that most of the volume of the Chagres would still have to use the canal to get to the sea: 'The de Lesseps idea of a still water canal is thus replaced by a regulated river.' To build this canal, it was estimated, would cost $250 million and take twelve to thirteen years.

It is testament to the confidence of the European engineers in the power of US government financial backing, and in emerging technology, that they should adopt a plan that had proved such a disaster a generation before. It might seem surprising that General Davis sided with the Europeans. After all, he had actually lived on the Isthmus and thus must have had the best understanding of how many times more difficult were the conditions in Panama compared to Northern Europe and the United States. Yet, for him, it was a question of national honour and of taking on and proving that the 'impossible' was no such thing. Had not a passageway through Suez been deemed unfeasible? Yet, he went on, sounding worryingly like de Lesseps in 1879, Suez was more than twice the length of the canal needed at Panama. 'Should the United States withdraw from the attempt to make at the American isthmus a channel as open, free and safe as already existing at Suez?' he asked. 'Should they climb over the hill or remove it? . . . I think that the dignity and power of this great nation require that we should treat this matter not in a provisional but in a final, masterful way.'

Virtually every element of this plan was sharply criticised by the proponents of the minority report, submitted to the ICC at the same time. Because of the immense depth of excavation needed at the Continental Divide, even with the steep sides the sea-level plan envisaged, the waterway at the bottom of the great gorge would be only 150 feet wide at its surface. Ships would be unable to pass each other, but would have to moor, as at Suez. Some eighteen streams or rivers, it was calculated, would pour their waters into this deep and narrow chasm, creating currents of some 2.6 miles an hour as well as eroding banks and depositing silt. Even without the cross currents, the 'narrow gorge' would be 'tortuous'. For nineteen miles, a large ship would have to be continuously changing direction in channel with a width only from one-quarter to one-fifth her own length. 'Such a waterway,' wrote one of the minority report authors, 'is far from meeting the conception of free and unobstructed passage popularly associated with a sea-level canal.' The danger of landslides – with hindsight the unconquerable obstacle to a sea-level plan – was alluded to, but not stressed. The nightmare of slides in the Cut was still largely to come for the Americans.

The minority report, largely the work of Joseph Ripley and distinguished US engineer Alfred Noble (who had helped build the Weitzel Lock on the Soo Canal), was in favour of a lock canal. To satisfy the requirements of the Spooner Act, the locks would be 900 feet long and 95 feet wide, big enough to handle 'the largest ships now existing or under construction' – the *Mauritania* and the ill-fated *Lusitania*, of the Cunard Line, both over 760 feet long with a beam of 88 feet. In comparison, Eiffel's locks had been under 600 feet long and about 60 feet wide. These new locks might be bigger than anything so far attempted, but they were not, Ripley and Noble argued, 'beyond the limits of prudent design'. The example of Ripley's Soo Canal, where a huge volume of traffic between Lake Huron and Lake Superior had been handled without mishap since the 1850s, gave them confidence that such locks could provide 'safe and efficient administration'.

The main difference between this lock-canal plan and everything that went before it was not just the scale of the locks, but the location of the 'controlling feature' of the scheme, the great dam for the Chagres. This the minority reported shifted from Bohío to Gatún. It had been accepted that neither offered ideal situations for a dam, with their bedrock in places far below sea level. But Gatún had several important advantages, in spite of the fact that the dam there would need to be immense – a mile and a half long and 100 feet high, an unprecedented size. Because Gatún was downstream of Bohío, and rivers tend to deposit large and coarse material upstream and finer and denser material near the river mouth, the alluvial deposits that sat above the bedrock at Gatún would, it was hoped, be less permeable. But more important was the site of the dam. A far bigger expanse of water would be created than by blocking the river at Bohío – a new lake of some 164 square miles was envisaged stretching all the way through the Cut, drowning several villages and settlements as well as much of the existing Panama Railroad. It would be, if completed, bigger than any man-made lake before. And this additional size was the key: not only would the lake provide simple navigation for a large part of the transcontinental route and, because of its size, nullify problems of silting and currents; it would also tame forever the volatile Chagres. Unlike the previously mooted Lake Bohío, Lake Gatún would be wide enough so that the greatest floods would only raise its level a few inches, easily coped with by a spillway, whereby the Chagres would resume its route to the sea at San Lorenzo. At the same time, the proposed lake would provide, even in the dry season, enough water for the huge locks for twenty-six transits a day, or some 30–40 million tons of traffic annually. When and if this limit was reached, further control could be imposed on the water supply by the construction of a second dam upriver above Gamboa.

The ICC spent just under a month considering these two different proposals, then, on 5 February, they opted to give their backing to the minority lock-canal plan. Stevens' influence seems to have been important. Although initially in favour of a sea-level canal, by October 1905, and the consulting board's trip to Panama, he had declared himself undecided. The following month, having, as he put it, carried out a 'personal study of the conditions', he was urging the ICC not to back the 'impractical futility' of a canal *à niveau*. According to Stevens, he also talked round President Roosevelt, during a trip to Washington in January 1906. So when the matter was handed over to Congress to decide, the pro-sea-level board majority report was accompanied by the ICC's decision for locks, as well as a letter from the President backing up this decision. Taft had also changed his mind since the year before. So there was a letter from him in the package as well, in which he upped the time and money estimate of the majority report considerably, as well as warning of 'the difficulties and dangers of navigation' the sea-level plan threatened. 'We may well concede that if we could have a sea-level canal with a prism from 300 to 400 feet wide,' he wrote, 'with the curves that must now exist reduced, it would be preferable to the plan of the minority. But the time and the cost of constructing such a canal are in effect prohibitary.'

To the frustration of everyone, especially those on the Isthmus, the decision was tied up in committee for several months. Almost anyone ever connected with the canal was wheeled out to give their opinion. Then, on 17 May the committee chose, by the margin of just one vote, to reject the advice of the ICC, Roosevelt and Taft, and recommend to the Senate that they adopt the sea-level plan.

Drastic action was called for. Stevens was summoned again from Panama, to endure once more the sea-crossing and the machinations of political Washington. He had seen the Chagres in flood that month, and was more convinced than ever that the river would wreck a sea-level canal within a year. He hammered away to the House Committee on Interstate and Foreign Commerce about the problem of the Chagres, and put together a compelling speech in favour of the lock canal to be made in the Senate. This was delivered on 19 June, by Philander Knox, previously Roosevelt's Attorney General and now a senator from Pittsburgh, where, happily for his constituents, the massive steel-lock structures would most likely be built.

But it was a good speech, well delivered, and just enough to do the job. Two days later the Senate voted by 36 to 31 to back Stevens' judgement and on 27 June the House followed suit. Thus only a handful of votes determined the United States' choice between a lock canal and a sea-level attempt that, in all likelihood, would have ended in failure.

So at last the decision was made, and the aimed-for canal had a definite shape for the first time. Starting at Limón Bay, a ship would take a short sea-level passage to Gatún, where it would find three tiers of double locks. These would raise the ship to the level of the new lake – 85 feet above sea level. The vessel would cross the lake, which continued, like the spout of a funnel, through the Culebra Cut to Pedro Miguel. There, a much smaller dam would be encountered, containing a single tier of locks that would lower the boat to a small, intermediate lake at 55 feet above sea level. This would continue through to the gap between Ancón and Sosa Hills, where another small dam would hold two locks to lower the ship back to sea level and out into the Pacific. As the locks were to be in tandem, simultaneous two-way traffic would be possible. With some modification, this was the plan followed to the very end. Thus in the place of de Lesseps' dream of an 'Ocean Bosporus' the Panama Canal would be instead a 'bridge of water' between the two oceans. Instead of requiring the moving of the mountain, the waterway would go over it. And with the drowning of much of the French diggings in the Chagres valley under the new lake, millions of cubic yards of excavation, for which so many engineers and workers had suffered and died, were at a cruel stroke rendered irrelevant.

As soon as the decision was announced, the critics swung into action. The majority report of the board of consulting engineers had judged that a dam at Gatún would be a 'vast and doubtful experiment'. 'It is nothing short of monstrous to jauntily rest this national enterprise upon untried methods vastly beyond the range of experience and past success,' they had argued. Subsequent criticism remained focused on the dam, a 'simply preposterous piece of work,' wrote one expert. In 1889 a large dam at Jonestown in Pennsylvania, similar in design to that mooted at Gatún, had collapsed; an entire city had been washed away and over two thousand lives lost. There were no proper foundations at the chosen site, argued a contributor to the *North American Review*: 'To base any scheme on a work like the Gatun dam, is to build a house on sand.'

Next in the firing line were the plans for the massive locks, the safety of which one engineer, Lindon Bates, called 'the greatest engineering conflict of the canal'. The terrible danger was that a ship would ram the Gatún lock gates and thus cause the entire lake to pour out through the breach. 'Every vessel in the waiting basin and every building and structure between Gatún and the sea in its path would be swept to utter annihilation,' wrote Bates in late June 1906. 'The damage to the canal and locks could not be repaired for years. To refill Lake Gatún would consume nearly a year of itself. The adoption of the lock flight arrangement, which puts so fearful a premium upon an accident, cannot be characterized as other than a most colossal and disastrous mistake.'

Just before the time of Bates' writing, there had been a bad accident on the

Manchester Ship Canal when a ship had failed to slow down and broken through a lock. As the *Manchester Guardian* reported on 22 June: 'In the Irlam lock the water is sixteen feet below the normal level; the muddy bottom is in many places exposed; and an abominable stench fills the air.' Bates concluded that 'accidents are therefore at the Isthmus certain and inevitable', particularly under tropical conditions where, he wrote, 'the vitality of men is reduced, alertness and initiative are at their lowest'.

In fact, on the Isthmus, Stevens and his engineers had shown considerable initiative. Gambling on the lock-canal version being adopted, they had laid plans to start right away. Twenty-four hours after the decision, work began on clearing the site at Gatún and laying rails to bring spoil from Culebra to start the process of building the biggest dam ever seen.

If the metaphorical clouds cleared when the final plan for the canal was at last decided, on the Isthmus the wet season in mid-1906 was all too real. 'Heavy rain day and night,' remembered one West Indian. Every worker recalls his clothes permanently soaked. 'Hard rains had set in by this time, writes Rose van Hardeveld. 'Everything smelled of mold and decay. Water fell from the sky in great drenching sheets. The house and everything in it was sticky and wet.' Her husband, working with his friend Jantje in the Cut, seldom had a dry shirt or a pair of dry shoes. Every night, exhausted, he would come home with mud and water squishing in his shoes and his shirt and trousers wringing wet. Rose's iron cook-stove and kerosene lamps were little help in getting the clothes dry.

'Never patient, Jan was now decidedly irritable,' remembers Rose. 'His thin face grew thinner, his prominent nose larger, it seemed to me. His cheeks were gaunt and hollow. He ate very little, and I felt sure he had malaria . . . All he thought of night and day was the Canal.' Then, in mid-May 1906, Jantje 'came bounding up the steps three at a time one evening, shouting that his wife and baby were coming on the boat tomorrow morning!'. The next day, before they had even had a chance to unpack, Jantje, carrying his baby boy on his shoulder, brought the new arrivals to meet Rose, Jan and the girls. It seems to have given Rose a great lift. 'I looked upon her at once as a close friend – closer than I could ever hope to feel to any of the native women,' she writes of Jantje's 'pretty young wife' Martina. The young couple, as they moved into Las Cascadas House Number Seven, were full of plans for saving their money, and moving to the United States, 'so that their boy might grow up in America'. Above all, 'they were very happy just to be together'.

All the while, everyday life seemed to be slowly improving for the white workforce, or at least for those who remained healthy. Las Cascadas was expanding as new homes were being erected nearby. There was ice aplenty, and better food was reaching the new commissaries. In late June, Jantje made a trip to Panama City and came back with 'something wonderful'. In a wholesale importers' he had found them unpacking the first consignment of Edison phonographs to arrive on the Isthmus. He bought two and half a dozen records for each. 'We had not realized how starved we were for music and entertainment until we heard the first strains of "Silver Threads Among the Gold" floating from the big tin horn,' remembers Rose. They all sat entranced, playing the records over and over, and then Jantje – or 'Teddy' as he was called by his wife (his middle name, fittingly, was Theodore) – took his son Jack in his arms and danced around to the music of 'Hungarian Rhapsody'. 'Supper, rain, canal, everything was forgotten for the time being,' Rose remembered fondly.

Even after the decision on the canal finally came through, Stevens still had the majority of his workforce assembling and repairing buildings – quarters, clubhouses, hotels, warehouses, schools, churches or commissaries. In two years, 85 million feet of board was used on new buildings, and by June 1906 over a thousand – nearly half – of the old French quarters were in use.

For John Meehan, who had arrived back in 1904, a turning point had been reached when, in late 1905, a new hotel opened which had different sections for those wearing coats and those not. 'The rule,' he wrote, 'marked the first definite break in the community of interest that had existed up to this time among construction men, engineers, artisans, and office men.'

In other ways, too, the white community became more stratified as the facilities improved. A policy was adopted by the new Head of the Quarters and Labor Department, Jackson Smith, whereby white workers were assigned homes exactly linked to their position in the canal hierarchy – one square foot of floor space for each dollar of monthly salary. This, according to Stevens, 'proved a strong incentive to encourage individual ambition. A promotion in rank meant not only a better wage, but more commodious living accommodations, and a certain rise in the social scale. Distinctive social lines were drawn on the Ithsmus,' he goes on, 'as sharply as they are elsewhere.'

Not everyone was happy. It was widely believed that Jackson 'Square-foot' Smith, as he became known, tended still to give the best accommodation to his own friends. Mary Chatfield was living in one of the resurrected French dwellings. She complained that the large verandas let the rain in, and, sleeping up near the roof, she would be woken by the storms, which 'sounded as though some one was throwing boulders and trying to tear the boards off of the roof'.

Her fiercest criticism was for the food served by the new ICC hotels. 'The meat served is almost always beef, and such beef! It does not taste like anything,' she wrote to her literary ladies in June. 'Tho' the waters abound in fish, there is *never* any fish served . . . the vegetables are all canned and very poor quality. The soup always tasteless as hot water.' She concludes that part of the problem must be widespread pilfering. Her letters do, however, give evidence of the increasing amount of organised activity available to the US workforce. For the 4 July celebrations, she reports, there were tug of wars, obstacle races, horse and mule races, pole vaulting and dancing competitions, with first prizes of $25. There were a couple of sour notes, however. The food served 'was *worse than usual*, which was only just possible', and another incident had upset her: 'A few colored people tried to watch the games at Cristóbal and were chased off by mounted policemen. A very unpleasant sight.'

Mary Chatfield was also less than impressed with the typical American attitude to the Panamanians, whom many referred to dismissively as 'Spiggoties', from the familiar cry of Panama City pedlars and pimps: 'Speak de English?' While working in the hydrography department, Chatfield actually had a Panamanian boss, a Mr Arango, the only local to occupy a senior position in the canal setup. 'I was angry at first to find that I had been placed under a Panamanian engineer,' she writes. 'But presently discovered him to be a gentleman, and an educated man, which I hear cannot be said of many from the States.'

In the many bars and gambling dens of the terminal cities there was constant tension between locals and Americans, particularly the seemingly ever-present US military personnel. In early June 1906, an incident in Colón's red-light district led to the arrest by Panamanian police of two US Marine Corps officers and a midshipman from a gunboat in the bay. They were subsequently 'severely manhandled' by the Panamanians. Magoon blamed both sides. The US citizenry encountered by the Panamanians were largely from the South, he explained in a letter to Taft on 5 June, 'and [made] no distinction between Panamanians and negroes'. The Latin Americans, for their part, were 'liable to these quick and furious exhibitions of uncontrollable rage'.

Aside from cultural or racial friction, there were also political and economic issues that were giving the locals cause for complaint. Panamanians remained wary of American intentions, particularly towards the anarchic terminal cities, seen by Zone authorities as a threat to the increasingly orderly nature of life in the US enclave. Local merchants, who had hoped for a return of the glory days of the de Lesseps era, were furious about the rapid expansion of ICC commissaries and restaurants.

The Americans, for their part, were concerned above all with political stability and the rule of law. The volatile history of the Isthmus had been a powerful argument against Panama being selected as the canal's location. The project's backers, such as Roosevelt and Cromwell, did not need telling how damaging to the canal effort would be headlines in US newspapers about political violence in Panama.

This was all about to get much more difficult. In July there were to be two important elections in the new Republic – for the municipal councils and the National Assembly. Tension between the two opposing parties, the incumbent Conservatives and the opposition Liberals, had been growing for some months. In October 1905, Magoon, at the request of Amador, had put marines and Zone police on alert when a Liberal rally in Panama City had threatened to turn violent. The following month, when Taft was on the Isthmus, the Liberals had presented him with a 'Memorial'. In it they asked whether during the forthcoming elections US forces would be used to 'guarantee public order and constitutional succession in the Republic'. If so, did this include supervising the polling stations and ensuring a fair election?

The role that the US would play was of vital importance to the Opposition. They knew that without American intervention in the polling process, the government, who controlled all the electoral machinery, would not allow themselves to lose a national vote, however small their real popular support. This was, of course, long established: the way real political change came to Panama was through intrigue or revolution, rarely the ballot box. So if the poll was going to go ahead without US supervision, wondered the Liberals, and thus deliver an inevitable Conservative victory, how would the US react if they took the traditional step and tried to gain power through a coup?

The response from Elihu Root, now US Secretary of State, communicated to Magoon at the beginning of December, was a careful exercise in tact. The United States earnestly wished for 'fair, free and honest' elections, he said, but would not take direct control of the voting process. Root knew how that would look to other already fearful Latin American countries and he had his hands full dealing with problems in Cuba. The United States, said Root, would exercise its rights to maintain order in the terminal cities and the Zone, or 'in that territory [in] which [disorder] can be prevented by the exercise of its treaty rights, and will not go beyond those treaty rights'.

The Liberals pronounced themselves satisfied, but interpreting Root's response to mean that US troops would not intervene in Panama's rural areas (in fact, treaty rights and the Constitution permitted intervention anywhere) they started preparing for an armed uprising in the countryside, their traditional stronghold.

Two days after the receipt of the Root reply, Amador reported to Magoon that armed bands were assembling in the interior, with red ribbons around their hats, the traditional symbol of revolution, warning that without US intervention he would be forced to re-form the army, something neither man wanted. Magoon reported to Washington that 'party feeling is very bitter, and serious disorder during the elections in June and July should constitute no cause for surprise'. The Conservatives were accused by their domestic enemies of being traitors and sellouts. The official Liberal newspaper the *Diaro de Panama* described the choice for the voters as between electing the Liberal Party or seeing Panama being annexed by the United States. In reply, a senior Conservative declared that he would sooner see the nation under US control than have it fall into the hands of the 'niggers'.

On the urging of Magoon, at the end of April came the clarification of Root's reply: the United States would move to 'suppress any insurrection in any part of the Republic'. The uprising was dead in the water, and the continuance of the rule of Amador and the Conservatives was assured. As Mallet explained to his superiors in London, because it was 'customary' for government candidates to win elections, the defeat of the Opposition was a safe prediction. At the same time he passed on a report from the *Star and Herald* that 'explicit directions have been given to the police to prevent by every means in their power the success of the Liberals, who, in a fair election, would overthrow the Amador government by one hundred to one'.

Thus by refusing to supervise the elections and at the same time banning revolution from the Isthmus, the United States' actions were decisive in maintaining in power an unrepresentative, undemocratic government. Mallet puts this down to American dislike for the Liberals' racially mixed constituency. Certainly, for now, the US felt more at ease with the Conservative faction. Taft had reported to Roosevelt after his trip to the Isthmus that the Liberals were much less trustworthy and that if they came to power they would bring an injection of unwanted 'Negro influence' into Panamanian politics.

Amador's party, as predicted, won both elections with ease, but not before requesting, and receiving, a cache of arms from the Americans, as well as the deployment of three hundred marines just outside Panama City. During the municipal voting on 24 June there was sporadic violence leading to four dead and over twenty injured. Widespread fraud was evident: thousands of Liberals arrived to vote only to find that their names had disappeared from the list. William Sands, the US chargé d'affaires, reported, 'The police [who owed their jobs to the ruling government] voted the first time in uniform and the second time in civilian clothing, returning again to the polls with their rifles "to preserve order".' There was a week until the National Assembly elections, during which Magoon hauled the party leaders before him and appealed for calm. With the marine presence and US gunboats in the harbours at both ends of the line, the election went off peacefully with the result never in doubt. Therefore, as in the defeat of the Huertas coup plot, Amador was in power thanks to American support, and the

United States found itself by the end of 1906, as much through events as by design, in almost complete control of Panamanian affairs.

As the painstaking work to prepare the 'canal-digging machine' continued, the leadership of the project sought to take control in other ways as well, to create the best, the most efficient environment for the enormous task undertaken. Stevens' argument with West Indian labour was not just that they were slow and incompetent but also that they were part of a 'prevailing clannishness' that needed to be broken up. In fact, the blacks workers – usually lumped together as 'Jamaicans' – were far from being the homogenous body that Stevens feared, consisting instead of a great variety of nationalities and cultures. And even among the British West Indians, as opposed to those from French-, Dutch- or Spanish-speaking areas, there was little fellow feeling among the nationals from the various islands; instead there was competitiveness and distrust. One of the complaints of the 'Jamaican Carpenter' about the ICC barracks was the mixing they had to put up with: 'There is no sense in putting so many different races together – Jamaicans and Bims [Barbadians] and Martiniques in the same room. It is not right.'

But to the American leadership they were all just a collective black mass, and one, furthermore, that felt itself indispensable to the canal effort. According to Stevens, 'some sort of hazy idea had gotten into their heads' that they 'controlled the labor market'. To put them in their place (as well as to find better-working men) Stevens decided to carry through his idea, mooted at the end of 1905, to bring in 'laborers of other races and different characteristics'. In February 1906, nearly three hundred Galicians and other Spaniards were shipped to Panama from Cuba, where they had been working on railway construction. Thus they brought the track-laying skills that were vital for Stevens' transport revolution on the Isthmus. The Chief Engineer monitored the new men carefully, deciding that 'one of them will do and is doing, as much work as three of our West Indian negro laborers'. So, although Karner was to keep his work going in Barbados as well as in other islands, in mid-1906 the ICC set up recruiting agencies in Madrid and Rome and started importing European labourers. Spaniards were the first preference, but Stevens had decided that he wanted at least 'three separate nationalities of laborers . . . so that none of them will get the idea that they are the sole source of supply on earth'.

Over the next two years, some 12,000 Europeans were brought in on ICC contracts, 8200 from Spain, 2000 from Italy (largely from impoverished Sicily and Sardinia) and 1100 from Greece, where another agency was set up in 1907. Typically, the men were contracted for three-year tours. Unlike the West Indians, the Europeans were expected to pay their own passage. The fare – a whopping

$45 – was deducted from their pay and there was no guaranteed repatriation. However, they were offered twenty cents an hour, as opposed to the ten-cent rate of the Karner contract. The reasoning went that it was worth paying twice as much for workers three times more productive.

For some in Madrid it was an illustration of how far the nation's fortunes had fallen that Spanish men were to return as lowly workers to a country once the crossroads of their great empire. The indignity was almost too much to bear: 'If America needed common laborers, let her seek it among her own people,' wrote one national newspaper. 'The American is too proud to work with his hands! He must work with his head, and Spain must be her hands! Spain refuses to be the hands of an American head!'

Nonetheless, there were plenty of takers for the chance of leaving Spain behind. The country had seen some of the worst anarchist violence in Europe, and, with industry and agriculture depressed, there was widespread unemployment and hardship, made worse by a string of influenza epidemics. There had been large-scale recent emigration from Spain to Cuba and elsewhere, and now, the stories went, the best money to be made was in Panama. Antonio Sanchez was different from the typical emigrant in that at thirty-five he was older than most, and had once been reasonably well off. He still owned a fruit and olive farm at Valero de la Sierra, in the province of Salamanca, but prices had fallen too low to make the business viable. When disease and famine carried away his wife and two daughters, Sanchez decided to leave his surviving nine-year-old son behind with grandparents and take his chances in Panama, where, he had heard, 'everything was gold and all things were as sweet as honey'. 'Everybody in his area was so scared of disease,' Sanchez's stepson explains. 'His farm was worthless; he just had to try his luck somewhere else. He had to leave.' With about a dozen friends from Salamanca, Sanchez sailed from the port of Vigo; he would never return to Spain or see his son again.

The first shipments from Spain seem to have contained a mixed bunch. According to an Italian doctor sent by his government to investigate conditions on the Isthmus, not all on board in the early days were 'gentlemen'. No checks were made on the applicants, 'and those of any colour or background were accepted. This explains why the first crossings . . . were characterised by continuous agitation and subordination, some of which had to be put down by force.' Among the Spaniards there seem to have been a number of men recently released from prison after a Royal Amnesty, as well as politically active anarchists. Certainly a Spanish anarchist newspaper was up and running in Panama before the end of the year. It may not have been quite what Stevens had envisaged.

Antonio Sanchez's first impression of Panama was that he had exchanged one site of '*peste*' for another. 'It was not a livable place,' he explained to his stepson years later. But the Americans, determined to avoid the problems they had suffered with the West Indians, had pulled out all the stops for the new arrivals. Relatively comfortable barracks had been built of similar size to the West Indians' but housing only twenty-five people each, rather than seventy-two. Castilian Spanish were carefully kept separate from Galicians as, Sanchez explained, 'they hated each other'. Special kitchens were constructed for them. Unlike the West Indians, the Europeans got chairs and tables. Most importantly, every effort had been made to provide familiar food such as potatoes and spicy Spanish sausage. They were even given wine at lunch in the European manner.

For now, all this looked justified, as the Americans assessed their new workers. 'Not only are they more than twice as efficient as the negroes, but they cope better with the climate,' gushed the 1906 ICC annual report. 'The Spaniard is certainly the more intelligent and better worker,' wrote a visiting journalist. 'We knew we had with us genuine workers,' remembers Rose van Hardeveld, whose husband's frustrating West Indians had been replaced by Spaniards. 'Stocky in build, with strong shoulders, they walked . . . with a purposeful stride . . . Jan was jubilant with the gangs . . . He had men he could understand and teach and like.'

Furthermore, the influx seemed to have fulfilled its other purpose as well: 'It did exactly what was expected in changing the self-confidence of the negroes,' Stevens later wrote. 'From an amusing but embarrassing attitude of self-complacency, they soon exhibited the aspect of men who were afraid of losing their jobs, and their value increased accordingly.'

According to Antonio Sanchez there was mutual respect and affection between the Americans and the Europeans during the construction period. Relations with the blacks, however, were strained from the outset. 'The Europeans hated them,' Sanchez remembered. It was partly the language problem: 'Every time one of them said something the other would take it as an insult, and vice versa. There were a lot of fights. With fists, shovels . . .' Stevens' plan to divide and rule the workforce seems to have succeeded.

In May 1906, an American journalist, told of the decision to recruit Europeans based on the success of the first shipment from Cuba in early February, went to investigate these paragons of efficiency himself. Assigned to track work in the Cut, the Spaniards had been quartered nearby in unscreened barracks close to marshland. 'Toward the end of the first fortnight, they began to fall ill,' the journalist discovered. After four weeks, 165 of the 270 had been hospitalised, over 60 per cent, 'practically all with malaria'.

During the headline-grabbing yellow fever epidemic of May to August 1905, 48 people had died of the disease. But during the same period twice as many had died of malaria, 49 from pneumonia, 57 from chronic diarrhoea and 46 from dysentery. The mortality rate for the year, not including accidents, was 24.3 per thousand. In 1906, this would jump to 39.29, the highest level of the US construction era. This is nothing like the 70-per-thousand rate suffered by the French during their *annus horribilis* of 1885, but it is still higher than anything under the New Company in the 1890s.

The dry start to the year gave no indication of what was to come. But with the onset of the rains in mid-May, and the transformation once again of the Isthmus into, as Mary Chatfield wrote to a friend, 'driving rain and muddy, muddy, much muddy mud', both malaria and pneumonia struck hard. In June, of the three hundred marines deployed near Panama City for the Panamanian election, more than half came down with malaria. By the end of the month, Ancón hospital was admitting seventy-five people a day with the disease. 'This rainy season has been a heavy trial on the canal builders, the railroad and the sanitarians,' read a despatch from Panama to the New York *Daily Herald*. 'There has been a riot of malaria, all departments being hampered by having so many men in the hospital.' In July, the black workers suddenly started dying from pneumonia at a rate of eighty a month. By November, there had been nearly four hundred fatalities from the disease, along with two hundred from malaria.

But the number of deaths from malaria does not tell the whole story. Although debilitating, the disease was rarely fatal, at least on its first attack, but in 1906 the cases that came to the attention of the medical system numbered nearly 22,000. Sanitary officer Joseph Le Prince estimated that an astonishing 80 per cent of the overall workforce was hospitalised at some point during the one year for malaria alone. The fallout rate of the Spanish pioneers from Cuba was not so bad after all.

This sort of rate of attrition meant that life in the field, out on the mosquito-ridden works, was a desperate, bewildering struggle. 'You turn up to work in the morning with a gang about 125 men and by Eleven clock you will find about 40 men all the others fall down with malaria,' remembered West Indian Rufus Forde. 'They spin all around like a top before they fall and that get you so frighten that at some times you don't come back after dinner.' Benjamin Jordan, the Barbadian who had lied about his age to get selected for an ICC contract, contracted malaria within weeks of arriving on the Isthmus. 'I can't describe them,' he says of the mosquitoes. 'I hear "woo" and they are into you.' Malaria, he says, 'took me at night . . . in the morning when I woke I couldn't get out of bed. But I did manage it, I got out, and my neighbour advised me to go to hospital. When I was discharged was deaf as bat . . . Malaria and the mosquito brother were top.'

A number of the West Indian accounts are full of praise for the hospital care they received once they had, almost inevitably, come down with one of the prevalent diseases. Jamaican James Williams, in his early teens, worked in a kitchen at San Pablo, on the banks of the Chagres River, 'where mosquitoes were frequent, especially at nights. Consequently I began to get fever.' The following day a doctor was visiting and someone told him that Williams was ill. 'The doctor immediately advanced to me and felt my pulse. I could remember he said to me "You are going to be sick, boy, go up to the hospital right a way." He further asked me, "Are you a God fearing man?" I replied, "Yes,"' recalls Williams. 'He said to me "You are going to die."'

Williams was put on a train to Ancón hospital, where, fearing malaria, he was given two-hourly doses of quinine and an ice bed bath. He had never been in hospital before, and it was a 'fretting' and alarming experience. The next day, parched with thirst, he drank a bowl of water left out near his bed. This turned out to be poison to kill flies and mosquitoes and brought on severe vomiting in young Williams. That night his blood was tested and he was shortly afterwards moved to the typhoid fever ward, where he made a slow but steady recovery. He remembers the staff who cared for him with great affection: 'I can truthfully say those American nurses – my own dear mother could not be more kind and tender to me.' After a couple of weeks, Williams was eating again and being given 'eggnog twice a day also real American Whiskey every day'.

But other accounts tell, as in the bad old French days, of men released from hospital before fully well or able to return to work – and thus qualify for further free hospital care. Other men were sacked when they started to look ill. St Lucian Charles Thomas worked at the iron foundry at La Boca: 'I was fired after two days,' he says. 'I remember the foreman call to me & said to me you are fired, you are looking tired. I was not exactly tired but I was feeling quite sick & just trying to make a week so I could get a commissary book for $2.50 to get something to eat and drink.' As the West Indians, unlike the white Americans, had no paid sick leave, some, unwilling or unable to forfeit their wages, would work on to the point of dropping. One account, redolent of the worst horror stories of the railroad or French era, explains how men suffering the dysentery which often hit those weakened by malaria would sometimes just disappear, never to make it to hospital (or on to the official casualty figures). Says Barbadian Clifford Hunt: 'Men in my gang, tell the Boss I am going out to ease my bowels and they die in the bush and nobody look for you.'

Pneumonia may have been the biggest single killer in 1906, but malaria, it was judged, offered the greatest threat to the success and efficiency of the project. Pneumonia was almost unknown among the white workforce, and dead black workers were easily replaced, such was the glut of labour in the islands. Malaria, on the other hand, also affected the whites, and, because rarely fatal, usually resulted in expensive hospital treatment for the black worker. Virtually nothing would be done to counter pneumonia, while the campaign against malaria was on an undreamt-of scale, far surpassing that against yellow fever.

'The question of controlling malaria appeared at first sight to be utterly hopeless,' wrote Joseph Le Prince. Part of the reason why malaria was a greater challenge than yellow fever was in the different nature of the two diseases. Those unlucky enough to contract yellow fever either survived and were free of the virus and immune forever or were dead. Either way, they were no longer a source for the infection. If someone caught malaria, on the other hand, they were far more likely to live, but the disease seldom went away for good. Usually the patient would remain both a recurrent sufferer and, for about three years, an ongoing source for the continued propagation of the bacterial parasite. Gorgas's very earliest tests in 1904 had shown that some 70 per cent of Panamanians carried the infection in some form. So the approach of keeping the mosquito away from the disease, successfully followed in the yellow fever campaign, was a non-starter.

The only point of attack had to be the *Anopheles* mosquito itself. To an extent, the species was the same everywhere, and as such, it was accepted, was going to be a much more formidable enemy than its yellow fever-carrying cousin, the fastidious, house-dwelling *Aëdes aegypti*. The *Anopheles* was, in contrast, omnipresent in the deepest bush as well as the backyard. For Gorgas eradicating the *Aëdes aegypti* was 'making war on the family cat', while a campaign against the malaria-carrying *Anopheles* was 'like fighting all the beasts of the jungle'.

July 1904 had seen Joseph Le Prince, one day off the boat from New York, poking about in hoofprints below Ancón Hill looking for *Anopheles* larvae, and thereafter, although the yellow fever mosquito enemy had first priority, investigations continued into the 'Isthmian *Anopheles*'. The researchers started by determining the local species most responsible for malaria transmission. Thousands of mosquitoes were captured and analysed, and their behaviour studied. Tests included getting human volunteers to sit in a mosquito-filled net. As Le Prince explains, 'Very patient negroes were necessary . . . Conditions soon became unbearable even to those persons who were accustomed to be bitten frequently.'

By the end of the first year, it had been established that the insect most responsible for malaria on the Isthmus was the 'white-footed' *Anopheles albimanus*. Unfortunately this was not only the most abundant, but also the species most determined to enter inhabited buildings. One of its tricks was to cling to dark clothing and thus gain entry to houses even if they were screened.

Thousands of eggs were collected, hatched and observed at every stage. Adult specimens were dyed using an atomiser so that tests could be conducted on their flying distances and habits. When it was established that the mosquito could not fly far without alighting on some sort of vegetation, work started on clearing 200-yard-wide areas around where people lived and worked. Tests showed that *Anopheles* preferred to rest on a dark surface on the leeward side of buildings, so black bands 2 foot wide were painted on sheltered walls at a convenient height for mosquito catchers to collect them. When it was noticed that certain species of spiders and lizards started congregating there to feed, these were bred in great numbers and released to wage war on the enemy.

Analysis of the larval stage showed that, disappointingly, it was far hardier than that of *Aëdes aegypti*, able to survive in water only a fraction of an inch deep, or even in mud once the puddle had dried. It had no particular preference for clean or dirty water and would still be alive after up to two hours under a film of oil. Nevertheless, the larval stage was still the mosquito's most vulnerable time, so the challenge was to deal with the breeding grounds.

It was a massive, almost hopeless, task. During the wet season, when at Culebra, for example, it averaged twenty-four rainy days a month, there was simply water everywhere. But even during the short dry season, there were swamps, springs or seepage outcrops near every settlement in the Zone. Fast-growing vegetation clogged streams, protecting the larvae from its natural predators and providing still pools for egg laying. The ongoing engineering work made it difficult, too. Badly placed spoil dumps blocked natural drainage, and excavations constantly filled with water. Every time a railroad tie was moved, it left an indent in the ground that could collect water and therefore mosquito larvae.

Thus Gorgas was never going to defeat malaria in the way he had yellow fever. But he believed he could control it by reducing the *Anopheles* population of the Zone. Swamps were drained using hundreds of miles of ditches, or filled with spoil from the works. Elsewhere, further natural predators were encouraged or introduced, including a top-feeding minnow from Barbados. According to Le Prince, 'larvae of dragon flies and water beetles were of great value'. But above all, vast quantities of poison and oil were deployed across the Isthmus. A special plant was built at Ancón to manufacture a larvicide consisting of carbolic acid, resin and caustic soda. Some two hundred barrels were applied monthly around the edges of pools and streams. Vegetation that clogged up running water was cleared by burning or with phenol or copper sulphate. To smother the 'wrigglers', crude oil, mixed with kerosene to increase its spreading qualities, was sprayed everywhere. At its peak in early 1907, the campaign was getting through sixty-five thousand gallons of crude in a month. Unsurprisingly, visitors to the Isthmus started commenting on the pervasive smell of petroleum.

Results would come, but for the men in the field, particularly those near the jungle, work at the end of 1906 meant swarms of mosquitoes. Some men took to rubbing exposed parts of their bodies with a mixture of kerosene and coconut oil, but they still got bitten and they still got malaria. The only treatment was quinine, either in a pill – 'the size of a quarter and twice as thick' – or as a sickeningly bitter liquid. Mallet reckoned that quinine was 'the cause of many break downs in the constitution, it ruins the stomach and digestive organs'. John Prescod, who arrived from Barbados in June 1906, described another nasty side-effect: 'Malaria fever have me so bad I has to drink plenty of quine tonic tell I heard singing in my ears murder murder going to quits drinking quine was getting me deaf.'

'The prevailing illness is malaria,' wrote Mary Chatfield in a letter home dated 30 June 1906. 'Many and many are the corpses I see carried past the office . . . the majority of the victims of malaria are the negro laborers.' 'I went to the Cristóbal dispensary this morning to get some tonic,' she wrote a month later. 'It was a pitiful sight to see the sick coloured laborers. Many of them were so weak they could not sit up while their medicine was being prepared, but lay on the benches and the floor.'

Albert Peters, who reached the Isthmus in August that year, 'eager for some adventure and experience', caught malaria within a week. He survived, but there was a daily reminder of those who did not. 'Every evening around 4:30,' he wrote, 'one could see No. 5 engine with a box car and the rough brown coffins stacked one upon the other bound for Mt. Hope [cemetery] which was called Monkey Hill in those days. The death rate was high . . . If you had a friend that you always see and missed him for a week or two, don't wonder, he's either in the hospital or at Monkey Hill resting in peace.'

'That's the reason we all used to go to Church more regular than today,' says Barbadian Amos Parks, 'because in those days, you see today and tomorrow you are a dead man. You had to pray everyday for God to carry you safe, and bring you back.'

Rose van Hardeveld had, from the outset, found, 'the horrible and unfamiliar noise at night' in Panama more nerve-racking than any other 'trials and tribulations'. As well as the strange, unearthly sounds made by alligators, bats, night birds or insects, 'the very worst of all was the wailing for the dead that came from the labor camp below us'. 'When one of their number died,' she continues, 'the friends and kindred of the deceased would gather in the room where the corpse lay. All night long they would drink rum and wail and sing Old English Gospel hymns in the flattest, most unmusical way imaginable . . . These tones would sway and swing in the air like the dance of witches.' It would leave her sleepless and 'utterly unnerved and filled with a vague, mounting dread'.

By July 1906, as malaria and pneumonia hit hard, 'The wailing and singing at the labor camp went on so often that there was hardly a night when the camp was silent . . . Slowly but surely my natural fortitude was giving way, and I was becoming a nervous, fearful woman. I believe it was the consciousness of what would happen to the children that kept me from going to pieces.'

But then their youngest daughter, Sister, fell seriously ill with a combination of malaria and dysentery. She became, Rose writers, 'a limp, feverish little bundle, crying night and day'. She was told to give her quinine, but the young girl could not keep it down. 'All the time I was becoming lower in spirits and less able to cope,' Rose remembered. 'The thought of putting my baby in a strange hospital was the last straw. That night I gave way to old-fashioned screaming hysterics, outside beside the roaring cataract. Poor little Janey clung to me, her frightened eyes searching mine for the cause of such carryings on! After that when I sat through the long nights, comforting my whimpering child in my arms, the howling and moaning from the labor camp no long grated so shatteringly on my nerves. I knew what it was to seek relief in wailing. Though for me, such yielding to hysterics was a matter for private shame, never to be regarded as an accepted social custom, I could concede to the black people whatever gratification they might find that way.'

Sister recovered and Rose found that she could now sleep through the nightly din from the labour camp. Martina Milliery's improved English meant that she was a real support and help for her friend. Rose's spirits were also lifted when she started helping out her husband by writing out Sunday passes for his team of Spanish workers. 'With this little job to do for my husband, for the Canal Commission, for President Teddy Roosevelt, and for my country, I was in my glory,' she writes. 'I sometime had difficulty writing those strange Spanish names, but still I liked doing it for these black-eyed and very deferential men.'

Most the new Spanish arrivals, like van Hardeveld and Jantje Milliery's gang, had been put to work on track construction and repair. The basic work on the Panama Railroad main line had been slow-going. It was not entirely double-tracked until well into 1907, but a maze of sidings, branch and service lines had been constructed – some 350 miles by June 1906. Most of this track, however, was in the Cut. Here, readying completion by late that year, in spite of all the prevailing sickness, was Stevens' great digging machine, perhaps his era's most important contribution to the American canal.

The lock-canal plan adopted that summer would still require, it was originally (under)estimated, the digging out and removal of more than 50 million cubic yards of rock and soil. For this to be achieved as quickly and cheaply as possible would require, first, the greatest possible number of steam shovels in operation in the space available. Thus the excavation was planned to proceed along a series of horizontal benches, or terraces across the valley in the making, each wide enough to carry two parallel railtracks. Therefore, in places, up to seven shovels could work on the same hillside almost stacked on top of each other up the terraced slope. Next, it was crucial that the shovels be working at maximum efficiency. Under normal conditions, tests had shown, it took about a minute and a half (seven bucketloads) for a shovel to fill a single spoil car; about forty-five minutes for an entire train. In the case of the smaller French trains and cars, it was much less. Thus, Stevens calculated, if it were never to be idle, each shovel required the service of a virtual conveyor belt of three to five entire spoil trains so that there would be at least one in attendance at all times. To run this sort of traffic in the narrow confines of the Cut required an enormous and highly intricate track system the like of which probably no one else in the United States had the expertise to design and build.

But Stevens went further. To put it simply, by starting the work at the two ends of the nine-mile Cut and working inwards towards its highest point, the site could, in the main, be organised so that there was a small but significant upward gradient on the terraces. This meant that empty spoil trains would be climbing up to their shovels, but then, when fully loaded, had a downhill journey to the dump sites. The scheme had the added advantage that water in the ditch, a constant annoyance for the French, naturally flowed away to both ends where it was easily disposed of with giant pumps. If Stevens' track system was fantastically skilful and intricate, like the assembly line in one of the new mechanised US factories, the use of the gradient – whereby nature was made a helper rather than an enemy – was engineering at its simplest and most brilliant.

A surprising number of the 'moving' parts of this system were still from the French era. At the end of 1906, over half the locomotives in the Cut were old Belgian machines, able, in dry weather, to pull about thirteen small 9-cubic-yard cars. But as fast as the new American models arrived they replaced the old plant. The US locomotives could haul four or five times the volume of dirt. In the same way, the new American spoil cars were also on a different scale, of a different age, immensely strong and able to hold at least three times more weight than the old French models. On one occasion a single rock weighing some 34 tons was loaded on to a single one of these new cars without mishap.

There was another important innovation: instead of individual cars self-tipping, when wet soil or clay stuck and had to be removed by shovel, the American cars were one sided only and linked together with panels, making a single long surface, like a giant conveyor belt. This not only meant a greater area was available, but also brought into play at the dump sites an ingenious invention. At the end of each of the new trains was a wagon holding what looked like a giant, one-sided snowplough, linked to the locomotive at the other end by a thick chain. When the spoil arrived at the dumping site – whether it be a marsh-fill, a dam or a causeway – the open side of the cars faced where the soil was required and the plough, its blade at an angle of about forty-five degrees, was pulled from one end of the 'belt' to the other, scraping the mud and rocks over the side. The empty spoil car departed, followed along the dumping site track by another specially adapted locomotive with arm-like blades at ground level. These flattened the soil, making room for the next load. When a new and firm terrace had been thus created, the rails were simply moved across to the edge and the process repeated. The contrast with the French period, when much of the spoil had to be unloaded by hand, is sharp. The saving in man-hours was immense.

The most labour-intensive aspect of the process was now moving track, either at the dump sites or in the Cut, where teams would have to update the intricate system as the site constantly changed shape and dimensions. Then, at the end of 1906, the general manager of the Railroad, W. G. Bierd, came up with an ingenious invention, a swinging boom mounted on a flat car that lifted extant track and moved it nearby without the need for disassembly. It was a slow process, but not nearly as slow as doing it by hand. Like the other innovations, it was just the sort of miracle machine that de Lesseps had hoped in vain would come to the rescue of his own canal effort.

As well as heavier, stronger and cleverer plant, the Railroad Era American canal builders had more useful experience than their French forebears. The railway boom in the US had provided an invaluable training ground for hundreds of American engineers, whose expertise was far in excess of anything available to the de Lesseps effort. It was not just Stevens and Bierd, but a host of switchmen, signalmen, locomotive drivers and mechanics, electrical engineers and railroad foremen. If transportation, the railway, was the key to building the canal, as Stevens had decided, then he had a depth of talent to call on.

The Americans also had the luxury of time. Stevens may have had a tight-fisted Congress and, in parts, suspicious domestic press to contend with, but that was nothing compared to the pressure on a private company watching the Bourse every day and with its life in the hands of volatile 'confidence' and the 'folly and gullibility of Capital'. It helped that Stevens was not easily thrown by advice or instructions from above. But, crucially, largely freed of direct money-raising concerns, he had the freedom to do what was right for the engineering of the canal rather than for its public relations. So instead of having to feed the *Bulletin* monthly excavation figures, he was able to concentrate on the painstaking, unglamorous preparatory work, without which the canal project would not have succeeded.

The long delay between the stopping of 'making the dirt fly' in August 1905 and the resumption of excavation in earnest at the beginning of 1907 turned out to be time very well spent. In spite of the terrible rates of sickness among the workforce, the digging machine was now ready, and excavation records were about to be shattered.

The massively increased traffic on the railroad, serving not just the engineering part of the project but the houses, shops and restaurants of the employees as well, did have its downside. There were virtually no roads on the Isthmus, and often the only way to get somewhere was to walk along the tracks. As the line got busier, so it became more dangerous, and railroad accidents started becoming an almost daily fact of life. Often, amid the shouts, blasts and din of the works, and partially deafened by the side-effects of quinine, men simply did not hear the danger in time. In mid-August, after 'a very pleasant day in Gorgona', the train carrying Mary Chatfield back to Panama City 'ran over a colored man, cutting off one leg far above his knee, and I think, killing him – I hope so – he was so mutilated. A fearful sight,' she wrote home. 'The school teacher at Cristóbal and a nurse from Ancón were in the car with me. The nurse went right out to see what she could do, but I sat still and shuddered.'

Although West Indians were most at risk, especially soon after they arrived, due to their lack of familiarity with locomotives and track, the danger affected everyone, particularly in the Cut, where a bewildering network of tracks, in constant use, now covered almost every flat surface. On 17 September, Jantje 'Teddy' Milliery had lunch at home as usual, before returning to his work site just below their House Number Seven. He stepped over the first two tracks across his route, but then, just as he reached a third track, he turned to wave to his wife and baby son watching him from the doorway of their house. At that moment he was hit by ICC Locomotive No. 215, an empty spoil train that was reversing without the customary lookout in place on the end car. According to his official file, his 'pelvis and both lower extremities [were] completely crushed'. 'I saw the accident and reached Jantje before he died,' Jan would tell Rose. 'I tried to tell him we would look after his wife and baby. I hope he understood me. He had such an awful time dying . . .'

Rose had been on a trip to Panama City with the girls buying, on Jantje's instruction, a new record for the phonograph. She got back to Las Cascadas at two o'clock. Jan was at the station with their West Indian maid-cum-childminder, Miriam. 'Send the children home with her and come with me,' Jan told his wife, and the two of them made their way to the hospital. In a small room on the upper floor sat a black woman holding Jantje's wailing baby, Jack. Martina was there in a hysterical state, begging to be allowed to see her husband, to kiss him one last time. To spare her, this was not allowed: a final note in his file reads, 'Remarks: Body too badly mangled to embalm.' Turning to Rose and Jan, she cried out, 'Oh, give me my Teddy, give me my Teddy.' 'Sorrowfully, Jan answered her in her own language, in words of one syllable. "Child, I can't." In her misery, she did not even seem to hear.'

During the weeks that followed, Rose spent her days with Martina, 'helping her in any way I could to bear up under her grief and start planning her life anew'. But her nights she spent 'pacing the floors of our little house atop the hill, wringing my hands and trying desperately, futilely, to unknot my nerves.' Martina was now no longer entitled to live in House Number Seven, and had to earn a living. For a couple of weeks she tried setting up a laundry, but then decided to return to Holland. 'I went with her to the Dutch Consul in Panama,' writes Rose, 'to arrange for passport and passage. After a last sad pilgrimage to the damp grave of her husband, she went back across the ocean, a mournful figure in black.'*

With Martina no longer relying on her, Rose began to break down again. She talked with Jan about quitting Panama. He had been badly shaken, too, but responded by 'hurling his energies with renewed determination into the job at hand'. Rose, however, found herself, she writes, 'drifting closer and closer to the yawning chasm of panic into which I had fallen once before, during the height of Sister's bout with the fever. And finally, much to my own disgust, I was put to bed with another spell of hysteria. The children haunted the bedside like frightened little shadows. I realized that I must pull myself together.'

Then there was a much-needed and very welcome boost: 'news came to us of the expected arrival, soon, of a visitor who – Jan triumphantly told us – had the welfare of all of us at heart: Theodore Roosevelt.' It appeared that the man for whom 'anything was possible' was coming to see his canal.

* Martina Milliery, née Korver, remarried and had another child in about 1912. This son emigrated to South Africa in 1930 but the rest of the family moved to Dutch-controlled Indonesia. During the Second World War they were interned by the Japanese. Jantje's son Jack died in the camp in 1945, shortly after his stepfather. Martina survived, but was blind from malnutrition. She returned to Holland and died in 1958, impoverished, in a home for the blind.

23

SEGREGATION

Roosevelt's visit to the Isthmus can only really be compared with the two much-celebrated visits of 'the Presiding Genius of the Nineteenth Century', Ferdinand de Lesseps. The authorities there had heard about the impending presidential descent back in July, and from that moment on thousands of workers had beavered away preparing the Isthmus for Roosevelt's November inspection. According to Mallet, Panama had never before been so thoroughly scrubbed, swept and cleaned. A wing of the new hotel, the Tivoli, was rapidly completed to house the honoured guest, and a new railway station built nearby. An elaborate schedule was prepared, replete with ceremonies, speeches and dinners.

Roosevelt sailed on 9 November on board the 16,000-ton *Louisiana*, the largest battleship of the now rapidly growing US fleet, with two cruisers in attendance. It was an unprecedented moment. Never before had a serving US President left the country. Of course, there were some on the Isthmus cynical about the 'momentous visit'. Mary Chatfield had written home in September: 'There is much talk about the anticipated visit of the president. All agree that if he wants to find out how things are he will have to come in disguise.'

To be fair to Roosevelt, from the moment of his arrival he went to great pains to throw the canal leadership on the Isthmus off balance, to dig beneath the prepared façade. Frank Maltby wrote that Roosevelt 'seemed obsessed with the idea that someone was trying to hide something from him'. For his landing on 15 November, schoolchildren had been lined up to sing 'The Star-Spangled Banner' and a cannon procured for an official salute. But the President arrived on shore an hour early, much to the consternation of the official greeting party, which included the President of Panama, who were still at breakfast in the Washington Hotel.

On the first of his three days in Panama, Roosevelt excused himself after lunch at the new Tivoli Hotel, saying he was retiring to his room. 'Instead,' writes Maltby, 'he bolted out the back door, rushed up the hill to Ancón Hospital and into the wards, where he began talking to the patients as to their treatment and care.' Thereafter, Roosevelt continually evaded his official schedule to drop in on messrooms and kitchens, to interview passing workers, or to ferret around in ICC barracks and lodgings. Canal officials were thrown into confusion. Mary Chatfield reports that 'When the president was at Cristóbal they were in a panic at the Cristóbal Hotel, hurrying off the filthy table clothes and replacing them with clean ones, fearful he might come bounding in.'

Roosevelt, in contrast to de Lesseps, deliberately came to Panama at the height of the rainy season. He wanted to see conditions at their worst. And it rained. On the second day of his visit, 3 inches fell in two hours, a new record even for Panama. Roosevelt took it all in his stride, rushing about or posing in a downpour sitting at the controls of one of the huge Bucyrus steam shovels, all the time making what a Washington *Post* headline called 'A Strenous Exhibition on the Isthmus'. 'He was intensely energetic,' remembered Frank Maltby. 'He seemed to be able to carry on a conversation with me and dictate a cablegram to his secretary at the same time. He revelled in the publicity and commotion his visit created. He would make a speech at the slightest opportunity and without any preliminaries.' Everywhere he went, addressing the white workers as 'the pick of American manhood', he urged them to 'play their part like men among men'.

After two days, Stevens was exhausted. 'I have blisters on both my feet and am worn out,' he told Maltby. 'Shonts is knocked out completely.' On the last day, Maltby showed the President the site of the controversial Gatún dam. To get a better overall view it was suggested they climb a nearby hill. 'We, together with three or four Secret Service men, charged up the hill as if we were taking a fort by storm,' Maltby reports.

On the evening of his departure, a mass reception for President Roosevelt was held in the great building that covered the largest wharf of the Commission at Cristóbal. Virtually the entire American canal force was present, crowding the immense structure, which was decorated with flags and lanterns. Roosevelt then made an impromptu speech, which captures the martial heroism of his vision of the great enterprise: 'Whoever you are, if you are doing your duty, the balance of the country is placed under obligation to you, just as it is to a soldier in a great war,' he proclaimed. 'As I have looked at you and seen you work, seen what you have done and are doing, I have felt just exactly as I would feel to see the big men of our country carrying on a great war.'

The visit went down well with the press at home, even among those papers most critical of the canal, and provided a significant morale boost on the Isthmus. Rose van Hardeveld remembers the effect it had on her. 'We saw him once, on the end of a train,' she writes. Jan had got hold of small flags for the children, and told them when the President would be passing their house, 'so we were standing on the steps. Mr. Rossevelt flashed us one of his well-known toothy smiles and waved his hat at the children as though he wanted to come up the hill and say "Hello!" I caught some of Jan's confidence in the man. Maybe this ditch will get dug after all, I thought. And I was more certain than ever that we ourselves would not leave until it was finished.' Two months later a visiting English journalist noted the 'energy' and 'optimistic spirit' of the Americans working on the canal. 'Every man,' he wrote, 'seems animated with the idea that he is doing a necessary part of the canal, and a feeling of pride prevails everywhere.'

Roosevelt reported back to Congress on 17 December. There was impressive progress to outline. In spite of the rainy season, and the thousands of men still employed putting up buildings, the month before his visit had seen a new record for excavation – 325,000 cubic yards. The long period of preparation was at last coming to an end, and the actual digging was underway. At Gatún over a hundred new borings had been made on the dam site and excavation had started on the lock basins. At Cristóbal he had seen the new bakery in action, churning out 24,000 loaves a day, as well as the nearly-completed coal depot and cold storage plant. The workforce had topped twenty thousand, and the supply from the West Indies and Spain seemed secure.

But the bulk of his message to Congress concerned what he must have judged to be the two interlinked problems that posed the greatest threat to the success of the canal: the high turnover of skilled men (still running at nearly 100 per cent a year) and negative publicity at home. A new reward for long service – the Roosevelt medal – was announced. And the canal effort was described in unmistakably patriotic terms – 'something which will redound immeasurably to the credit of America'. The sanitation effort of Gorgas was praised to the rooftops. Among the Americans, including dependants, there had not been a single death from disease in three months, a very impressive record, Roosevelt pointed out, even by mainland United States standards. On numerous occasions direct reference was made to Poultney Bigelow and his report rubbished, which gives an indication of the great and lasting effect it had had. It was almost as though Roosevelt went to Panama specifically to erase the stain that his law-school classmate had put on the enterprise. It was simply unpatriotic to criticise the canal effort, the President exclaimed. For detractors, he said, he felt 'the heartiest contempt and indignation; because, in a spirit of wanton dishonesty and malice, they are trying to interfere with, and hamper the execution of, the greatest work of the kind ever attempted, and are seeking to bring to naught the efforts of their countrymen to put to the credit of America one of the giant feats of the ages.'

However, Roosevelt himself had seen that all was not rosy on the Isthmus. On his return to Washington he wrote to Shonts: 'The least satisfactory feature of the entire work to my mind was the arrangement for feeding the Negroes. Those cooking sheds with their muddy floors and with the unclean pot which each man had in which he cooked everything, are certainly not what they should be.' And while the health of the white workers was indeed impressive, 'the very large sick rates among the negroes, compared with the whites' was alarming. In fact, as Roosevelt's own figures acknowledge, the black workers were three times as likely to die of disease. In the ten months of 1906 before the presidential visit, thirty-four white workers had died, compared with nearly seven hundred West Indians. Roosevelt suggested that 'a resolute effort should be made to teach the negro some of the principles of personal hygiene'.

Gorgas had concluded, in his health report of July 1906, that the black workers were dying at three times the rate of the whites because, contrary to the earlier belief, their race could not stand the climate as well as their American employers. 'We do not agree with the doctor,' countered the Colón *Independent* angrily. 'The higher death rate is, in our opinion, due to circumstances. The white employers are better housed, better paid, and therefore live better; they do the bossing while the blacks do the actual labor, such as work in mud, and water and rain. Change conditions with the two races and see if there would not be triple the amount of deaths.'

Mary Chatfield, soon to leave the Isthmus having, she felt, 'done one citizen's duty towards the building of the Panama Canal', gives a vivid description of what conditions were like for the black workers. Riding on a train near Mount Hope, she spotted some labourers' quarters near the railroad – 'wretched little houses rest on stilts, and now during the rainy season the water is constantly on the level with the floors . . . When I left the office at 5 o'clock the negro laborers were returning to their quarters and were getting their suppers on on little charcoal braziers outdoors. It was a sad sight to me . . . How could they get their suppers with rain pouring in torrents for an hour? They are not allowed to cook in their quarters for fear of fire and no covered place is provided for them in which to cook, so these poor men exist under difficulties. Consider the brilliant criticisms some of the authors of magazine articles make on "The lazy, worthless, Jamaican laborer." They sleep all night on a strip of canvas but little wider than their bodies, they must get up in the morning and cook their breakfasts out of doors in the tropical rain or shine, as it happens. They must be at work at 7 o'clock A.M., they get their noon meals under the same weather conditions as other meals. They receive the splendid wage of 10 cents, U.S. currency, an hour. They are obliged to pay at the government commissary as high prices for food as are charged at the grocery stores in the City of New York, whose proprietors have high rents to pay and are doing business for profit.'

The West Indians, often unfamiliar with modern machinery, and given the most dangerous jobs, were also suffering from accidents at twice the rate of the white employees. At Ancón hospital, detailed autopsies were carried out on West Indians, which included the measurement of brain weight, skull thickness and skull shape. The doctors' conclusion was that the large number of accidents befalling the black workers indicated 'a striking lack of appreciation for a dangerous environment [in] the negro's mental processes'.

If accidents and disease were deemed to be the West Indians' own fault, or as a result of their inherent weaknesses, this reflects deeply held ideas about race. These, in turn, would shape every aspect of life in the Canal Zone, and nowhere more so than in the division of the workforce into the Gold and Silver Rolls, described by one canal historian as a 'notorious' example of 'racial and ethnic discrimination by the U.S. Government'. Harry Franck, a travel writer who worked in the Zone as a policeman for three months in 1912, remembers his surprise at seeing notices everywhere stipulating whether a shop, railway car, toilet or drinking fountain was for Gold or Silver Roll employees. But he quickly worked it out. 'The ICC has very dextrously dodged the necessity of lining the Zone with the offensive signs "Black" and "White",' he writes. 'Hence the line has been drawn between "Gold" and "Silver" employees. The first division, paid in gold coin, is made up, with a few exceptions, of white American citizens. To the second belong any of the darker shade, and all common laborers of whatever color, these receiving their wages in Panamanian silver. 'Tis a deep and sharp-drawn line.' For Franck, there was little doubt as to the model being followed. 'Panama is below the Mason and Dixon Line,' he concluded.

It has often been noted that US imperialist expansion went hand in hand with rising racism. Influential thinkers such as Alfred Mahan and politicians such as Senator Albert Beveridge of Indiana had used a social Darwinian doctrine of Anglo-Saxon superiority and the 'civilizing mission' to justify US imperialism in the Philippines, Hawaii and Cuba. It was not long before people started applying this theory to race issues closer to home. 'If the stronger and cleverer race is free to impose its will upon "new-caught sullen people" on the other side of the globe,' asked *Atlantic Monthly*, 'why not in South Carolina and Mississippi?'

Indeed, the closing years of the nineteenth century saw the abandonment of the Southern blacks by Northern liberals, and as the 'white man's burden' was shouldered overseas, the Southern states began a process of disenfranchisement and officially sanctioned discrimination against their black populations. In 1896 Louisiana had contained 130,000 black voters. Four years later, there were only 13,000. And what became known as Jim Crow laws spread across the South, officially segregating whites and blacks, with the best facilities always reserved for the former. What had previously been unspoken and unenforced was, by the time of the beginning of the US canal effort, rigid and backed up by the law.

This system – with the euphemisms 'Gold' for Anglo-Saxon whites and 'Silver' for everyone else – was imported into the Canal Zone in Panama by the US authorities and would survive in various forms for nearly a hundred years. But it was not imposed, as is often believed, en bloc, but was rather a gradual and complex process that parallels the other ways in which the Commission sought to impose itself on the lives of the canal builders of all backgrounds. It started with the decision taken at the outset of the project to pay some workers in US gold currency and others with local silver money. Attached to the Gold Roll from the beginning were privileges such as paid sick leave and holidays and better accommodation (basically the generous deal needed to lure workers from the US). Who got what was decided by an amalgam of precedents – the PRR had always paid its US workers in gold and the rest in silver, while the French companies had paid almost everyone in local currency but had divided its workers from all backgrounds into skilled and unskilled grades. The early Gold–Silver system merged these two approaches (a US government report in 1908 would describe the distinction in terms of skills, but noted that the Gold Roll was 'nearly all Americans'). Either way, white American citizens in the Zone, all in theory skilled workers, were almost always on the Gold Roll, and as the vast majority of the earliest unskilled workers were West Indians, the terms Gold and Silver quickly took on racial connotations.

Initially, however, it was not that simple – a relatively large number of West Indians, approaching a thousand, were put on the Gold Roll as skilled workers. These included foremen, office clerks and teachers. This was considered a good way to co-opt potential leaders of the 'Jamaican' community, and also to incentivise workers to train in useful skills and thus gain promotion to the higher-status Gold Roll.

Then, with the arrival of Stevens and the building of Commission hotels, restaurants, additional hospital facilities and shops, it was discovered that by limiting access to parts of these establishments to Gold Roll employees, it was possible to keep undesirables away from the elite white sector of the workforce. Supposedly, it all began with a pay car. When two separate windows were used, one marked 'Silver' the other 'Gold', it was found to provide 'the solution to troubles growing out of the intermingling of the races'. Thereafter this was widely adopted, and no commissary or post office was built without separate sections for Gold and Silver. In everything, there was a premium service for the Gold employees.

But with the distinction now being used to prevent 'intermingling of the races' on the Zone, the blacks on the Gold Roll presented a problem. In September 1905

Stevens closed the door to the Gold Roll for West Indians by ending both direct recruitment to the Gold Roll and promotion from the Silver Roll. At the end of the following year he started removing blacks from the Gold Roll, even if they were skilled and valuable employees. There was the occasional protest. The manager of the commissary at Cristóbal wrote to Stevens, 'It would, I think, be very impolitic to separate all of the Commissary employees by color putting all the colored men on the silver roll. They would naturally feel it to be in a measure a humiliation. We have a number of colored men in charge of Departments . . . We also have two or three colored clerks in our Shipping office, who are very valuable men and draw larger salaries than some of our white clerks.' Nevertheless by mid-1907 only a tiny handful of blacks, mainly postmasters and teachers, remained on the Gold Roll, and they would be gone by the following year.

The arrival of the Spanish and other southern European workers from mid-1906 onwards might have had an unsettling effect on this rapidly solidifying racial system. But although Latins were thought higher up the evolutionary pecking order than the blacks, they were certainly beneath the Anglo-Saxons and were in colouring, it was suggested, somewhere in between white and black. Thus they formed an intermediate layer – paid in Silver, but with better food, accommodation and general treatment along with some Gold Roll privileges.

The education system in the Zone provides a microcosm of the development of this system of inequality based on race. Some of the earliest Canal Zone schools had a mixed intake of West Indians, Panamanians and a few whites. As more families came out from the United States and the West Indies to live, the classrooms were segregated, and then the white and black children were separated out into entirely different schools. Light Panamanian children from good families as well as the children of white European labourers enrolled in the white schools, the latter only under sufferance.

The white schools, housed in new buildings and well staffed and equipped, performed at a level at least equal to that back at home in the States. The non-white schools, however, were less than second class. In 1909 there were about seventeen children per teacher in the white schools; in the others, it was 115 pupils per teacher, an astonishing disparity. Furthermore, the black schools were usually housed in dilapidated buildings, staffed by less-trained teachers and had to make do with textbooks discarded by the white schools. There was no question of pretending to provide separate but equal facilities.

The West Indian children were taught American history, discipline, orations, manners, the three Rs and subjects such as carpentry and gardening that would equip them for unskilled work on the Zone. In 1911 a secondary school was opened for white children, but for the black students there were only advanced classes in agriculture, sewing and domestic service.

This official sanction of racism nourished and legitimised racist behaviour on a day-to-day basis. Harry Franck commented that a 'new amalgamated' national 'type' was being created in the Zone: 'Any northerner can say "nigger" as glibly as a Carolinian, and growl if any of them steps on his shadow,' he wrote. So prevalent were the attitudes associated with 'South of the Mason–Dixon line' that newcomers assumed that most of the Americans were Southerners, although in fact Northerners were in the majority. Even the nursing staff, who mostly cared very well for their black patients, were not immune to prejudice. Among three nurses arriving in November 1905 was Miss Emma M. Jeffries, a black American. On the steamer from the States Miss Jeffries, according to the Colón *Independent*, 'was made to feel the prejudice against her color, as one of the white nurses refused to occupy the same state room with her'. It got worse when she was taken to Ancón hospital. 'Miss Jeffries was informed at the nurses' reception room that she had made a great mistake in coming here, as all of the other nurses were white and had decided to go on strike if forced to work with a Negro. They even refused to sit with her at the same table for meals.' Jeffries returned to New York in disgust.

Others also found the way colour dominated life in the Zone too odious to cope with. 'My father read of Panama and thought it a wonderful place to come to because he saw progress in Panama,' an Antiguan lady told a researcher in the 1970s. But he did not work for long in the Zone. 'He just could not take it – the life was so different. We were not accustomed to be told so much about your colour or to have to think about it often, black and white. He couldn't stand it so he left the Canal Zone and came to Panama [City].'

For black labourers out on the works, 'some of the foremen were very polite, while some were very rough and impolite', as one West Indian recalled. Edward White, from Jamaica, remembered being very lonely when he first reached Panama, but found himself made to feel part of a family by his American foreman and timekeeper. 'The lonely feeling started to leave me, as these men treated me like their own. Mr Arthur, Mr Chambers, and I were so knitted together, I felt as if I was their own son.'

This tone tends to be the exception in the West Indian accounts, however. Most are at best mixed about their treatment. Jeremiah Waisome was born in Nicaragua, but had lived in Panama since a baby. When he was twelve or thirteen, proud of his ability to read and write, he applied for work on the canal: 'Unknown to my mother one morning instead of going to school, I went to Balboa to look myself a job. I approach a boss one morning for a waterboy job. "Good Morning, boss," I said. "Good morning, boy," he retorted. At this time he was chewing a big wad of tobacco. I ask him if he needs a water boy, he said yes. He ask me "What is your name?" I told him. Then I noticed that my name did not spell correctly, so I said, "Excuse me, boss, my name do not spell that way." He gave me a cow look, and spit and big splash, and look back at me and said: "You little nigger! You need a job?" "Yes, sir." "You never try to dictate to a white man."'

An American journalist sympathetic to the US canal authorities reported in 1906 that he had 'often seen the threat of the slave-driver in the foreman's eye – the menace of brute force'. Occasionally, this was more than a threat. 'Among the white employees on the "gold roll" some times an employee would use his hands or foot on one of the "silver employees",' admitted a steam shovel engineer. On 23 March 1906, the Colón *Independent* ran a story about how a man at Bas Matachín Machine Shop 'by the name of Bryan was thoroughly clubbed and kicked by Master Mechanic Cummings because he refused to lift up a bucket of metal which was beyond his strength'. When the accusation was taken to court, it was the victim Bryan in the dock, with his attacker Cummings demanding that the West Indian be punished for insolence.

The very worst foremen were dismissed, and treatment improved as the Americans learnt that shouting and hitting were not the best ways to get results from their gangs. However, actual physical aggression against the blacks continued. After the death of Jantje Milliery, Rose and Jan van Hardeveld had made a new best friend, Charles Swinehart, the mining engineer given a job through his father's connections with the local Republican Party at Steamboat Springs, Colorado. According to Rose, Swinehart was very much 'the he-man type' and one evening at dinner a 'troublesome' West Indian discharged from his gang 'elected to place himself under the veranda and shout abuses at the house and Americans in general. He cursed and swore, while everyone at the table tried to act as though nothing were happening. Suddenly Charley, his lips set and his face white, politely excused himself. He left the table, went into the bedroom, and then we heard him go down the steps. In a moment there was silence below. We heard the young man coming back up the steps. He entered the bedroom, came out, and reseated himself at the table. The conversation and the meal continued. A few days later Charley was summoned to appear before the judge at Empire to answer the charge of knocking a British subject over the head with the butt of a revolver. He pleaded guilty, and was fined twenty-five dollars. "Was it worth the money?" asked the judge with a twinkle in his eye. "Yes, indeed, sir," answered the aggressor gravely.'

A journalist visiting the Isthmus in 1908 was advised that 'it cost twenty-five dollars to lick a Jamaican negro and if I did it be sure and get my money's worth'.

The standard response of the black worker to bullying or abuse, according to virtually every American account of the construction period, was to 'straighten himself up and say to the foreman "I wish you to understand, sir, that I am a

British subject, and if we can not arrange this matter amicably we will talk to our Consul about it"'. In fact, Mallet had his work cut out caring for those denied wages or hospital care and utterly desperate. There was no way he could deal with complaints from over twenty thousand British citizens on the Isthmus, as he frequently pointed out to the unresponsive Foreign Office. Nevertheless, pride in being British seems to have sustained the self-respect of the West Indian workers in often very difficult circumstances. Guyanese novelist Eric Walrond, who moved to Panama as a fourteen-year-old in 1911, would write in 1935, by that stage a committed Garveyite, that the West Indians had 'developed an excessive regard for the English'. But in his 1926 short story 'Panama Gold', the protagonist, returned from the Isthmus to Barbados, triumphantly explains how he came to be given compensation for a lost leg: '"Pay me," I says, "or I'll stick de British bulldog on all yo' Omericans!" . . . Man, I wuz ready to stick Nelson heself 'pon dem. . . . I let dem understand quick enough dat I wuz a Englishman and not a bleddy American nigger!'

'As British subjects,' William Karner wrote of the Barbadians he recruited, 'they think they are close to royalty and quite superior to white laborers from the United States.' In fact, the West Indians *did* consider themselves superior to Americans. After all the British Empire was still the most powerful in the world, as they would point out, and they were as much a part of it as anyone. The Americans thought this was hilarious.

In other ways, too, the West Indians resisted the Commission's attempts to dehumanise and control them. In early 1907 there were nearly 12,500 workers in the ICC's austere, heavily regimented military-style barracks. Two years later there were fewer than 3500. The others preferred to pay the exorbitant rents of the terminal cities or simply put up a hut of flattened tin cans and old dynamite boxes in the bush. Either way, the move reclaimed independence and dignity. In the same way the attendance at the ICC-run kitchens collapsed, with 80 per cent making their own arrangements by the end of 1909.

In the workplace there was little point in complaining. 'You couldn't talk back,' remembered Constantine Parkinson. 'It would get you fired if you talked back.' Young Jules LeCurrieux, who had done a variety of jobs since starting work on dynamiting Gold Hill, protested on behalf of his gang when they found work unloading cement impossible because of the choking dust. He was promptly fired. So the workers simply voted with their feet, walking away from the worst jobs or the worst bosses. On other occasions, as in the French days, they would move about the line looking for the best pay or to be with their friends, taking on a new name each time so that they could be re-employed. In both cases it was bad news for the efficiency of the canal effort: the dispersal of the workforce in 'private' accommodation made the control of malaria and other diseases much more difficult, and the moving about of the workers from job to job caused frequent delays to the construction programme.

The Spanish workers were always treated better than the West Indians, but by the beginning of 1907 they too were beginning to cause difficulties for the authorities. For one thing, their impressive initial energy and zeal had not lasted. If they did not succumb to disease, they soon adjusted their work rate to a more realistic tropical pace. By the middle of 1907 a divisional engineer at Culebra was even requesting that his Spanish workers be replaced by West Indians. The Europeans, he argued, were 'little better than the West Indian negro', and as they were paid twice as much they were a waste of money. Even Stevens was forced to admit that while the introduction of the Europeans might have improved the work rate of the blacks, 'the efficiency of the Spaniards did not hold up to the standard first developed'.

For another thing, they were breaking their contracts and leaving in large numbers, mainly to move on to better-paid railway or mining work in South America. The Chilean consul was among those actively recruiting among the ICC's Spaniards, to the fury of the American authorities. Consul Mallet estimated that nearly half of those recruited during 1906 were gone by the beginning of the following year. The main impetus was money – although the Spaniards accepted that they were well paid by the Commission, the cost of living in Panama was such that they would struggle to earn the steamer fare home, let alone the riches they had anticipated. Antonio Sanchez tells of how his group 'were deeply disappointed when they realized they would not be able to save enough money for the return trip to the land of their birth'. And if any of them 'became a victim of misfortune', he says, they were in real trouble.

In late January 1907, the thousand or more Spaniards working in the Cut went on strike demanding an increase in pay from $1.60 to $2.50 a day. The West Indian workers were not supportive, however, and had to be protected by the police. After a tense standoff, fighting erupted that led to several deaths and serious injuries among Spaniards and Zone police. The strike's ringleaders were rounded up and the protest broken. Stevens later ascribed the violence of the repression to the need to give a 'severe lesson' to prevent future demands endangering the project. But clashes between Spaniards and police continued for the rest of the year.

Around this time letters and articles started appearing in Madrid newspapers reporting that all was not well on the Isthmus for the expatriate workers. A letter from three workers printed in *El Socialista* at the end of December 1906, complains about the high expenses in Panama, the retention of a proportion of their wages to repay their outward fare, and poor food and accommodation. Furthermore, they wrote, 'People are falling ill the whole time . . . Many are leaving.' They end by warning others not to be deceived by the 'siren songs'. The hospitals were no good, a subsequent letter from a Spaniard on the Isthmus declares. You get 'handed over to the care of a black, who, as you'd expect, doesn't bother himself with any such a thing, and he who gets better, gets better, and he who doesn't, dies'. A Spanish journalist sent out to Panama noted, 'The labourers' lives are not highly valued, so there are frequent accidents.'

Towards the end of 1906 worrying news also began to reach Italy about the fate of the thousand or so workers recruited to work in Panama. It was said they had to labour eight hours a day in a swamp with water up to their knees, under the sun in torrid heat, exposed to torrential rain and suffering from dreadful illnesses. A Naples paper claimed that most of the workers had died, and there were thousands of corpses on the streets. In both countries, the governments began to come under pressure to prevent further migration to the Isthmus.

To Stevens this was simply petulance. 'My own private opinion,' he wrote to Shonts in mid-January 1907, 'is that no European nation is favourable to the building of the Panama Canal: that they do not want it built; will do anything they can possibly short of open hostility in the shape of force to prevent the consummation of the project, and will, if the movement of laborers from their countries assumes large proportions, take steps directly or indirectly, to prevent such movements.' Stevens, however, was about to become yesterday's man.

The Chief Engineer was in Washington in December 1906 and those who saw him were shocked at how weary and sour he had become. It appears he had fallen out with Gorgas, whose starring role in Roosevelt's Congressional Message would have irked Stevens. The following month ICC chairman Theodore Shonts resigned to take up a lucrative post in New York, about which the President could have no complaint. Shonts had told Roosevelt that he would quit the project once the preparatory phase was completed. However, this resignation cannot but have pleased Stevens. Relations between the two men, never good, had deteriorated of late, and Shonts' departure also cleared the way for Stevens to take absolute control over the project, on the Isthmus at least, as he had requested for so long.

In other ways Stevens had no cause for gloom. The heavy rains experienced by Roosevelt had continued, leading to flooding of the works in December, but in January, with the return of dry weather and the deployment of no less than sixty-three Bucyrus shovels, over half a million cubic yards had been excavated. This monthly figure would grow steadily thereafter, proving that Stevens' machine was working well.

But at the end of January, Stevens sat down and wrote an extraordinary letter to Roosevelt. Six pages long, it revealed the depths of his exhaustion and bitterness. Although he appreciated the support the President had given him, Stevens wrote, he had never sought the Panama job and did not like it. The 'honour' of being the canal's builder meant nothing to him. He had been endlessly attacked by 'enemies in the rear'. Even the level of his salary had been questioned, when, in fact, he could have returned to the States and secured any of a number of far more lucrative and less stressful jobs, some of which, he wrote, 'I would prefer to hold, if you pardon my candor, than the Presidency of the United States.'

Roosevelt received the letter on 12 February. He did not 'pardon the candor'. Only two months before, he had told the canal workforce that they were like an army in the field. Now their general, to whom Roosevelt had given almost unqualified backing, was looking to desert his troops in a most unmartial way. The letter was sent on to Taft with a note from the President attached: 'Stevens must get out at once.' Then he telegraphed Stevens to tell him that his resignation had been accepted, effective 1 April.

Stevens never spoke or wrote about his real reasons for quitting, leaving the field open for a miasma of speculation. He had fallen out with the President, it was alleged; he had found that the Gatún dam plan was unworkable; he had discovered something about the role of Cromwell in the sale of the New Company so explosive that it would 'blow up the Republican Party'.

Many felt that he had not actually meant to resign, but was either letting off steam or flexing his muscles. Mallet reported to London that Stevens' resignation was never formally tendered and that 'an immoderate amount of adulation over the success of Mr Stevens' organization and management led him to imagine his services were indispensable to the successful prosecution of the works'.

Stevens, like Shonts, had secured on his hiring the promise that he would be allowed to leave the project once it was up and running. His career before and after Panama shows a succession of departures to take on new challenges, and perhaps Stevens felt that his job was done on the Isthmus. By his own reckoning he handed on a 'well-planned and well-built machine'. Whoever came after him would merely have to 'turn the crank', he said. But perhaps he also realised that the nature of the task had fundamentally changed with the firm adoption of the lock-canal plan. From being an unprecedented but essentially low-tech canal, it had become an equally huge, but also technically complex project. There is little doubt that Stevens was the best man to design and build the transportation system for the excavation of the canal, but he had little experience of hydraulics, lock design or dam construction. Perhaps he understood that it was time for a man with different skills to step up to the plate.

But probably the greatest factor leading to Stevens' departure was mental and physical exhaustion. Stevens once said to Maltby, 'I know you pretty well now and without raising the question of your competence, if you were chief engineer you wouldn't last thirty minutes.' Working twelve to fourteen hours a day, suffering from insomnia, endlessly dragged to Washington to be hauled before 'idiotic' Congressmen, he had had enough.

On the Isthmus the news came as a severe blow – 'astonishing' wrote the *Star and Herald* on 28 February. Over the following weeks the paper traces the surprise, sadness and then anger of the canal workforce. 'Unless this step has been forced upon Mr. Stevens, a supposition which is scarcely likely,' the paper wrote on 2 March, 'his action in retiring from the canal work looks suspiciously like an abandonment of a trust, and unless it be his desire to lay himself open to the same scathing rebuke which was heaped on Mr. Wallace his obvious course is to at once withdraw his resignation . . . we think that in his place a strong sense of loyalty, we might even say of devotion to an ideal, should have outweighed mere personal considerations.' No one really knew what these 'personal considerations' were. When asked, Stevens merely growled back, 'Don't talk, dig.'

A petition was organised, begging him to stay, and promising to work even harder for him in the future, but to no avail. After numerous farewell functions the Chief Engineer sailed from the Isthmus for the last time at noon on Sunday 1 April. There was a huge crowd on the wharf to see him off. Other vessels in the harbour, reported the *Star and Herald*, 'whistled their salutes, the crowd waved hats and handkerchiefs, and many shed tears while the I.C.C. band played Auld Lang Syne. Mr. Stevens stood at the rail, and as long as he could be recognized his face was pale and sad.'

Roosevelt had rated Stevens highly – he was his sort of 'strenuous man' – and he was grieved as well as angered by his departure. He also knew full well that the canal would never be built if it kept losing its Chief Engineers. So he now decided to place the work 'in the charge of men who will stay on the job until I get tired of having them there, or till I say they may abandon it . . . I shall turn it over to the army.'

24

'THE ARMY OF PANAMA'

To an extent, then, the project had come full circle. After all, the military needs of the United States had been of primary importance in starting the American canal. It was as a conduit for sea power that the canal's supporters had successfully sold the idea to the US Congress and public. But the all-new Isthmian Canal Commission, ordered to take over on 1 April 1907, was not entirely military. Its new chairman was Lieutenant Colonel (later Major General) George Washington Goethals, one of the army's finest engineers, with particular expertise in lock construction. There were two further Engineer Corps officers, a Navy man and Colonel Gorgas was given a seat on the Commission for the first time. But there were also two civilians – an ex-senator from Kentucky and Jackson 'Square-foot' Smith, like Gorgas promoted to the Commission. Furthermore, the military men were detached from their usual chains of command, reporting to Goethals, who himself dealt directly with Taft. All the Commission members were required to live on the Isthmus, where they would work as heads of department.

There was no question, however, of the seven Commissioners having equal say, as envisaged by the Spooner Act. Each man was summoned before Roosevelt and told in no uncertain terms that there would be only one boss. 'Colonel Goethals here is to be chairman,' said the President. 'He is to have complete authority. If at any time you do not agree with his policies, do not bother to tell me about it – your disagreement with him will constitute your resignation.' As well as chairman, Goethals was appointed Chief Engineer, head of the PRR, and would wield total control over the government of the Canal Zone. The new arrangement made Goethals, in the words of his biographer, the 'most absolute despot in the world . . . [who] could command the removal of a mountain from the landscape, or of a man from his dominions, or of a salt-cellar from that man's table'. It was the 'one-man proposition' demanded by Stevens, and there would be only one end in view. As Goethals himself explained: 'It was asserted that the Department of Government, generally, regarded the construction of the canal as of secondary importance and seemed to consider that the main purpose and object of the work on the Isthmus was to set up a model of American government in the heart of Central America as an object lesson to the South and Central American republics.' Governor Magoon had left the Isthmus the previous September to help out with the crisis in Cuba and was now told that he would not be returning to Panama. Henceforth, as Goethals wrote, 'everything should be subordinated to the construction of the canal, even the government'.

Goethals himself told a New York friend that his taking the job was 'a case of just plain straight duty. I am ordered down – there was no alternative.' He landed on the Isthmus in mid-March, for a two-week handover period with John Stevens. It was an awkward time for the new man. Stevens' popularity was everywhere apparent, along with deep unease among the civilian engineers about the nature of the new army regime and the inevitable changes in personnel that the new leadership would bring. On 18 March there was a reception in Goethals' honour at Corozel. Stevens was not there, but every time his name was mentioned in a speech a loud cheer rang out. When it was Goethals' turn to speak, he tried to reassure the men. There would be no military uniforms or saluting on the Isthmus, he said. 'I expect to be chief of the division of engineers, while the heads of the various departments are going to be the Colonels, the foremen are going to be the captains, and the men who do the labour are going to be the privates . . . I am no longer a commander in the United States Army. I now consider that I am commanding the Army of Panama, and the enemy we are going to combat is the Culebra Cut and the locks and dams at both ends of the canal, and any man here on the work who does his duty will never have any cause to complain of militarism.'

Goethals had been on the Isthmus before. In November 1905, early in the Stevens regime, he had accompanied Taft to Panama as part of a group of army experts looking at the fortification requirements for the canal. At the time he had commented on the chaos and hysteria, but now, starting to look around, he was agreeably surprised. 'The magnitude of the work grows and grows on me; it seems to get bigger all the time,' he wrote to his son on 17 March. 'But Mr. Stevens has perfected such an organization so far as the RR [railroad] part of the proposition is concerned, that there is nothing left for us to do but to just have the organization continue in the good work it has done and is doing.' The Stevens system was operating well. In March over 800,000 cubic yards had been excavated, and the following month would see this rise again to nearly 900,000, with five hundred trainloads of spoil being dumped every day. Eighty per cent of the necessary plant was in place, and, although there were still nearly four thousand men employed on building work, 70 per cent of the required Gold Roll accommodation was completed. About a fifth of the workforce of nearly thirty thousand was off sick at any one time, but infection rates for malaria were falling as Gorgas's two thousand sanitarians continued and extended their campaign against the *Anopheles* mosquito, meticulously draining hundreds of square miles of swampland.

Goethals' main concern was for the more technical parts of the project, the locks and dams, areas outside Stevens' expertise. 'The hydraulic part of the propositions is not so good and is a way behind,' he wrote to his son on 22 March. Goethals quickly judged that some of the department heads did not have the necessary experience for the new tasks ahead. Maltby, for instance, although 'an excellent man at dredging', 'had no work on foundations and locks and is therefore of no account'. Reorganisation was needed 'and yet not to demoralize the other branches of the work we have to be careful in making changes'. Such was the new man's confidence in Stevens' system for the Cut that he judged that the canal's completion date now depended not on *la grande tranchée* as everybody had always assumed, but on the creation of Gatún Dam and Lake, and the necessary prior relocation of large parts of the railway. That said, excavation in the Cut was still in its infancy – it had been widened by over 100 feet but hardly lowered at all. The terrible setbacks that would accompany deeper excavation were still, for now, in the future and unanticipated.

Goethals took official charge on 1 April 1907. He immediately threw himself into the job, spending the mornings on office work and the afternoons inspecting the line, propelled along the railroad in a gasoline-driven railway car, known as the 'brain wagon' or the 'Yellow Peril'. He would frequently dismount to talk to a foreman or manager. He was, a contemporary writes, 'a tall, long-legged man with a rounded, bronzed face and snow-white hair. His moustache was also white, but stained with nicotine, for he smoked many cigarettes . . . He wore civilian clothes with the usual awkwardness of a man who has spent most of his lifetime clothed in the uniform of his country.' Every day the role of 'Czar of Panama', which included leadership of civil government, courts, schools, post offices, the police, and the battalion of US marines in addition to the canal work, seemed to grow in size. 'The strenuous existence of the past seems like mere child's play to the 5 past days,' he wrote to his son on 4 April.

There was no mass clearout of the Railroad Era men, but changes were inevitable. Soon after Goethals' arrival Frank Maltby left Panama, although he would return as a private contractor later. The excellent head of the railway, W. G. Bierd, the inventor of the track-shifter, resigned as well, supposedly because of ill health, but, Goethals noticed, he cropped up soon after working for Stevens in the latter's new job as head of the New Haven Railroad. Then, at the beginning of May, the steam-shovel men decided to test out their new boss, requesting a steep pay increase with the threat of a strike if their demands were not met. As one of the new Commissioners, Major William Sibert, commented, 'the President, in his talks, praised the men for their patriotism and enlarged upon their hardships to such an extent that . . . after the visit the steam-shovel men asked for a rise in wages'.

Goethals, still finding his feet and assessing the extent of his power on the Isthmus, acted cautiously, referring the matter to Taft, who was paying a visit to the Isthmus. Taft heard the men's demands, and then consulted with Roosevelt in Washington. 'Things are unsettled here,' wrote Goethals in a private letter on 7 May. In Washington, it was decided that as the steam-shovel men, on $210 a month, were already the best paid of the mechanics, their hoped-for $300 a month could not be granted. Instead Taft offered a 5 per cent pay increase. But the steam-shovel men remained determined. The increase was rejected and the men came out. The next day all but thirteen of the sixty-eight shovels were idle. It was the most serious strike on the canal so far.

But Goethals did not panic, even as the stoppage continued, reducing excavation to a quarter of its previous level. Handling it slowly, he gradually recruited strike-breakers until he had replaced the original workers. When the strikers gave in and asked for their jobs back, they were told that they would have to start again at the most junior level. By July, all the shovels were back in action, and excavation was once again at full pelt, with just over a million cubic yards extracted, a new record. It was an unmistakable victory for Goethals, and for the rest of the construction period there would be no stoppage on anything like the same scale. And the 'Czar of Panama' had not even had to use the most potent weapon at his disposal. Under the original terms of Roosevelt's Executive Order setting up the first Commission, the chairman had the right to expel from the Zone anybody, who, in his opinion, 'was not necessary to the work of building the canal, or was objectionable for any reason'. With his power enhanced by his defeat of the steam-shovel men, Goethals dealt ruthlessly with a small stoppage in November by boiler-makers at two large machine shops. Replacement workers were quickly in place and the strikers found themselves deported on steamers back to the United States. Thereafter, his response to any strike threat was simple: be back at work tomorrow morning or be expelled instantly from the country.

The European Silver Roll workers were dealt with even more firmly. They were continuing to leave the Isthmus for better work opportunities elsewhere, thus depriving the ICC of the repayment of their steamer fare. In response, Goethals banned the solicitation of labour within the Zone, and placed guards at the ports to prevent contracted workers leaving. 'I have no complaint of any kind against the Isthmian Canal Commission,' stated Spanish worker F. Olario when hauled off a Chile-bound steamer in May 1907. 'I was always well treated, liked the wages I used to get, but could not understand the orders of the foreman, and besides, I was most of the time sick, out of the four months that I have been a laborer on the Isthmus.'

Early in his tenure, Goethals had set up a routine that he would hear complaints from Gold Roll employees every Sunday morning, just as Stevens had done. Silver Roll employees had the same chance to air their grievances, but only with Joseph Bucklin Bishop, the unctuous Secretary to the Commission, who, among other things, acted as Roosevelt's eyes and ears on the Isthmus. Bishop recruited a Spanish-speaking Italian, Joseph Garibaldi, grandson of the famous independence leader, to deal with the southern Europeans. In late 1907 Roosevelt appointed a Commission to report on labour conditions on the Isthmus. Its conclusions came with an attached letter from Garibaldi explaining the genesis of the problems the Spanish labour seemed to be presenting. 'Supposed ill treatment has been the source of a great deal of trouble along the Canal Zone, and in most cases it was simply due to a misunderstanding between the men and the employees in charge, because of different languages,' he wrote. 'It is natural for a foreman to want his work to go on as rapidly as possible, and consequently giving orders in English to Italians and Spaniards, expecting to be obeyed at once.' As William Sands would write: 'Few or none of the gold-payroll employees bothered to learn enough of the languages and dialects of the silver-payroll people to discover the real causes of their dissatisfaction.' 'In some cases the laborer,' Garibaldi continues, 'failing to understand the order and not complying with it immediately, has been discharged. If the discharged man resented this action and made some comment in his native language, accompanying his remarks with gestures – as most Europeans do – the foreman, failing to understand the man, and thinking himself insulted, would in some cases use violence. The result of this would be a strike by the whole gang, and sometimes by the entire camp.'

Garibaldi tactfully points out that the foreman 'was not always to blame, this owing to the rather turbulent character of the imported laborer', but in July and August 1907 such small-scale strikes were happening all along the line among the European labourers. The ICC response was to try to identify the 'ring-leaders' and swiftly deport them as 'professional agitators'.

Although the measure brought success on the Isthmus – strikes fell away from November 1907 onwards – the reluctant returnees further fuelled the clamour in the Spanish press to outlaw the ICC recruitment agents. In May 1907 the Spanish Welfare Society of Panama, founded in 1885, had sent a report on conditions for the Spanish workers to Madrid. A copy was leaked to *El Liberal* newspaper and caused a sensation. 'In Panama, Spaniards are treated as of less importance than negroes,' the report had alleged. 'For any trivial incident, they are condemned to forced labour on bread and water and without any payment.

And to make the punishment more horrible, they put a chain 4 metres long on the right foot, with an enormous iron ball at the other end, as if they had committed a serious crime.' In October 1907 a new outrage came to the attention of the Spanish press. The celebrations for the feast of Sta Maria del Pilar by workers from Aragon had got out of control, resulting in a savage attack by the police. 'Without doubt, the workers were rather merry, something very natural in the circumstances,' conceded *El Socialista*, 'but that was sufficient for the Yankee police to attack the workers, treating them as virtually slaves. The Spanish workers received such barbarous and cruel treatment that many had to be taken to hospital, some in a serious condition.'

The same month the Spanish government started raising objections to ICC recruitment in their country, and with the arrival of a Liberal government in Madrid in early 1909 the representatives of the canal were finally banned from Spain. The Italian government followed suit, even though their official investigator had found very few of his countrymen still at work on the Isthmus when he visited in late 1908.

Goethals was unconcerned. The attitude was: we do not want you anyway. 'At the present time all of our superintendents and foremen are unanimously of the opinion that the efficiency of our 20-cent (40 cents silver) contract labor is much less now than it was a year ago,' he wrote to the Spanish chargé d'affaires in Panama City. 'In addition, several instances have been reported to me which indicate that the conduct of our contract laborers, as a whole, verges on insubordination; that the orders of foremen and others in authority are not received with respect and executed as the necessities of the work require.' Thereafter, although there were still some 2500 Spaniards working on the canal at the end of the construction period, the numbers steadily dwindled. For various reasons, Stevens' experiment had failed. In the main, then, it would still be British West Indians who would do the bulk of the work building the American Panama Canal.

In spite of the problems from the Spanish, the work continued steadily through 1907. Surveying parties were hacking through the jungle to map the contours of the new lake basin. In July work started on digging the lock basins on the Pacific side, and August saw a new, fresh record for excavation. By the end of the year, the workforce had grown by 15,000 to nearly 46,000, twice the peak number under de Lesseps. The year delivered a total excavation figure of nearly 16 million cubic yards, more than the entire American total up to December 1906.

Goethals divided the work into three divisions, as the French had done. The Atlantic Division stretched from Limón Bay to Gatún. To protect the entrance of the canal from 'northers', the French had dug their canal in the shelter of the bay's eastern shore. The Americans opted to head directly into the centre of the bay, and protect the entrance to the waterway from storms and silting through the construction of breakwaters out into the harbour. While these were being planned, dredges were scooping and sucking out a channel 41 feet deep and 500 feet wide from deep water three and a half miles offshore to the site of the planned dam, three and a half miles inland. This key structure was also the responsibility of this division. By the end of the year the dam site was clear of vegetation and the lock-basin excavation proceeding well.

The Central Division ran from Gatún to Pedro Miguel and included the preparation of the new lake basin as well as the excavation in the Cut. The Pacific Division ran from deep water in Panama Bay up the valley of the Río Grande to the foot of the mountains of the Continental Divide at Pedro Miguel. As on the Atlantic side, dredges and steam shovels worked their way upwards from the coast.

Inevitably, there were some changes to the original plan. After 2 million cubic yards of spoil had already been removed from the site of the Sosa locks, it was decided in December 1907, for a variety of reasons, to move them three miles inland to Miraflores. For one thing, Miraflores offered a more stable site for locks and dam, but, most importantly, it was safe from naval bombardment.

In line with the military requirements of the canal, and prompted by the US Navy, the width of the locks was increased from 100 to 110 feet, in part because of the extra compartments around naval vessels' hulls needed to combat the new threat of submarine attack. The largest navy battleship on the drawing boards, the *Pennsylvanian*, had a beam of 98 feet. (The *Titanic*, then under construction, was 94 feet wide.) The locks designed by Eiffel, when de Lesseps finally conceded to the lock plan, were little more than half this size. The width of the rest of the canal was increased as well from 200 to 300 feet at the bottom, making it four times as broad as the projected French canal. It was becoming increasingly apparent that the de Lesseps canal, had it been completed, would have been almost immediately obsolete.

These changes obviously increased the massive excavation still ahead for Goethals and his army regime. But for now this held no fear. By the beginning of 1908, the majority of the workforce was at last engaged in actual excavation, rather than building or sanitation work. 'The biggest boss is King Yardage,' wrote an American journalist who visited in February 1908. 'A toiling, moiling, delving potentate to whom all make obeisance, and who imperiously demands results every minute of the day.' And the results were spectacular. In 1908, 37 million cubic yards were removed, more than double the previous record year and about half of what the two French companies had achieved in seventeen years. The era of the 'solid inevitability' of the American canal seemed at last to have arrived.

*

But the turnover of skilled American staff, still running at a rate of nearly 100 per cent a year, remained a concern of the canal leadership. During 1907, more than three thousand new skilled workers had to be recruited in the States to keep up a Gold Roll force that in the middle of the year numbered only 4400. The response was to accelerate the process, started by Stevens and Magoon, of providing for the white workers every possible convenience and luxury.

Each morning a supply train of twelve cars left Cristóbal for the line, containing five of ice and cold storage provisions, two of bread, one of vegetables and four of staple commissary supplies. From April 1908 the bakery started producing pies and pastry in huge quantities, and the cold storage facility was expanded to include an ice-cream factory and a coffee grinding plant. Laundries and drying rooms were constructed for the Gold Roll employees. More and more stores were opened, to the dismay of the Panamanian merchants and delight of house-and-home runners like Rose van Hardeveld. When a commissary at last opened in Las Cascadas the vegetables might have been thin on the ground, she reports, but the staples were plentiful. And if you got there at eight in the morning, as most tried to do, you might even find something new to break the monotony. Rose felt that a corner had been turned: 'I realized . . . that the last vestige of fear and uncertainty seemed to have left us when our children were able to buy ice cream cones and soda pop at the clubhouse . . . we now felt thoroughly at home, truly, now, a transplanted bit of the United States.' Jessie Murdoch, the Ancón nurse who had arrived back in 1904, expressed a similar sentiment. By mid-1908, she writes, 'we were surrounded by all the modern comforts and conveniences. Telephones buzzed, electric lights were flashed on, and we recognized ourselves as a part of an ideal community.'

For Rose van Hardeveld, even more important than home comforts was the growing number of families on the Zone. Roosevelt's visit had helped improve the image of Panama, and the ICC offered strong inducements, mainly in the form of superior housing for married workers. By May 1908 there were well over a thousand families on the Zone, and a riot of weddings. On one steamer ten brides arrived from the United States and were all married on the dock within twelve minutes of disembarking. The bachelors on the Isthmus who could not persuade their sweethearts to join them had to look closer to home. This meant the nurses of Ancón hospital, who consequently could take their pick.

Once assigned married quarters, the young couples found that virtually everything was provided free by the ICC, including rent, light, janitor service, ice, distilled water and fuel, as well as hospital and medical care. All the bride and groom need buy was bedclothes and china. As the Zone policeman Harry Franck pointed out, 'It is doubtful, to be sure, whether one-fourth of the "Zoners" of any class ever lived as well before or since. The shovelman's wife who gives five-o'clock teas and keeps two servants will find life different when the canal is opened and she moves back to the smoky little factory cottage and learns again to do her own washing.'

In the summer of 1908 the van Hardevelds were told that a new house was ready for them nearby. Before moving out of House Number One, they took a holiday back in Nebraska, returning in November with a new addition to the family, a son. Their new dwelling was 'one of the brand new cottages over the hill . . . painted battleship grey'. Although the mould and insects soon moved in as well, Rose professed herself very pleased. The house had modern plumbing and electric light, and was 'clean and comfortable, just about the type of home a man in the States would try to provide for his family'.

There were now nearly forty families in Las Cascadas, a far cry from how it was when they first arrived. Families had been encouraged, of course, to give stability to the workforce, and as a way of keeping the men on the straight and narrow. According to Rose, this was working. With the arrival of the wives, 'attendance in the saloons fell off to a considerable degree, and normal social patterns became possible'.

'Our friendships with neighbours deepened,' Rose writes. 'We drew together in a sort of compact clique. How we worried together and laughed together.' The main meeting place was their old House Number One. This had been taken over by their friend Charley Swinehart. His father had died, and so his mother and two teenage sisters had come out to join him and his brother in Panama. Dakota, or 'Cote', Swinehart seems to have become something of a matriarch of the Las Cascadas community. Rose describes her as 'a fragile little person who suffered a great deal from the heat and humidity, but who maintained a cheerful outlook and a brisk efficiency that inspired and reassured us younger women'. The vast majority of the families were very young – they only had, on average, one child each. 'The presence of an older woman in our midst,' says Rose, 'lent a feeling of stability to our otherwise youthful community.'

Jan remained obsessed with the canal and would spend his evenings talking to Charley Swinehart about yardage excavated, the best dynamiting techniques and the challenges still ahead. The shared canal-building task – vast, historic, epic – united and inspired many of the Americans on the Zone. 'Nothing else seemed quite so important as this immense project moving gradually and steadily to completion,' writes Rose. 'Nearly all the women and children felt the same way . . . This was our life. All other things were subordinate.'

But not everyone was so motivated. According to an official report it was 'not until the business depression . . . in the United States, in the winter of 1907–8, [that] was there a lessening of the numbers leaving the Isthmus for the States'. Even in 1909–10 the turnover of skilled workers was nearly 60 per cent. 'Anyone who stays here through a year of it becomes depressed,' wrote an engineer on the project, 'and visions of the home country, with its bracing weather, its familiar scenes and its fond ties, begin to float out on the curling wreaths of smoke from pipe or cigarette.' A journalist who visited in early 1909 found a few Americans who unreservedly loved the country and climate, but in most he discovered 'a certain pathetic note of exile from all that is dear'.

To address this homesickness, it was decided to try to keep the men occupied as much as possible. Two and a half million dollars were allocated each year to entertainments and recreation, some $750 per white employee. Churches and Sunday schools were constructed, and more playing fields laid out. Most important, however, were the Gold Roll clubs, run by the YMCA. By late 1907, there were four in operation, at Cristóbal, Culebra, Empire and Gorgona. Each had bowling alleys, a billiard room, a library and a gymnasium. They also provided the location or focus for a bewildering array of organised activities: sponsored hikes and horse rides through the jungle, amateur theatricals, boat trips to Portobelo, athletics competitions, sight-seeing trips on labour trains to the Cut or the locks areas. Lecturers and professional entertainers were also brought in. There were numerous clubs for games including chess, chequers and bridge. Orchestras, bands and glee clubs were formed, and lessons offered in everything from Spanish to First Aid to Bible study. Over two thousand books were provided in the libraries, where more than eighty US newspapers and periodicals were also available. When the clubs were inspected in early 1908, the visitors were impressed, commending the clubs 'without reservation'. 'They fill a necessary place in the somewhat artificial life on the canal zone,' it was concluded, 'where a body of loyal Americans, far removed from the uplifting influence of home and friends, are performing with genuine enthusiasm a work of great importance to their country, in a climate demoralizing to the white man.'

The white community also had its own ICC-produced newspaper, the *Canal Record*, first published in September 1907, and free to anyone on the Gold Roll. It was determined that this should not replicate the French *Bulletin* – praise of department heads was expressly forbidden – but the *Record* charted the excavation and building work week by week, keeping the community abreast of progress and making them feel involved. By printing the excavation figures of particular divisions or even steam shovels, the paper helped fuel competition among the shovel men and train drivers, thereby increasing productivity. But the *Record* was also the social 'notice board' of the Gold Roll Americans, and as such offers a fascinating glimpse of community life. One issue in mid-September 1907 mentions a new baseball team organised at Culebra composed entirely of men from Georgia. There is notice about a forthcoming entertainment 'to be furnished by Sidney Landon, character delineator'. The results of a recent bowling competition between teams from Empire and Culebra are printed. Chess, chequers and billiard tournaments were, it appears, in progress at two of the clubs.

'Zonians' seem to have been, on the whole, great 'joiners'. By this time there was a plethora of societies, many based around place of origin or trade. Thus there was the Ohio Club, and the Order of Railway Conductors, Brotherhood of Locomotive Engineers, and so on. Around half of the Americans were members of one of the Masonic Lodges on the Isthmus, and there were a large number of social clubs. The same issue of the *Canal Record* includes a notice about the 'Improved Order of Red Men'. This now has seven hundred members, the item reports, with 'tribes' at all the main sites along the line. The meeting is at 7.30 on Friday night and you must come in full Indian outfit, with war bonnets, war paint and all.

It was all wonderfully wholesome, just as American domestic opinion demanded. To many it seemed that the impossible had been achieved – proper society had been created two thousand miles from home in the middle of jungle and depraved natives. In early 1909 Rose van Hardeveld's family moved to Empire. She was impressed. There was 'a really active American community . . . Here were nicely dressed, pretty young teachers and office workers. Clean, fine-looking, bronze-faced young chaps escorted them in the evenings to a dance or to the band concerts.'

One such office worker was Courtney Lindsay, whose long and detailed letters home have survived and offer a picture of everyday life on the Isthmus during the Goethals era. Lindsay arrived at the beginning of June 1907, about a month after Goethals took over. He was just short of his twentieth birthday. He had been working in his home town of Savannah, Georgia, for the local railway company when he met someone recently returned from the Isthmus on holiday. 'He says it is not home, but on the order of a boarding school,' he wrote to his mother about the encounter. 'The fare is not Delmonico's, but he says it is eatable and that if you want you can save half your salary.'

Thus encouraged, Lindsay secured a position in Panama paying $125 a month. His mother was a friend of Major David Gaillard, ICC Commissioner and head of the Central Division, so this might have helped. Lindsay's job was the same as Mary Chatfield's first position – stenographer in Arango's department of meteorology and river hydraulics, based in Panama City. So, like Chatfield, Lindsay would be part of the 'B-echelon' of canal personnel, working far from the construction and excavation 'front line'. He wouldn't even see the canal, apart from the view from the train, until eighteen months after his arrival. His letters show none of the heroic motivation of someone like Jan van Hardeveld.

His first impressions were favourable, however. 'Every day I am better pleased that I came,' he wrote to his mother a week after his landing. He had quickly judged the ICC-provided food – 'things are not always very clean' – and made alternative arrangements, eating lunch at the house of a Jamaican woman. 'I have adapted myself pretty well to the climate and conditions,' he wrote home a week later. He had even put on weight, and was, he reported, taking three grains of quinine every morning.

He was also agreeably surprised by the social life in the city. His boss tended to hand on invitations to gala occasions to his employees. 'The Tivoli is giving a reception and dance tonight to the Vice-President,' he wrote home excitedly at the beginning of July. 'So I am having my dress suit pressed for the occasion. This is the second time I had used it in the month I have been here. I never wore it once in Savannah.' His younger sister wrote to him, asking about the pineapples and her brother's romantic prospects. Pineapples are only fifteen cents 'spickity', he wrote back. 'Yes, there are a great many American ladies, not so many girls. This is a very "marrying" place, and nearly all the good looking girls are Mrs. There are a few exception among the nurses however.'

But it did not take long for Lindsay to adopt some of Chatfield's cynicism about the actual work. Less than a month in, he wrote that as the department boss was away, 'things have already begun to slack up. This job is like plenty of Gov't places in the States. There are one and a half men to do one man's work.' His immediate superior was an Englishman, Vince, 'who seems to have caught the "manana fever".' He was dismissive, too, of the endless Congressional committees visiting the Isthmus, assessing the works while being hosted to a round of dances and parties. 'Now what can they tell about it?' he asked in a letter in November. 'Seems to me it is a trip on Uncle Sam.' The following month he reported that 'the novelty has worn off and nothing ever happens'. There was a friend of his, Hugh Wills, due out soon to join him, but for now he felt homesick, left out of his life at home ('I'm doomed to bachelor hood'), and sad about being away for Christmas.

His first Christmas Day on the Isthmus turned out to be all right. He went fishing in the Bay of Panama, had dinner at the Tivoli and then went to a party at the Corrozal Club, where there was singing, stories, music and boxing bouts. Soon after, he took a week's sick leave at the sanatorium on beautiful Taboga. All Gold Roll employees were entitled to fifteen days' paid sick leave per six months work. For many, this was just a nice extra holiday. Lindsay says that while he was on Taboga 'I've never felt better in my life'. In the New Year his hometown friend arrived and Lindsay began to feel settled in. His bachelor residence was refurbished and electric lighting installed. There were trips to Portobelo and to Old Panama, the city up the coast destroyed by Sir Henry Morgan back in 1671, now a picturesque ruin. In March he reported that a show he had attended, the 'Empire Lady Minstrels' was 'the best amateur entertainment I've seen in a long while'.

All the while he was learning Spanish, and was proud to report that he could now say, 'I have neither one nor the other but I have the trunk which the sailor from the ship of the Captain gave me.'

In August 1908 he returned home for his annual leave, and when he came back he found himself posted to Culebra. It was a bit of a comedown after Panama City. 'Nothing ever happens here,' he complained in a letter to his father. 'The only thing worth mentioning since I came on the 20th of last month has been the repair of the YMCA phonograph.' But he soon adjusted. 'Am beginning to like Culebra better,' he wrote home the following month. 'The YMCA is an oasis in the desert. I enjoy the bowling, indoor baseball and gymnasium very much.' In January, he reports, he 'broke into Culebra society' by attending a dance at the club. 'The hall is small, and the floor not near so good. Still it is rather livelier than the Tivoli, – it's more like a country town where everybody knows everybody else.'

After leave in June 1909, taken in nearby Costa Rica, Lindsay was moved again, this time to the quartermaster's department. His bosses were awful – 'Pyne stupid and Saville unpleasant . . . he must have what you call a "bullying" disposition' – and he was compelled to board at the ICC hotel. 'It is pretty bad at present,' he reported of the food. 'They have a formidable bill of fare for each meal, but if you get three dishes you really like or want you are lucky. "You couldn't buy the same meal in the States for the price" is the howl of the Subsistence people, but who would want the same meal?'

His only health scare came at the end of this year. He spent Christmas in hospital, thinking he had malaria. But no 'bugs' were found in his blood, and although he was given quinine 'steadily', he was discharged after a week. The following year saw the young American acting in a farce called *Facing the Music* and learning bridge; there were moments when he really felt in love with Panama. 'It seems to me,' he wrote to his mother in October 1910, 'that a dry season night down here, with a moon, is about as near perfection as this world ever gets.'

The following year, it would get even better. In May his friend Hugh became engaged to one of the Ancón nurses, a Miss Dequine. Before the end of the year, Lindsay had followed suit, having met an English nurse, Olive. 'She's just about the nicest thing in the girl line there is,' he told his parents. As soon as family quarters became available, they married at the small Colón Episcopal Church, Christchurch-by-the-Sea, built back in the 1860s by the Railroad.

The majority of Gold Roll employees accepted the way in which the canal authorities dominated and organised their lives. Visiting journalists, however, were fascinated by the white society that had been created. Everything, all the essentials of life, were supplied by the 'state'. What was this system? they asked. Was it some form of 'military paternalism'? Or 'welfare socialism'? Certainly life on the Zone had little in common with the ideas of the capitalist democracy at home. No one was allowed to own meaningful property or vote for the Zone government. And it seemed to work. 'The commissary is an assured success,' reported one journalist in early 1909. 'It has shown the absurdity of the ancient superstition that organized society, the state, cannot attend to the needs of a people as economically and with as efficient service as can an individual or a corporation.'

American journalist Arthur Bullard, who had been in Barbados to watch Karner recruit labourers, actually tracked down a member of the US Socialist Party among the Gold Roll workforce. He asked him whether he was living in his ideal state. 'First of all, there ain't any democracy down here,' Bullard was told. 'It's a Bureaucracy that's got Russia backed off the map . . . Government ownership don't mean anything to us working men unless we own the Government. We don't here – this is the sort of thing Bismarck dreamed of.'

The concentration of power in one man, the 'Czar of Panama' George Goethals, had resulted in, one journalist wrote, 'the establishment of an autocratic form of government for the Canal Zone . . . not in accord with the principles of democracy'. Zone policeman Harry Franck described the regime as 'enlightened despotism'. According to him, it seems to have succeeded through the character of Goethals himself, whom he describes as 'an Omnipotent, Omniscient, Omnipresent ruler'. This is echoed in many other accounts. Rose van Hardeveld also calls Goethals 'omnipresent'. 'The Old Man' as he became known, 'was so constantly on the job that we never thought of him as being at home or eating or sleeping'. 'Goethals dominates over everybody and everything,' Mallet reported to London. To anyone in Panama, this was unmistakable. 'You can't realize what the Chief Engineer is until you live on the Isthmus,' wrote Courtney Lindsay home. 'His power is as near absolute as any man's can get.'

Goethals himself was uneasy about the nature of his regime but believed it was only way to get the canal built. But there is more than a hint of Big Brother in some of his methods. Officials, known as 'spotters', toured the works and the terminal cities in disguise, not only to punish 'loafing', but also to weed out potential troublemakers. The spotters had the power to deport 'undesirables'. 'The system is one that would be very repugnant to Englishmen,' reported a London journalist. 'Employés dismissed are given no notice nor granted any compensation.' 'The men complain of the savage rigor with which even petty misdemeanours are punished all along the Zone,' wrote an American journalist in an article published in April 1909. Two men who attacked the Zone government in a New Orleans newspaper made the mistake of remaining on the Isthmus. They were arrested, prosecuted for criminal libel and imprisoned. A watching journalist described the sentences as 'judicial terrorism . . . the kind of justice dealt out in some, if not most, of the courts is not the sort that would be tolerated long in a democracy'.

But on the whole, more subtle pressures were sufficient to keep criticism at bay. Those who complained were dismissed as 'kickers'. As one perceptive American journalist explained: 'there has grown up in Panama circles somewhat of a tendency to monopolise patriotism, and identity it with official designs, means, methods, and management. Dissent or a different viewpoint is too often hailed with cries of "Enemies of the Canal."' To criticise the leadership, then, was un-American, as many Zonians believed. One of the first resolutions of the new Women's Club in Panama was 'That every club-woman in the Canal Zone constitute herself a committee of one to foster favourable instead of adverse criticism of the conditions of the Zone and of the Isthmus of Panama.'

Goethals sat at the top of a rigid hierarchical structure. There was a racial 'ladder', of course, with the Americans and the hundred or so British at the top; next came the Panamanian and Spanish 'almost-whites'; at the bottom were the blacks, with the West Indians beneath the locals in status. But within the small white community there was also an obvious pecking order. 'Caste lines are as sharply drawn as in India,' wrote Harry Franck. 'The Brahmins are the "gold" employees . . . But – and herein we out-Hindu the Hindus – the Brahmin caste itself is divided and subdivided into infinitesimal gradations. Every rank and shade of man has a different salary, and exactly in accordance with that salary he is housed, furnished, and treated down to the least item, – number of electric lights, candle-power, size of bed, size of bookcase.' Such differences in status felt immensely important. 'D, who is a quartermaster at $225, may be on "How-are-you-old-man?" terms with G, who is a station agent and draws $175. But Mrs D never thinks of calling on Mrs. G socially,' Franck continues.

Diplomat William Sands would soon be leaving the Isthmus. He had offended Mrs Amador by mentioning her wayward son, who had got himself into difficulties with a woman of sensational reputation in the United States while serving as ambassador there. When Sands returned from his holiday in early 1907 he was refused re-entry into the country as *persona non grata*. It is unlikely that he would have minded much. He hated Panama, or, more exactly, the American community there. 'It was . . . a narrow ribbon of standardized buildings and standardized men working at standardized jobs,' he later wrote of the Zone. 'Its people had been hammered into a highly disciplined civilian army, into a mechanical state with rigid hierarchies of labor, by processes which at that time seemed utterly un-American.' Everyone behaved and looked the same – 'six gallon hat, khaki shirt and breeches, and protective leggings' – 'to the vast bewilderment of the native, who had previously thought of us as a race of extreme individualists and superlatively free men'.

But if the Zone had become a 'drearily efficient state', the terminal cities remained, in comparison, anarchic and chaotic. Panama City had more than two hundred bars, Colón 131, including forty in one street. Every Saturday night special trains were laid on, bringing in hundreds of canal workers. For Harry Franck, Panama and Colón acted 'as a sort of safety valve, where a man can . . . blow off steam; get rid of the bad internal vapors that might cause explosion in a ventless society'.

It was, then, a less than ideal setting for the fostering of understanding and respect between Americans and their Panamanian hosts. The Americans 'blowing off steam' seemed to the locals loud, rude and drunk. Harry Franck describes an American 'type' 'grown so painfully prevalent': 'a chestless youth . . . whose proofs of manhood are cigarettes and impudence and discordant noise, and whose national superiority is demonstrated by the maltreating of all other races'. Clashes continued, often centred on the notorious Cocoa Grove brothel in Colón. In September 1908, in Panama City, an American was killed and another seriously wounded in an altercation between citizens and US sailors. The Zone authorities demanded and received permission to patrol the terminal cities to prevent fighting reoccurring, and the Panamanian government, protesting that the fault was with the drunkenness of the Americans, was forced to pay an indemnity of nearly half a million dollars to the US.

Many Americans considered the Panamanians backward, deluded about their own importance and lacking in gratitude for everything the United States had done for them. In March 1908 an article appeared in the Philadelphia *Saturday Evening Post* entitled 'Life in Spigotty Land: the Cohorts of King Yardage'. Its author, Samuel G. Blythe, describes the 'scores of clear-eyed, broad-shouldered and hard-headed Americans who are carrying out their part of the work without hope of fame, but as Americans, doing an American job in an American way'. Little heed was paid to the Panamanians, he wrote, 'who are funny little people, vainglorious and thinking they achieved their own liberty, have an idea the canal is being dug for their especial benefit'. In fact, the truth was 'The republic was made for our convenience and it is held up by the scruff of its neck by this Government.'

Indeed, neither of the Republic's political parties could afford to antagonise Uncle Sam. The Conservatives, many still favouring annexation by the US, saw American power as the best defence of their interests. The Liberals needed the Americans as well. Unless the forthcoming 1908 presidential elections were supervised by the US, there would be an inevitable repeat of the fixed municipal and National Assembly votes of two years previously. The government candidate, backed by the outgoing Amador, was Ricardo Arias, brother of Tomás, and an ultra rightist. The Liberal Party, thinking its own candidate would never stand a chance, backed Obaldía for the presidency. Obaldía had been a popular vice-president and was also close to the Americans, having served as ambassador in Washington.

It proved a shrewd move. Taft considered Arias corrupt and utterly lacking in scruple, while Obaldía was seen as the best guarantee of a smooth succession. Predicting that Arias, with Amador's help, would so blatantly fix the vote that an uprising would result, he decided to acquiescence in Liberal demands that the United States supervise the election. In the meantime, William Nelson Cromwell popped up again helping run Obaldía's campaign. Realising he was fatally out of favour with the Americans, Arias withdrew and Obaldía was elected unopposed.

Clearly the Liberal Party no longer seemed a threat to American interests in Panama. Taft affirmed, 'We have such control in Panama that no Government elected by them will feel a desire to antagonise the American Government.' Indebted to the US for its 1908 election victory, the Party under Obaldía's presidency would make no trouble for Goethals. But then, at the beginning of 1910, the President died. His successor as first designate was Carlos Mendoza, who had been a leader of the revolution in Colón back in November 1903. According to Mallet, Mendoza was 'extremely tactful and friendly towards everybody', but for the Americans there was a problem. Mendoza was a mulatto. Not only would having a non-white Panamanian President contrast a little too sharply with the racial hierarchy in the Zone, but Mendoza, according to Sands' replacement as US chargé d'affaires, possessed 'a racial inability to refrain long from abuse of power'.

Mallet reports that on the prompting of Goethals, a junior officer in the US legation, Richard O. Marsh, 'put it about to the most notorious babblers in the city that the United States Government would regard the election of Senor Mendoza as unconstitutional, and that if the National Assembly persisted in his election a military occupation of Panama would be the inevitable result'. Mendoza formally withdrew his candidature and an elderly white Liberal patrician was installed as President. 'It is really farcical to talk of Panama as an independent state,' wrote Mallet to the Foreign Office. 'It is really simply an annex of the Canal Zone.'

The United States' grip on the Republic would last until the end of the construction period (and, of course, beyond). Mallet reported in 1913, after the election the previous year of Belisario Porras, the erstwhile 'notorious hater of foreigners', that it was now impossible to be President without being 'docile to American wishes'.

In the United States, the end of 1908 saw a presidential election campaign between Taft and William Jennings Bryan. Roosevelt had publicly backed his Secretary of

War. But then, a month before the vote, Panama was once again front-page news. The story had been reignited in September the previous year when Cromwell's claim for payment from the New Company, being arbitrated in Paris, was leaked to the New York press. Undoubtedly the lawyer had put a favourable gloss on his description of work performed for his client, but the extent of influence he claimed in the heart of US government was deeply unsettling. Then, in October 1908, the New York *World* published a story that accused Taft's influential brother Charles, and Roosevelt's brother-in-law, Douglas Robinson, of being members and beneficiaries of the syndicate supposedly set up to profit from the sale of the New Company to the US government. Further allegations were made during the campaign as the Democrats saw a way to attack the Republicans. Cromwell, the *World* claimed, was 'practically the Secretary of War as far as the Panama Canal was concerned' and that his 'law offices at No. 41 Wall Street were even regarded by many as the real executive offices of the Panama Canal'.

Roosevelt, furious that what he considered to be the greatest foreign policy achievement of his administration was once again mired in scandal, brought a prosecution for criminal libel against Joseph Pulitzer, owner of the *World*, two of his editors and two publishers of the Indianapolis *News*, which had picked up the story. To prepare his defence, Pulitzer sent two of his best investigative reporters to Washington, Paris, Panama and Bogotá to get the 'Untold Story of Panama'. Followed everywhere by secret service agents, they found obstruction at every turn. In Paris they were told that the details of shareholders were in a sealed vault. When their lawyers eventually got this overturned, they found the records virtually non-existent. The paper's British counsel commented, 'I have never known in my lengthy experience in company matters any public corporation, much less one of such vast importance, having so completely disappeared and removed all traces of its existence as the New Panama Canal Company.' In Panama, the journalists found vital cable evidence destroyed and the 'revolutionaries' – unwilling to lose the trust and support of the United States – good at keeping political secrets. They denied everything, even meeting Cromwell.

But by now the cases had become more about freedom of speech and federal versus state government than specific allegations. When the trial started in 1909, it was deemed unconstitutional for the government to 'drag citizens from distant States to the capital to be tried'. Judge Anderson dismissed the case, and the evidence of the syndicate and of United States collusion in the 'revolution' was never put to the test. The judge did, however, have one final comment to make on the case: 'There are many peculiar circumstances about the Panama canal business,' he said. 'Rather suddenly it became known that it could be procured for $40,000,000. There were a number of people who thought there was something not just exactly right about that transaction, and I will say for myself that I have a curiosity to know what the real truth was . . . I am suspicious about it now.'

On the Isthmus, however, the 'Army of Panama', now numbering nearly fifty thousand, kept up 1908's high excavation total during the following year, as fatalities, particularly from disease, continued to fall. The largest cause of death, for the first time, was in 1909 from accidents on the works. On the Atlantic Division huge amounts of silt and sand were being sucked from the channel from coast to deep water, the old French canal from the bay to the site of the Gatún dam and locks had been effectively redredged to carry materials to the construction side and old French-era dredges, many twenty-five years old, were making progress in the channel from the bay to the locks.

On the Pacific Division there had also been steady progress, working in from the sea, shifting material with dredge, shovel or hydraulic jet. In the harbour, an 11,000-foot-long breakwater was under construction to guard against submarines, and to prevent the channel from silting up. It was proving slow work as the dumped rock either disappeared into the mud of the bay or pushed up other sandbanks nearby. Eventually over ten times the originally estimated quantity of spoil would be required. But in the Central Division, the infamous Culebra Cut, *la grande tranchée*, had shown itself once more to be the biggest challenge of all.

25

'HELL'S GORGE'

In Culebra, the mountains were on the move. Work on deepening the Cut had begun in earnest with the end of the steam shovelmen's strike in July 1907. But as the gorge grew in size, it was as if the land was fighting back. On the night of 2 October, after particularly heavy rain, a great mass of earth and rock plunged down into the Cut from the slope just south of Gold Hill at Cuçaracha. Two steam shovels were overturned and nearly buried, track and piping carrying water and compressed air was destroyed, and the drainage system wrecked. Horrified engineers then noticed that the slide was continuing. An area of about fifty acres continued to move for the next ten days, sliding into the canal prism at a speed of about 14 feet a day. Gaillard, the engineer in charge of the Cut, described it as being 'a tropical glacier – of mud instead of ice'. Goethals reported that 'it required night and day work to save our equipment'. By the time the equilibrium of that particular part of the mountain had been restored and the movement stopped, over half a million cubic yards had entered the Cut. In his end-of-year report for the Foreign Office, Mallet wrote, 'the magnitude of the task is much greater than was at first thought'. 'There is less disposition,' he went on, 'to under-rate the French failure.' Certainly Goethals quickly revised his earlier view – the job in the Cut was now deemed the 'most formidable of the canal enterprise'.

The first great Cucaracha slide was just the beginning. As the ditch was lowered foot by foot, there followed numerous similar 'gravity slides'. In many places along the walls of the Cut, a layer of semi-porous clay sat on top of a stratum of impervious rock. Rainwater seeped through the clay to form a soapy, greasy layer on top of the harder rock. When this rock sloped towards the Cut, there would come a point when the friction between the two layers became so reduced by the water on the join that the top layer slipped into the Cut, 'like snow off a roof'.

The following year, 1908, saw slides of this type at Paraíso, near Gold Hill and at Culebra.

But such was the baffling geology of the Cut, with rocks of all different types in bewildering combinations, that gravity slides, most usual during the wet season, were not the only problem. Some of the strata, previously long-buried, reacted to the air in a way that caused them to become unstable and unable to support material lying above. Other harder rock, depending on its lines of fracture, would collapse into the Cut when its lateral support was removed, bringing down upper layers as well.

Spaniard Antonio Sanchez, who worked in the Cut for four and a half years, told of a curse going back to French times. The ground itself, he said, would take revenge against those who sought to 'dissect nature's creation'. Most vivid in his memory was the shrill sound of whistles from an accident site. If they were not at work, foremen would appear at their camp to demand that they come to help dig out men and equipment buried by slides. But while they were attempting to rescue the buried men, they too were subject to the danger of further slides. Sometimes the deep mud they worked in would prevent them from getting out of the way quickly enough if there was another earth movement. The majority of the time, he reports, they would only manage to dig out disfigured and broken bodies.

Most of the slides, however, were slower, although more substantial than these sorts of avalanches. But even if a slope moved towards the Cut at only a few inches a day, it still required the re-laying of miles and miles of track. And as soon as a slide was cleared, an engineer remembers, 'the old hill politely slid back again, completely filling the canal'. As a West Indian worker put it, 'Today you dig and tomorrow it slides.'

The slides made Culebra an unpredictable enemy for the 'Army of Panama'. The deep gorge, wrote a senior US administrator, 'was a land of the fantastic and the unexpected. No one could say when the sun went down at night what the condition of the Cut would be when the sun arose the next morning. The work of months or even years might be blotted out by an avalanche of earth.' At the end of 1907 Goethals had to refuse to set a completion date for the canal as, he said, 'The difficulties we are liable to encounter are unknown to ourselves and uncertain.' Another senior engineer confessed that it was impossible to plan for the final shape of the ditch as 'This material has or will ultimately make its own design as to slopes.' In other ways as well, the Cut was 'fantastic and unexpected'. Such was the geological chaos of the ground that dynamiters and steam-shovel operatives, as Goethals explained, 'found themselves handling hard rock one hour, while the next hour they might be working in earth or clay'. In some places the downward pressure of the unsupported rock faces would push up the comparatively soft strata of the canal floor, sometimes as much as 30 feet. On once occasion, Gaillard himself was standing at the bottom of the Cut when the ground he was standing on rose 6 feet in five minutes. Just as uncanny for the diggers were the cracks that appeared at the bottom of the Cut, spewing out stinking sulphurous fumes or boiling water. Blasts hot enough to char wood were emitted from the ground, caused by the oxidation of iron pyrites in the soil or by the vaporisation of water in the intense heat of friction as the despoiled ground writhed and slipped.

To prevent the slides Goethals tried all the techniques then available. Shovels worked at the top of the slopes to reduce the weight of material pushing down on the lower levels. Long 'nails' were driven into the sides of the trench to bind the porous layer to the rock beneath; slopes were plastered in concrete. Such measures were being used with success at the time by the British in Hong Kong, but the scale and complexity of the Cut would doom all to failure. To keep water from the slopes, large diversion channels were built near the crests to carry away moisture that might otherwise saturate the sides of the Cut. But this approach failed as well. The only option left was simply to dig it all out again.

Unlike the French, Goethals had the muscle to do it. The massive steam shovels, by now personalised with female names, were removing huge amounts of dirt. On average over the construction period they dug out a million cubic yards each, a testament to their sound construction and efficient maintenance. In the peak month of March 1909, there were sixty-eight shovels at work in the Cut excavating an astonishing 2 million cubic yards. In the same month seven hundred thousand pounds of powder was exploded. Thanks to the system set up by Stevens, 160 trains, carefully controlled from the construction headquarters at Culebra, ran in and out of the gorge every day, pulling thousands of flat cars to and from the dumps; in the Cut's nine miles there were now seventy-six miles of construction track.

The work never stopped. At the end of the day, the track shifters, dynamite gangs and steam-shovel operators were replaced by coaling trains and maintenance crews who worked through the night so that nothing should delay progress the next day. And although about a quarter of the effort involved digging out material from slides, the Cut still got deeper and deeper and more and more spectacular.

The Cut was the 'special wonder of the canal', 'one of the great spectacles of the ages'. Over 70 per cent of the vast total canal excavation came from its nine miles. For the increasing number of tourists, gazing down into the great man-made canyon from its edge high above, it was an inspiring sight. 'The Cut is a tremendous demonstration of human and mechanical energy,' wrote a British visitor. 'It is simply the transformation of a mountain into a valley.' It was more than 'heroic human endeavour,' said another. It was a 'geological event'. The scale was overwhelming. 'From the crest,' wrote an American tourist, 'you looked down upon a mighty rift in the earth's crust, at the base of which pygmy engines and antlike forms were rushing to and fro without seeming plan or reason. Through the murky atmosphere strange sounds rose up and smote the ear of the onlooker with resounding clamor.' These included the 'strident clink, clink, clink of the drills . . . the shrill whistles of the locomotives . . . the constant and uninterrupted rumble' of the ever-moving dirt trains, the 'clanking of chains' of the shovels, 'the cries of men, and the booming of blasts. Collectively the sounds were harsh, deafening, brutal such as we might fancy would arise from hell.'

William Baxter started work as an official guide in 1911. In that year there were 15,000 tourists. 'They are generally comfortable men and women of 50 or more,' he wrote. The English tended to wear cork hats, though 'some American men dress as if for a trip through the jungle when they go out on a sightseeing train. Most women wear heavy ugly shoes. All tourists carry umbrellas.' 'Patriotic tourists, or perhaps it would be better to say "chauvinistic tourists," are rather common,' he continues. 'They have two great topics: "The French Failure" and "The Cost". It is futile to explain to them that a private company of Americans would have failed as the French company did, under the same conditions. "We have done it, and they failed," is always the answer.'

For the 'antlike' figures working far down below, the Cut was known as 'Hell's Gorge'. The noise alone is hard to imagine. On a typical day there would be more than three hundred rock drills in operation as well as the steam shovels, trains and the blasting of some six hundred holes, with all the booming and crashing reflected and amplified by the walls of the 'big ditch'. But that was only a part of it. As a Barbadian dynamite carrier, Arnold Small remembered, 'There was no shelter from the sun or the rain. There were no trees, then, just a bare place. When the sun shine, you get it, when it rain fall, you get it. When the wind blow, you get it.' Ten feet of rain fell in the Cut during 1909, converting it to a muddy nightmare. The Barbadian John Prescod was working at Bas Obispo 'at the steam shovel in mud and water. One pair of boots last me one day. In the afternoon walk to the camp barefoot.' 'I had never saw so much rain in all my life as I see in the Cut,' says another digger. 'You had to work all through the rain, I remember when I was in the drilling gang, the boss allway say keep the drills agoing so as to keep your body warm sometimes, you are so cold that your teeth keep rocking together, in the morning you had to put your clothes on damp no sun to dry them.'

In the dry season the perpetual wet was replaced by a 120-degree heat, and clouds of choking dust. 'For the first couple of days or weeks, you are always out of breath,' says Arnold Small.

Harry Franck vividly remembered the day he spent during 1912 traversing the Cut enrolling workers in a census he was conducting. 'The different levels varied from ten to twenty feet one above the other, each with a railroad on it, back and forth along which incessantly rumbled and screeched dirt-trains full or empty, halting before the steam-shovels, that shivered and spouted thick black smoke as they ate away the rocky hills and cast them in great giant handfuls on the train of one-sided flat cars that moved forward bit by bit at the flourish of the conductor's yellow flag. Steam-shovels that seemed human in all except their mammoth fearless strength tore up the solid rock with snorts of rage and the panting of industry, now and then flinging some troublesome, stubborn boulder angrily upon the cars . . . Each was run by two white Americans . . . the craneman far out on the shovel arm, the engineer within the machine itself with a labyrinth of levels demanding his unbroken attention. Then there was of course a gang of negroes, firemen and the like, attached to each shovel.'

All around, scores of drills were 'pounding and grinding and jamming holes in the living rock'. Anywhere near them was 'such a roaring and jangling that I must bellow at the top of my voice to be heard at all. The entire gamut of sound-waves surrounds and enfolds me.' There were gangs everywhere, on the floor of the canal and on the terraces and 'stretching away in either direction till those far off look like upright bands of the leaf-cutting ants of Panamanian jungles'. And over it all hung heavy clouds of coal dust from the trains and shovels.

With so many men and machines crowded into this narrow space, almost nowhere was work more dangerous or life cheaper. 'There were so many engines at a time in the Cut,' remembers Rufus Forde, 'that most every month, a man lost his leg or badly damage. When any thing like that happen one engineer will turn to next engineer, one just grease the wheel. In those days a fowl life was more valuable than our lives.' One Panama-born West Indian remembered seeing a man cut cleanly in two, with his legs carried away by the train, which did not bother to stop. 'Billy had been the engineer. He will stop his train on the tracks for a horse or a cow, but not for a human. Those were his words always.'

Tales of serious danger from accidents dominate the accounts of the West Indian workers. One remembered seeing a Spanish track layer hit by a locomotive and pushed for about 20 feet along the track. 'He was still alive, with mostly all his skin was stripped off like a piece of ham bone. All I could hear him say was "Mi madre, mi madre!"' Harry Franck reckoned you needed 'eyes and ears both in front and behind, not merely for trains but for a hundred hidden and unknown dangers to keep the nerves taut'. Scores of men were killed by being hit by the swinging boom of a steam shovel. Jan van Hardeveld narrowly escaped being crushed trying to right an overturned shovel, but soon after had his leg badly injured by a flying spike maul. Antonio Sanchez worked several months of

310 hours, the overtime being night work on track relocation when 'the mud and slime was always present as well as the danger of the various spoil trains and rolling stock in dark and rainy nights'. In March 1909 he was disabled for three months when his foot was crushed by the wheel of a train.

Even worse than the traffic and machinery was the vast amount of dynamite being deployed to break up the rock so that it could be handled by the shovels. Accidents were numerous. Goethals blamed the incompetence of the workers, but some of the explosives became unstable from exposure to the Panamanian climate. On other occasions the subterranean heat in the Cut ignited the charges before the men were safely clear. Once, a premature explosion was caused by a bolt of lightning during a storm, killing seven men. The most common danger, however, was when excavating machinery hit unexploded charges. 'It was a very awful sight to see how they dig out the bodies,' remembered Constantine Parkinson, 'but it did not mean nothing in construction days people get killed and injured almost every day and all the bosses want is to get the canal built.'

The worst such accident occurred in December 1908, at Bas Obispo at the north end of the Cut. 'Preparation was made to shoot down a high Hill in the centre of the waterway on Sunday A.M.,' says Jamaican Z. McKenzie. 'Unfortunately on Saturday about 12.30P.M. the blast went off. I just leave the Gang to eat my lunch. I ran to the Spot & Saw what happened. Oh, it was a day of Sorrow for the living.' The accidental ignition of 22 tons of dynamite, in two separate explosions, was heard three miles away and left 60 injured and 23 men dead, 17 West Indians, 3 Spaniards and 3 Americans. West Indian Amos Clarke remembers seeing 'flesh hanging on the faraway trees. It was something terrible an awful to look at.'

Antonio Sanchez described going to work every day in the Cut as like 'going to a battlefield . . . we had to sweat and be brave'. Even at a supposedly safe distance from the great explosions, bits of rock would be flung into the air for hundreds of feet. 'Many times the rocks would hit labourers with such impact that they would fall unconscious on the spot,' he said. 'As there were no other means for our protection, we used our shovels to cover our heads from the impact of the flying projectiles.' Harry Franck noticed that the track switchmen, or 'switcheroos', built sheet-iron wigwams, not as shelter from the sun, but as protection from flying rocks.

John Prescod was in a drilling gang near Empire in mid-April 1913. In one 'difficult place', at the bottom of a steep and unstable cliff, it was impossible to set up the drills due to rocks falling from above. His foreman was Charley Swinehart, the friend of the van Hardevelds. 'General foreman came to spot,' Prescod wrote, 'say your all don't started up yet no boss rock falling down un us. Say if I go up and set up a drill God dam it I going to fire the whole bunch of you I am sorry to say sad accident occurred. Rock fall from the bank knock Mr. Swinehart down in the canal Put him on a flatcar rush him to Ancón Hospital die the same day.' According to his official record, thirty-two-year-old Swinehart, an old-timer having been in Panama since April 1905, died of a fractured skull. Rose van Hardeveld says that he was still breathing as the hospital car rushed him to Ancón, but he passed away before his mother could reach his bedside. 'Two days later, all of us who had become such close friends gathered in the hospital chapel to weep with the bereaved mother and sisters,' Rose writes. The surviving family returned, 'broken-hearted', to the States.

To stop the flooding Chagres from flowing into the Cut as the trench deepened, a huge earth dike was built across the north end at Gamboa. But as in French times, frequent flash floods caused delays and damage to equipment. Still the work was pushed on, even when, in 1910, the Cucaracha slide started up again. By 1912, it had deposited over 3 million cubic yards into the canal prism. And now the other side of the trench had come to life. At Culebra a huge crack appeared about 100 yards away from the crest of the Cut. The new clubhouse was disassembled and moved away, as were some thirty other buildings in the town. But still the crack widened as the edge started to slip inexorably downwards. Eventually seventy-five acres of what had been the town fell away into the canal. The mass dumped was twice that of the Cucaracha slide.

Goethals simply ordered it dug out again, but Gaillard was distraught. Then in January 1913 Cucaracha slid again, this time completely blocking the end of the Cut. For Gaillard, this seems to have been something of a final straw. He appeared to suffer a breakdown and left the Isthmus. Back in the United States he was diagnosed with a brain tumour and died before the end of the year. Everyone assumed that it had been the pressure of digging the Cut that had killed him. His fellow engineer and friend William Sibert wrote of his death: 'at the end of long years of patient, exacting work, of terrific responsibility, the tragic end has come . . . just as much a direct result of the struggle itself as if it were the work of a hostile bullet.'

Even as the shovels lowered the floor of the canal scoop by scoop, there were many who believed that the Cut could never be finished, that it would continue to fight back until it had defeated its despoiler. For their lock-canal plan the French had estimated that 23 million cubic yards had to be moved from the Cut. The Americans initially upped this to 53 million cubic yards, but the estimate rose to 78 in 1908, 84 in 1909, 89 in 1911, then to 100 million in 1913. This was partly because of the widening of the bottom width but also due to breaks and slides. George Martin, who had been on the Isthmus since 1909, remembers when working in the Cut in 1911 his bosses' 'encouraging talk. "Boys, are you saving your money? It won't be long now, we will see water into the Cut." But we just take it for a joke,' he writes. 'I personally would say to my fellow men, that could never happen. My children would come and have children, and their children would come and do the same, before you would see water in the Cut, and most all of us agree on the same.'

George Martin, an apprentice carpenter, had been eighteen when he arrived in Panama from Barbados. He had heard, he writes, 'A voice from a great people' inviting him to help build the Panama Canal. His first job was on the relocation of the Panama Railroad, identified by Goethals as a priority. The line had to be moved to high ground to avoid the area of the proposed Lake Gatún. The job was far bigger than the original railroad construction of 1850–55, requiring either fills, cuts or bridges for most of its length. The work took Martin deep into the jungle, where spiders and snakes abounded, as well as what he called the 'Goosyana Fly' – 'when he stings, he leaves worms in the flesh'. There were also swarms of *Anopheles* mosquitoes. 'The fever lashed good and plenty,' Martin writes. 'In those days you watch men shake, you think they would shake to pieces.' As it was impractical to carry out anti-mosquito measures in the temporary camps along the new line, Gorgas's department was reduced to catching them in traps, one of which netted 1800 specimens in a week. But Martin had a good boss who 'did not order or compelled, he only pleaded, so we obeyed'. Martin worked on a culvert, and, when it was completed, on laying the tracks on it. 'We took the spiking of the rail, to the pulling, like a merry-go-round,' says Martin. 'This were a sight to watch us work along this line; the work was hard but we did it cheerful . . . Every man with an iron bar about five feet long, one would sing, and while he sings, you watch the track line move. The white bosses stand off and laugh. The Songster had a song, goes this way, he would sing the first part, and we comes in with the second part, it goes:

> Nattie oh, Nattie O – first
> 2nd Gone to Colón
> Nattie O Nattie O
> 2nd Gone to Colón
> 1st Nattie buy sweet powder
> 2nd Powder her –––– – – you know
> 1st Nattie buy sweet powder
> 2nd Powder her ––––– – same

And so he would sing this song over and over, gentlemen watch track line move, the work appeared sweet, the white foremen enjoyed the singing they laugh and did laughed.'

Martin's is perhaps the most positive of all the hundred or so accounts in the 'Competition for the Best True Stories of Life and Work . . .' collected in 1963 (he received the second prize of $30). He writes of the incessant rain and constantly wet clothes but adds, 'We worked joyfully in these days.' The food he was given was good, and the box cars they were first billeted in were 'like palaces'. Martin worked for the canal for nearly forty-six years until the mid-fifties. Looking back at the construction days he sees them as a time of excitement, and also comparatively low prices. '$2.50 [commissary] book was plenty in those days,' he writes. 'Construction days were better days, never to be seen again, the money was paid small, but we live big.' On one occasion he was accidentally given two $5.00 books for the price of one. 'What to do with $10 in those days? . . . I bought a ham, at that time it look as big as I were . . . real lean, I took ham to work every day in order to have it finish, my associate and I ate ham for days. I don't think about ham these days it's too high in price, now it is for the other fellow.' Best of all though was 'our ice-cream, I am saying here it was refreshing. We worked hard, but cheerful, I can assure you,' he goes on. 'Our boss never had any worries, he only says what he wanted, and it was done.'

Mallet reported in early 1913 that the West Indians had become a 'fine body of disciplined and skilled workmen'. Many of the Americans were beginning to agree. The secretary to the Commission, Joseph Bishop, would later write that the work of the West Indian artisans 'proved very satisfactory'. Overall, he continued, the West Indians were 'quiet, usually honest, as a general rule well and respectfully spoken, demonstrating an aptitude to learn the rudiments of the various sorts of work for which they were contracted'. Another engineer declared that the West Indian labourer had lived down his bad reputation and developed into a good workman, 'and pretty certain always to make a fair return to the United States on the money it paid him in wages . . . The American republic always must stand indebted to these easy-going, carefree black men who supplied the brawn to break the giant back of Culebra.' Unlike the Spaniards, wrote Harry Franck, 'the negroes from the British West Indies . . . could almost invariably read and write; many of those shovelling in the "cut" have been trained in trigonometry'. (Not that any of this was reflected in their pay. A black West Indian would have to be skilled and to have served for a number of years to earn as much as the minimum for a European labourer – US$0.20 an hour. By September 1909 fewer than a thousand had qualified.)

As a rule, the West Indians were sober, industrious and religious. Harry Franck remembers frequently coming across 'young negro men of the age and type that in white skins would have been loafing on pool-room corners, reading to themselves in loud and solemn voices from the Bible'. 'What was the black culture that the West Indians brought to Panama?' asks poet, social historian and 'silverman' descendant Carlos Russell. 'An amalgam of European customs and ideals with a decidedly British (Anglo-Saxon) veneer imposed on a fragmented African base, weakened, but not eradicated by centuries of slavery. It was a culture with a distinct penchant for things and ways British. Proper spoken English, conservative dress, a black suit, stiff collar and tie in the tropics, proper deportment and a loyalty and dedication to a job which demanded much and paid little.' And life and culture were changing as the construction years passed. 'Now here comes a little improvement,' writes Jules LeCurrieux. 'The West Indian Negro woman began to immigrate here, then the poor old bastards found themselves wives of their tribes and began to live like human beings and not beasts, or slaves, they found someone to cook them a decent meal, to wash their clothes, some one to be a companion.' Although the authorities approved of the increasing arrival of West Indian women and children, as they had for the whites, there was precious little ICC-provided accommodation for families, so most lived in expensive rented flats in the terminal cities. Harry Franck, while taking the census of the Zone in 1912, went into many of these. 'They lived chiefly in windowless, six-by-eight rooms,' he writes, 'always a cheap, dirty calico curtain dividing the three-foot parlour in front from the five-foot bedroom behind, the former cluttered with a van-load of useless junk . . . a black baby squirming naked in a basket of rags . . . Every inch of the walls was "decorated" . . . With pages of illustrated magazines or newspapers . . . Outside, before each room, a tin fireplace for cooking precariously bestrided the veranda rail.'

The black workforce were not only ruthlessly excluded from Gold Roll facilities, but also virtually nothing was provided for their amusement, edification or recreation. In one year, the ICC actually spent fourteen times more per person on 'extras' for the white personnel than for the blacks, who, of course, made up the large majority of the employed numbers. So the West Indians were largely left to their own devices.

The church provided the centre for the developing West Indian communities. 'The men are kept hard at work full six days a week,' wrote a visiting American journalist. 'On Sunday morning every religious community is busy – you would think a great revival was in progress.' By 1910, nearly forty 'black' churches were in operation in the Zone, almost all established without any material assistance from the ICC. The majority were Anglican but there were also Methodists, Baptists, Pentecostals and Episcopalians, as well as more 'charismatic' congregations. For one of the leading black Panamanian historians, the church was 'a forum for expression on many issues. It preserved the self-respect of the workers, and stimulated their pride.' Others are more cynical, seeing the white-dominated Anglican Church in particular as a tool to 'tame them and provide a relief valve'. The loud singing, extravagant dress and general exuberance of the low-church black congregations was commented on by Americans, with approval as well as condescension. As far as they were concerned, emotion and energies were being worked out that could otherwise be directed against them.

But the Christian churches of various denominations did not have a monopoly on the spirituality of the imported black workers. As in the islands, their influence was leavened by other traditions that took in 'obeah', or sorcery, herbal medicine and rituals of spirit possession, all of which survived from their African inheritance.

Other carry-overs from Africa via the Caribbean islands were the Mutual and Friendly societies, designed to protect those injured or bereaved. As during the French era, 'Burial clubs' or 'su-sus' became widespread in Panama, whereby small sums were deposited every month against the cost of one's funeral. This was not only because of the high death rate, but also because of the social importance of funerals to the West Indian community. Within the separate island communities – Barbadians tended to stick with Barbadians and Jamaicans with Jamaicans – there was in general a high degree of communal and interdependent living. Harry Franck comments that while West Indians seemed to know everything about their neighbours down to the most intimate detail, the Americans he came across would often not even be aware of their neighbours' names.

In the absence of any ICC-organised activities, the West Indians put together their own cricket teams, as well as card or domino-playing circles. Clothes and music were also important. The Caribbean people, says one 'digger' descendant, 'were the unquestioned leaders of glamour and glitter'. 'Let me tell you,' says West Indian Benjamin Jordan. 'To see people at night. Saturday night they have dances in different places. People put up huts and have dance parties and the rest of it. On Saturday night, it was a joyful time in Culebra. Liquor was common at that time. You give a dollar or a dollar and a half for a quart. That did a lot. To see those people dancing and making merry . . . Boy!' Another West Indian remembers, 'the elegant quadrille dances, men and women graciously moving though the many fancy figures' to the music of 'Calypsos, mambos and meringues'.

For policeman Harry Franck Saturday nights, when the men had just been paid, was 'the vortex of trouble on the Isthmus'. On one occasion he went into 'the rough and tumble' of New Gatún, where he encountered 'such a singing, howling, swarming multitude as is rivalled almost nowhere else, except it be on Broadway at the passing of the old year'. With a colleague, he went into one of the bars, or, more exactly, the white side of it. 'Beyond the lattice-work that is the "color line" in Zone dispensaries,' he writes, 'West Indians were dancing wild, crowded "hoe-downs" and "shuffles" amid much howling and more liquidation; on our side a few Spanish laborers quietly sipped their liquor.'

Indeed, the 'Silver and Gold' distinction made no exception for Saturday nights or any other time or place in the Zone. Away from the works, there was next to no mixing between the races. In fact, the system was ever tightened. In February 1908, Taft, to please the unions as he prepared his bid for the presidency, declared that only US citizens (and, after a protest, Panamanians) could be on the Gold Roll. The last of the West Indians were demoted, but the order also further complicated the tortuous euphemism of the Gold/Silver distinction. For one thing, there were a number of black US citizens working for the ICC. Some five hundred had been recruited during the early years, almost all on to the Silver Roll. It had been hoped that they would be good at 'managing' the West Indians. But the authorities soon realised they had made a mistake – the American blacks were far less malleable and passive than the imported 'third-country' workforce. As a West Indian writes, 'The majority of them were employed as team drivers, and when delivering goods would refuse to unload same, claiming they were no labourers, they were team drivers. They also were tutoring the other employees to act accordingly. In the view of the fact, they were sent back home.' 'We had colored Americans working, good men, skilful men,' remembered another West Indian. 'But they can't pull with the White Americans always a fight and trouble.' Furthermore, the small number on Gold Roll contracts vigorously protested when, as blacks, they were refused service at Gold Commissaries, or ordered to take off their hats 'although no such requirement is made of white employees'. Consequently, to preserve the colour criterion that underpinned the Gold/Silver distinction, no more US blacks were given Gold Roll contracts, and by February 1909 only one such employee remained, a Henry Williams.

So by 1908 there were hardly any non-white Gold employees left, but when Taft made his nationality order as far as Gold and Silver was concerned, it opened a can of worms for the authorities. Soon after, Goethals received a petition from ten Silver Roll American blacks complaining that as US citizens they were entitled to the lavish Gold Roll privileges. After much discussion among the top brass, a compromise was decided on whereby the men were given home and sick leave entitlements, were paid in gold, but were still officially Silver Roll and thus excluded from clubhouses and commissaries.

Another protester, Henry W. Scott, proved more of a handful. In early January 1910, he complained that he had been denied a foreman position in the Pacific Division because, he was told 'that he was colored and not eligible for employment on the Gold Roll'. Utterly fed up, he now wanted a job in Panama City, where the 'prejudice feeling' did not exist 'as on the Canal Zone'. 'My father having performed distinguished service in the Civil War. And on behalf of the twelve million American colored people and taxpayers, I most respectfully make application for one of the above places,' he ended. But he had also been in touch with his senator back in the States, Wesley L. Jones, who, it turned out, had fought with his father in the Civil War. Jones wrote to Goethals, 'Mr Scott is part colored but I understand he is a young man of splendid ability.'

Goethals had to tread carefully. Not only was the racial segregation of the Zone legally dubious, but also the Republican Party had sold itself as the defender of the blacks. He wrote back to the senator on 24 January 1910: 'The fact of his being an American citizen does not entitle him to employment on the gold roll, as employment on the gold and silver rolls, respectively, depends entirely upon the class of work our employees do and not upon their nationality or color.'

Cases like this highlight the confusions, contradictions and hypocrisy of the Gold/Silver distinction. Happily for the authorities, the number of these awkward American blacks continued to dwindle. In July 1912 there were only sixty-nine at work for the ICC. The following year, only fifteen remained and this threat to the 'logic' of the system was almost gone.

Later generations of Panamanian Antilleans, when looking at the actions of the American blacks or the Spaniards, would accuse their West Indian 'silvermen' fathers and grandfathers of passivity during the construction period in the face of poor working conditions and the discriminatory policies of the canal authorities. Certainly, there would be no organised West Indian labour resistance until well after the canal was finished and the workforce much depleted. But many factors weighed against concerted labour action during the construction period. For one thing, there was still a surplus of labour on the islands. At the end of 1907 over two thousand men were actually laid off at the completion of the building work, and by the end of 1909 Karner could pack up his recruiting operation in Barbados as labour needs were more than being met by independent emigration. The result was a pool of some five thousand unemployed and usually desperate West Indians living in Colón or elsewhere from which the ICC could draw as and when it liked. As the *Star and Herald* noted, every man had 'the knowledge that there are ten hungry applicants for each vacancy who will like the conditions well enough'. Furthermore, there was little tradition of organised labour in the islands, and the West Indian community was, for now, divided by loyalty to individual islands.

The West Indians were also kept in line by the vigorous and often violent efforts of the Zone police to punish even the mildest infringement. If this was not enough to curtail organised action, there was the daily struggle to make ends meet when often three-quarters of a wage would have to be spent on rent. There was also the exhausting fight against malaria, pneumonia, other diseases and the effects of accidents. The West Indian accounts nearly all tell of at least one stay in hospital, often many more. By 1914, Gorgas's sanitation squads had drained more than a hundred square miles of swamp through the building of nearly two thousand miles of ditches and drains. But although infection and mortality rates kept on falling, malaria and its recurring symptoms of agonising fever and shaking, followed by mind-numbing lethargy, would continue to affect many of the inhabitants of the Zone. In 1914, nearly half the workforce, over 24,000 people, were admitted to hospital at some time during the year for a variety of illnesses or accidents.

The two simplest explanations for the lack of West Indian protest or action against the ICC's regime were provided by a 104-year-old resident of Colón, Mr Foster Burns. First, he said plainly, the men wanted and needed the money. However it looks now, work on the canal was the best get-rich (or at least stop-being-hungry) scheme on offer. Second, the men were so busy working that they had no time or energy left for anything else. This certainly rings true. Although most American employees – with the exception of a few foremen and doctors – were limited to an eight-hour day by a law of Congress, Stevens had secured a special clause exempting 'alien' labour on the canal from this stipulation, so the West Indians worked ten hours a day for six days a week. And this was before overtime (often compulsory and unpaid) and the journey to and from work.

The actual labour was usually backbreaking, and frequently carried out, of course, in very difficult conditions. There were few cushy jobs available to the black workers. Albert Peters tells of how being assigned to a dredge involved having to continually dive into the muddy, slimy water to free the suction pump. In the burning heat or ankle-deep mud of 'Hell's Gorge', thousands carried heavy and dangerous dynamite boxes, manned the largest and most violent drills ever seen, and stoked suffocating, red-hot steam furnaces on shovels or locomotives. And from 1909 onwards, an increasing proportion of the labourers were working on the giant locks – the 'concrete cathedrals' – being built at either end of the central 'bridge of water'. Here the work was, if anything, even more dangerous and unpleasant than in the Cut.

26

'LORD HOW PIERCING!'

It had been the 'monstrous experiment' of the plan for the Gatún dam that had attracted most criticism in the US. In fact, there was nothing unprecedented about the dam apart from its scale. More justified were worries about its siting – over two deep geological gorges filled with dubious alluvial material. The plan was to block the Chagres valley at Gatún with an enormous, essentially earth structure one and a half miles long. The height was adjusted in the course of the building, but ended up being 115 feet above sea level, 30 feet above the planned level of the lake at 85 feet above sea level. At its top it was to be 100 feet wide, then 500 feet at the lake's surface and almost half a mile wide at the bottom. Between the two gorges across the dam site was a hill of solid rock, which would serve to anchor the structure and to provide the site for the spillway, through which the waters of the Chagres would be funnelled and thence flow to the sea at San Lorenzo. Consisting of a convex wall 800 feet long and nearly 100 feet high, the spillway was to be the new gaoler of the formidable Chagres River. For the planned locks to work properly, it was important that the level of the new river should not deviate from its 85-foot level by more than 2 feet. Within the wall of the spillway were a series of openings with massive steel gates. When all the gates were lifted, the spillway could cope with a discharge of 182,000 feet a second, judged to be the worst that the Chagres could throw at it. Next to the spillway was planned a hydroelectric station, to generate enough electrical current from the falling water to supply all the canal's energy needs. The triple-tiered double locks would be built at the dam's eastern end.

The dam itself was well suited to the height and length requirements of the chosen site, but more importantly was to be constructed largely of the materials excavated from elsewhere along the canal line. A dam in concrete would have been equally feasible, but a great deal more expensive. Earth dams are not complex

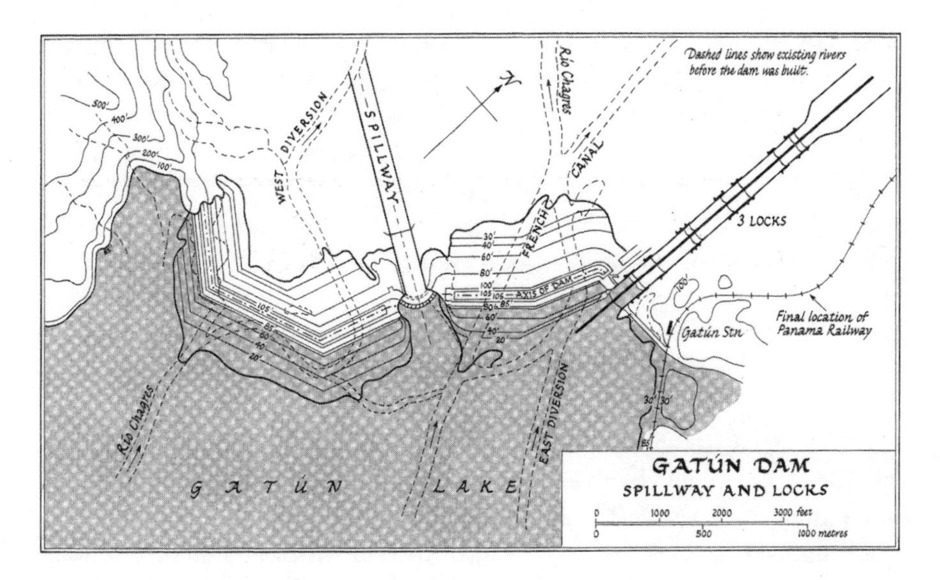

structures and are sometimes also referred to as gravity dams, because that is exactly how they perform; the resultant lake is held back purely by the huge mass of the dam and thus the friction at the base between the dam structure and the existing ground on which it is founded. The principle of earth dams is based on the fact that most clays are impervious. To construct such a dam, one constructs in layers. First, outer walls of hard rock are laid down on both the upstream and downstream faces. This is to hold the earth and clay filler in the construction phase, but also to protect against scour in the permanent case (from the lake on the upstream face and storm water on the downstream). The next layer inside both faces is then normally constructed of soil and smaller rocks/stones, and then the central core is composed exclusively of appropriate clay. The clay can either be placed 'dry' and compacted using rollers, or liquefied and pumped in. As the slurry drains and dries it will harden and become impervious.

At first all went well. To prevent organic matter forming a potentially porous layer under the dam, the six-hundred-acre site had to be meticulously cleared of its rich cover of tropical jungle. This task was completed by the end of 1907 and the dam site was barren. It reminded one visitor of the scene of a battle in some unimaginable war to come: 'an ugly denuded waste of land . . . stubble was everywhere, and standing out like pockmarks were hundreds of black ash heaps where the greenery had been burned. Across this soggy wasteland . . . a dredge in the Chagres sucked mud from one place and vomited it into another; and dynamite crews sent up enormous geysers of rock and water. Men in gangs of forty to a hundred swarmed about the valley, all in the blue shits and khaki trousers of the Zone Commissary, while the air was filled with the babel of more than twenty languages.' As well as the dam site, it was necessary to clear all the large trees from the future route of the shipping lane through Lake Gatún. Barbadian Edgar Simmons described this process: up to fifteen holes were cut in the trunks of the trees, which were then stuffed with explosives: 'Three sticks of dynamite, with a cap and coil, about 18 inches long, and covered with mud. So all are set for evening,' he writes. 'After the 5:15 passenger train pass for Panama, we start lighting. Some of us has up to 65 or 72 holes to light . . . Nine of us start out, each one with two sticks of fire in our hand, running and lighting, at the same time trying to clear ourselves before the first set begin bursting on us. Then it's like Hell . . . it was something to watch seeing the pieces of trees flying in the air.'

In 1907 the Chagres flowed through four routes to the sea: its old river bed, the French canal and the two diversion channels the French had built either side of their waterway. All four flowed through the site of the new dam. The Western Diversion was the lowest and this was widened and deepened, while the other three channels were dammed without any complications. The plan was to start work on the eastern end of the dam and the spillway, and then close the West Diversion when it was time for the river to be diverted through the spillway.

Shovels started cutting a channel through Spillway Hill while suction dredges thoroughly cleaned out the old river and canal beds. Then two parallel lines of trestles were erected at 30 feet above sea level along the upriver and downriver faces of the dam some 1200 metres apart. On these were laid railtracks to carry cars of rocks to create the two stone 'toes' that would hold the innards of the dam in place. Locomotives from Culebra or the Mindi Hills started arriving regularly and slowly the rock walls began to crawl across the wide valley. Meanwhile a suction dredge began pumping in between the walls a mixture of clay and water sucked up from the bed of the old French canal behind the dam site. Towards the end of 1908 a solid wall of rock, 60 feet high, had been laid out along the extremities of the dam.

Back in the States, the doubters were still vociferous, and now Bunau-Varilla had added his comments: water would seep through, emptying the lake, he said; the dam should have been located at Bohío (as per the French lock plan), where there was better bedrock; Gatún dam would develop fissures, and would be damaged by even a light earthquake. Then, at end of November came the news that everybody had been dreading. 'Collapse of the Gatún Dam' read one headline in the US. 'Chagres River plunges through Gap in Isthmian Wall.' It turned out that an American journalist in Panama had noticed that a large section of the southern 'toe', some 200 feet, had slipped by 20 feet where it crossed the line of the old French canal. Goethals denied any serious problem and was backed up by John Stevens, who wrote an article denouncing 'the outbreak of yellow fever journalism in regard to the Gatun Dam'.

But with their predictions of doom seemingly confirmed the press kept the story running. The *New York Times*, a long-time enemy of Roosevelt and his Panama route, declared that the canal had to be begun anew at Nicaragua. Taft was quickly despatched to Panama with some experts in tow to find out the truth about the dam. He confirmed Goethals' line that the slip was not serious, and rounded on critics of the great project, accusing them of creating a 'fire in the rear . . . calculated to break down the nervous system of those persons on the Isthmus working day and night, tooth and toenail to build the greatest enterprise of two centuries'.

The damage to the 'toe' was repaired, and early in 1909 the first concrete was poured at the spillway site. Throughout the year, the massive dam structure began to take shape, rising to block the valley. By the end of 1910 four dredges and more than ten locomotives were at work, adding between them over a million cubic yards a month to the mass of the dam. As the new chasm was being created in the Cut, much of the spoil removed ended up as the new mountain at Gatún. The peak year of building was 1911, when there were two thousand men at work and over a hundred trainloads of rock and earth were being dumped every day. There were further alarms: in October 1911, 1000 feet of the dam's eastern end settled a few feet, and in August the following year about 800 feet of the crest of the dam dropped 20 feet. On other occasions, like the walls of the Cut the dam writhed and twisted as it sought a new equilibrium, wrecking track and threatening the lives of the locomotive drivers. But these were largely movements within the fill, rather than a slip of the entire structure, and as the clay centre of the dam dried out and hardened so the period of greatest danger passed.

With the Panama Railroad relocated on to higher ground, it was time to start filling the new lake. At the end of April, preparations were complete for the closing off of the West Diversion. The engineers knew that the final taming of the Chagres River would not be easy. For one thing, delays meant that the river level was 7 feet above its lowest dry-season level, and more rains were on the way. Furthermore, the West Diversion ran directly over a deep gorge, with mud to great depths. It was also about 20 feet lower than the bottom of the spillway, through which the river was now to be directed. This meant that the river would need to rise by this height before it would begin to find the new outlet.

Huge trestles were driven into the mud of the channel, and hundreds of flat cars, loaded with rock from Culebra, were readied nearby. The plan was simply to unload the rock into the current faster than it could be washed away.

Starting at either side, rocks were dumped from the trestles around the clock for four days. Initially, all was fine, but as the channel contracted the flow picked up power and pace. When it had shrunk to 80 feet wide and 6 feet deep, as fast as the spoil was dumped it was washed away. Even rocks weighing a ton were lifted up and carried off by the powerful Chagres current. A great mass of old twisted French rails was brought up to the site and thrown off the trestle into the open section. As predicted, the rails snarled on the trestle piles, forming a web that started to catch and hold the bigger rocks. But this also transferred huge pressure on to the trestle itself, which groaned and trembled, then broke, with parts starting to move downstream. The water rushed through again, and parts of the surviving dam slid and collapsed.

But the battle was not lost. Men clambered out on to the trestles to repair them, more cars were brought up frantically to dump rock, and a dredge started pumping clay upstream of the dam breach to reduce the force on the trestles. Finally the dam held and as the water at last found the outlet of the heavily regulated spillway, the destructive force of the Chagres, the bane of the French effort, was gone forever.

With the West Diversion closed, the water was allowed to back up and the lake began to rise at 2 inches a day. Eventually the water would cover 164 square miles of jungle, as well as several villages and much of the digging of the French. Some villagers refused to move out and had to be forcibly evicted. For many of them the drowning of the Chagres valley was like a biblical flood. A way of life based around the river, which long predated Columbus' first voyage, was coming to an end. 'It is not hard to realize why the bush native does not love the American,' wrote Harry Franck, who took a trip out on the growing lake in a police launch in mid-1912. 'Suppose a throng of unsympathetic foreigners suddenly appeared resolved to turn all the world you knew into a lake, just because that absurd outside world wanted to float steamers you never knew the use of, from somewhere you never heard of, to somewhere you did not know.'

Franck's boat trip had some of the dreamlike qualities of Bunau-Varilla's a generation before as the then Acting Chief Engineer surveyed the flooded works and steered clear of the tarantula-laden treetops. Franck's party followed the line of a submerged river as it provided a path through the treetops. 'Splendid royal palms stood up to their necks in the water . . . gasping their last,' Franck writes. 'Great mango-trees laden with fruit were descending into the flood . . . once we passed acres upon acres of big, cod-like fish floating dead upon the water among the branches and the forest rubbish. It seems the lake in rising spread over some poisonous mineral in the soil. But life there was none, except the rampant green dying plant life in every direction to the horizon.'

But then they came to a small island, once the top of a hill, 'congested with game'. Some Americans would take to boats to 'hunt' these animals, while others took the unique opportunity to collect rare treetop-living orchids. But there was work to do as well. 'The water brought everything to the top, except stones,' says

George Martin, who was working at the lake between his railroad relocation job and his time in the Cut. His task was to collect the 'branches and forest rubbish' from the water's surface to prevent it clogging up the spillway or locks. 'Some of us were placed in death-traps,' he says, 'that is, we were to seize two stumps or body of trees that had float on the water and nail them together like rafts, and whether you could swim or not, two of us would have to get on this and go all in the water on the surface and clean it of all the debris that had float on the water . . . the thing that had me so scared was that I could swim but like lead.' As much of the land being covered was deep swamp, decaying tree trunks and vegetation from the bottom rose up forming bizarre floating 'islands', consisting of grass and small trees. On a couple, deer were even found. The 'islands' had to be towed away by tug lest they cause a blockage in the locks.

The first task in constructing the locks had been to excavate to bedrock the basins in which they were to be built. For Gatún, this meant digging a hole a mile and a quarter long, 200 yards wide and 50 feet deep – the removal of nearly 7 million cubic yards of rock and silt. Most difficult was the lowest of the three basins, whose bottom was 66 feet below sea level, and whose sides constantly slipped. As the engineer in charge wrote, 'No one expected on returning to work in the morning to find things as they were left the evening before.' Even though there were temporary dams at both ends of the lock basins, there was a constant danger of seepage and floods. American Harry Cole worked on the locks on the Pacific side – one pair at Pedro Miguel, two at Miraflores. In the lowest basin on the Miraflores locks – 'an extremely dangerous place to work' – he experienced the same problems. 'Sometimes, in the rainy season,' he writes, 'even the small rivers became large rivers and often overflowed and inundated much of the construction work, putting even our drainage pumps out of action. Canal banks would break loose, cover up lock wall foundations, fill up culverts, submerge railroad service tracks and cause weeks of delay to clean up and put the work in order again. Those were our heart-breaking times.'

As excavation continued on the lower basins, in late 1909 actual construction started on the completed upper basin of the Gatún locks. The lock components of the 1906 plan had attracted almost as much criticism as the dam. Certainly, they were a gamble. Nothing of the sort even remotely as large had been built before, and the choice of concrete as the primary building material, mainly to save time, was even more daring. Today, concrete technology is an industry in its own right, as it can be a more difficult material to work with than it first seems. In the early 1900s little was known about the properties of concrete and the basic materials (sand, gravel, cement and water) were simply thrown together in a fairly fixed proportion, mixed, and poured. That the locks at the Panama Canal are still working nearly a hundred years later without any major problem in the concrete is testament to the standard of design and construction employed.

The structural design of the locks was unremarkable. They were to be constructed of 'mass concrete', that is, there was to be little use of steel re-inforcement bars. This type of structure, like the Gatún dam, relies mainly on the enormous mass of material present to withstand the loads imposed upon it. In fact, the side walls of the locks are structurally similar to miniature dams. The massive width of the bottom of the wall, some 50 feet, creates an equally large frictional resistance to the force of the water pushing against it when the lock is full. The wall becomes thinner and thinner as the height increases and the applied hydraulic force reduces.

The key to building a major concrete structure was the same as that for the excavation works at the Cut: creating an organised and balanced production line. In this, Goethals was equally as successful at the locks as Stevens had been in the Culebra Cut. To build any concrete structure, the wet concrete must be poured into giant 'moulds', referred to as shutters. Traditionally, these are made of wood. In massive structures such as the Panama locks, using wood becomes extremely wasteful, time-consuming and thus expensive, as the moulds can only be used two or three time before being scrapped. Goethals therefore opted to use shutters made of steel, which has a very high initial cost, but a relatively low total lifetime cost. Such a shutter weighed many hundreds of tonnes and constituted a major feat of precise manufacture in itself. These main shutters were then mounted on rails, running parallel to the walls, so that they could be easily moved.

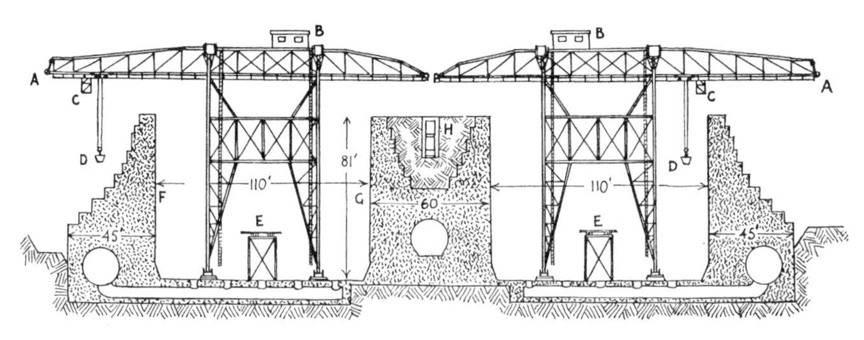

A Travelling trolleys
B Machinery houses containing electrical operating apparatus
C Operators' cabs
D 2-cubic yard bucket for placing concrete
E Double-track trestle from which the concrete buckets are served
F Side walls of locks, showing side wall and floor culverts
G Centre wall and centre wall culvert
H Operating tunnel

Chamber-cranes at Miraflores

An enormous amount of concrete was required. For the Gatún locks alone, 2 million cubic yards would be poured. The cement came from the States, about five million bags or barrels of it. Obviously it made sense to source the stone and sand from nearer to the site. The Commission sent a party to try to secure sand from San Blas, which, although ninety miles from Colón, had the best quality for cement. But this area was inhabited by Cuna Indians, who had long memories: 'They did not look with favour on visits from the white men, whom they suspected were searching not for sand but for gold,' reported one of the engineers. The Americans were led before a seventy-year-old local chief, 'seated on a block of timber, and he motioned his visitors to a seat on the sand at his feet. An air of great solemnity surrounded the proceedings.' The visitors outlined their plans to connect the two oceans by a canal, and how this would bring a better price for the Indians' coconuts and ivory. 'The Chief listened,' the account continues, 'but when the story was finished he said that God had given the Indians their country, the land and the water, and that which God had given to the Indians they would neither sell or give to the white man.' That was his final word, and permission to anchor for the night was only granted on the condition that the Americans left the next morning and did not return. So the rock for the gravel, including old stones from the fortress of San Felipe de Todo Fierro, ended up coming from Portobelo, where a crushing plant was built, and the beach at Nombre de Dios was stripped of its sand, all to feed the voracious appetite of the Gatún locks. On the Pacific side suitable rock and sand were found much closer to the lock sites, but to supply Gatún required barges negotiating the dangers of a sea journey of between twenty and forty miles. Any stoppages from bad weather would upset the brilliant mechanised system Goethals had now established to deliver the mixed concrete to where it was needed.

Once the barges from up the coast arrived at Colón, they made their way up the newly redredged old French canal to Gatún, where unloading facilities and a mixing plant had been established. The concrete was assembled by the use of an ingenious system of small electric cars carrying buckets passing under hoppers containing sand, crushed stone or cement. These would end up above one of the eight giant mixers, underneath which other similar cars waited for the finished concrete.

Spanning the site were four giant, 800-foot-long cableways, supported on either side of the giant basin by towers eighty-five foot high. As each pair of buckets was filled with 6 tons of liquid concrete, the car moved off and was then picked up by a hook on the cableway and taken off to the site. At the same time, an empty skip returning from the site was set down on the railcar.

The system worked brilliantly. In one year a million cubic yards of concrete were delivered to be moulded and set. The walls were built in 36-foot sections all the way to the top, which took about a week. Then the scaffolding, shutters and cableway towers, all rail-mounted, were simply shifted along to the next section.

On the Pacific side concrete was batched on site and lifted to required position using a tower crane. Here even more concrete was used for the single lock at Pedro Miguel and the double tier at Miraflores. A small dam was also required at Miraflores in conjunction with the locks, forming the tiny Miraflores lake, but this structure presented no serious difficulties.

Barbadian George Martin pops up in yet another job, this time unhooking concrete buckets from a crane at Miraflores. It was his least happy posting. Hit on the body by one of the buckets, he was in hospital for six days. The *Canal Record* from this time contains a number of reports of men falling from the high walls of the lock chambers. Typical is this from 1913: 'Joseph Seeley, a West Indian laborer employed at Miraflores, fell from a ladder yesterday morning while working on the wall of the west chamber of the lock. Instantaneous death resulted. It is presumed that he was overtaken by an attack of vertigo.'

The lock basins are the construction marvel of the canal. They are massive, a hundred thousand tons of concrete each, 1000 feet long and more than 100 wide. Seen full of water they are impressive enough, but those who stood on the empty floor and gazed upwards at the sheer, featureless walls, more than 80 feet high, taller than a six-storey building, were awestruck. 'These locks are more than just tons of concrete,' commented one such visitor, 'they are the gate to that pathway of which Columbus dreamed and for which Hudson died. They are the answer of courage and faith to doubt and unbelief. In them are the blood and sinew of a great and hopeful nation, the fulfilment of ancient ideals and the promise of a larger growth to come . . . I left with the feeling that follows a service in a great cathedral.'

The mechanical marvel of the canal was the machinery of the locks. For one thing, the entire system was powered by electricity (supplied from the hydroplant next to the spillway), a great innovation at a time when steam- and horse-powered systems were the norm, and the electrification of factories was in its infancy. The power that did the hard work of lifting or lowering ships was, of course, gravity, the filling of a lock from water from the lake or a higher lock or the release of water to further down the system. Water was drained or admitted through tunnels 18 feet in diameter, built lengthwise within the centre and wide walls of the locks. Running perpendicular to these were smaller tunnels or culverts under the floors of the locks, fourteen per chamber. Each had five, evenly spaced openings in to the floor, so that when water was admitted the turbulence would be minimised. The main culverts had large steel gates as valves. The principle was ancient: to fill a lock the valve at the lower end would be closed and the upper one opened, and vice versa to drain it. Each lock had a lift of an unprecedented 28 feet, and the system aimed to raise or lower a ship this distance in just fifteen minutes.

Worries voiced in the US press prompted the adoption of a series of stringent safety measures to avert the worst-case scenario of a ship smashing through the upper lock causing the lake to pour through the breach. No ship would pass through the locks under its own steam. Instead the movement of the ships would be controlled by a group of powerful electrical cars, mounted on rails at the lock's edge, each in control through a windlass, a steel cable, attached to the ship. Each lock had a double gate and a chain ran in front of the mouth of the top lock to catch and slow any vessel approaching in a dangerous way. Failing that, an emergency dam could be lowered across the channel within minutes.

All of the mechanisms for the lock operation were controlled from one position, where a working replica of the locks had been made, with the controls for each machine adjacent to its model. In addition, a system of safeguards meant that it was impossible, for instance, to open a culvert valve if the relevant lock gates were not firmly shut.

The manufacture of the lock gates was the only substantial part of the canal work that was subcontracted. A Pittsburg firm, McClintic-Marshall, started assembly and installation in May 1911 at Gatún, August at Pedro Miguel and at Miraflores in September 1912. The lock gates were designed to close into a flattened V. When open they slotted into a recess in the lock wall. Each individual gate was 65 feet wide and 7 feet thick and varied in height from 47 to 82 feet depending on its position. The highest were the bottom gates at Miraflores, which had to cope with the Pacific tides. The gates were opened and closed by giant wheels within the lock gates. These needed to be hugely powerful machines, but the gates weighed less than they looked. They were hollow, consisting of steel plates riveted on to a steel frame, and watertight so relatively buoyant. At their base were rollers, which ran on enormous steel plates embedded into the floor of the lock.

Few ships at that time needed anything like the space the giant locks provided, so intermediate gates were installed so that vessels less than 600 feet long could pass through quicker and with less expenditure of water. So in all there were forty-six pairs constructed and installed. The great steel gates were one of the most spectacular sights on the Isthmus, particularly while the lock chambers were still empty and their huge size could be appreciated.

However, the work on the gates also figures, along with the explosion at Bas Obispo, as the worst memory of the construction period for the West Indian workers. The money was good – McClintic-Marshall paid US$0.25 an hour, more than double the basic labouring wage – but the work was among the most dangerous anywhere. As a West Indian work song from the time had it: 'Yuh gets more money for that job than working in the cut/But it all depends muh honey on if yuh don't get hut/ For if you ever get a drop yuh surely have to die/For dem gates lawd gad gal is seventy five feet high.'

In mid-1912, when work on the gates was well advanced, Harry Franck went to Gatún locks looking for a man accused of theft. 'I found myself racing across the narrow plank bridges above the yawning gulf of the locks, with far below tiny men and toy trains, now in and out among the cathedral-like flying buttresses, under the giant arches past staring signs of "DANGER" on every hand . . . I descended to the very floor of the locks, far below the earth, and tramped the long half-miles of the three flights between soaring concrete walls . . . On them resounded the roar of the compressed air riveters and all the way up the sheer faces, growing smaller and smaller as they neared the sky, were McClintic-Marshall men driving into place red-hot rivets, thrown at them viciously by negroes at the forges.'

This was the worst job of all. The riveters, using heavy pneumatic hammers, worked on scaffolding that normally consisted of chains running down the face of the gate to which hooked planks were attached one above the other. There were four men per plank, and any sudden movement could unbalance and unhook the platform; if one came down from above, it usually took a few lower ones with it.

Eustace Tabois remembers a particularly gruesome accident at Gatún. 'We were working there one Sunday,' he says. 'I was inside the gate, right at one of the portholes, and I just happened to look out. I saw a shadow come down like that. And when I look out I saw this man down below. The scaffold break away with him and he went right down and the plank down there with him a spike went right though his head. Kill him dead, kill him dead. Man used to die every day.' 'Scuffles would break away,' says James Ashby, 'hurting many and sometimes killing men instantly. Lord how piercing!' As another West Indian commented, 'The family of those men working on those locks were always fearful as to who may be next to fall.'

It was the same on the Pacific side. Jamaican Nehemiah Douglas was working on the giant gates at Miraflores when the cable holding his plank broke, 'kill[ing] some men, on the spot. The amount of blood that flowed gave the appearance of a little gully,' he later wrote, 'and when I saw what appeared to be an island of blood, I got nervous, I think, because how I got down, I do not know; but I got down and ran like never run before, straight home in Paraiso.' Many stayed away after witnessing incidents like this, even if they had pay to collect, but Douglas returned and was soon after hit by a crane and received a fractured skull.

Nevertheless, with the money that could be earned, men still 'poured like sand' to go and work on the locks. However many were injured or killed, 'there were always others to take their places without hesitation'. This became more urgent as the project progressed. It was obvious that the well-paid work was not going to be around for that much longer. The canal project was now nearing its triumphant conclusion.

27

THE LAND DIVIDED, THE WORLD UNITED

During the four last years of construction, nearly 100,000 American and European visitors arrived in Panama to view work on the canal. Interest was so great that steamship lines diverted vessels from other routes to the Caribbean. At Ancón the Isthmian Canal Commission created a tourist station with a lecture room, relief maps and models of the locks. Tourists could also visit the work site by taking a special train whose open sightseeing cars had been converted from Panama Railroad flat cars.

Back in the States, Tin Pan Alley was busy churning out hits such as 'Where the Oceans Meet in Panama (That's Where I'll Meet You)' and 'The Pamela', and the press eagerly reported, alongside sombre news of rising tensions in Europe, each and every landmark and breakthrough.

Work on the west breakwater from Toro Point in Limón Bay started in 1910. As the local rock was too soft to withstand the power of the 'northers', 12- to 18-ton rocks were blasted from the quarry in Portobelo and shipped by sea to be dumped by barges along the outside of the breakwater line. Then a series of creosoted piles were driven in to form a trestle on the inland side. Next, local stone was loaded into flat cars, pushed out along the trestle, and unloaded by the plough method on the inside of the line of hard rock. Work progressed steadily and Colón became a safe harbour for the first time. The east breakwater caused more problems, mainly because the French had dredged deep channels across its line, but was completed in 1915.

The digging inland did not all go smoothly. At Mindi huge amounts of explosives were required to break up the rock, and there were constant problems from flooding. However by the summer of 1912 only a small belt of land 1000 feet wide separated the open Atlantic channel from the site of the Gatún locks. By

August 1913 it was the same at the other end of the canal, with just a dike holding the waters of the Pacific back from the Miraflores locks.

Construction had also got underway for the running of the canal post-completion. A new Administration Building was built at Balboa, as La Boca had been renamed, while on Toro Point and Margarita Island in Limón Bay and on the islands of Perico, Flamenco and Naos in the Bay of Panama 16-inch guns were installed, the largest then in the possession of the United States military. Elsewhere, smaller bore artillery was sited, along with mortar batteries, to repel a land-based attack. In all, some $15 million was spent on fortifying the canal from attack by sea or by land.

At 4.30 p.m. on 20 May 1913, shovels No. 222 and No. 230 met 'nose to nose' at the centre of the Cut. At 40 feet above sea level, the Cut had reached its full planned depth. The mighty shovels whirred and clanked to a stop and the men cheered and threw their hats in the air. Whistles were sounded all along the gorge and photographers recorded the event. Now all the shovels could concentrate on removing the massive slides from the eastern end of the Cut.

Further red-letter days followed thick and fast. On 27 June 1913 the last of the Gatún dam spillway gates was closed, allowing the lake to rise to its full height. On 31 August, a Sunday, so that, Goethals suggested, 'everyone could join the fun', a huge charge was exploded in the dike between the Miraflores lock and the Pacific channel. The hundreds of spectators then settled back to wait for high tide at noon. The Pacific tide at Panama comes in fast. Soon it was pouring over the breach in the dike, and then rushing in, clearing the dam debris as it went. Then the waters of the Pacific were lapping at the closed gates of the Miraflores lock. Two days later the same operation was completed for the barrier below the Gatún locks. Steadily, waters from the two oceans were moving inland to meet each other.

The locks were all completed by the end of the following month, and the Gatún lock had been successfully tested. But in the Cut the battle against the slides was not going well. The slopes at Cucaracha and Culebra had been cut back to as gentle as one in five, but the movement of earth and rocks into the canal prism continued. In October it was decided to flood the Cut and complete the excavation 'in the wet' with dredges. In September, the last steam shovel was moved out, the rails lifted and carried away, and old ties piled up and burnt.

The huge earthen dike at Gamboa, which separated the Cut from the rising lake, had six large pipes in its base. These were now opened, and water started to pour into the gorge. Everyone who had worked on the canal had imagined and pictured in his mind's eye the appearance of the Cut when full of water. Now it was happening. Then, on 10 October 1913, in a stunt dreamed up by a newspaperman, President Woodrow Wilson pressed a button in Washington and relayed by telegraph from Washington to New York, to Galveston to Panama, the signal that blew the centre of the dike to complete the flooding of the Cut, and, it was hoped, blast a passage through the slides at the eastern end. Just before the signal was sent, Rose van Hardeveld remembers, 'There was a reverent silence' among the hundreds gathered to witness the spectacle. 'No one spoke at all.' Then, 'there was a low rumble, a dull muffled B-O-O-M! A triple column shot high in the center, turned, and fell gracefully to both sides like a fountain. From the multitude came a spontaneous long, loud roar of such joy and relief that I felt sure I would remember the sound all my life. As the water poured out of the lake into the Cut, hats came off.' Rose and the children could see their father Jan shaking hands with his boss. Both men were crying.

As the fountain of earth cleared, the spectators could see a great wave sweep down the channel. At its crest rode two small launches and a native dugout, in fierce competition to see who would be the first to get to the Pedro Miguel locks at the other end of the Cut. But they were in for a disappointment. At the mighty Cucaracha slide the water was stopped. The frustration was immense. Massive amounts of dynamite were used and dredges were brought up from the Atlantic through the Gatún locks and across the lake to work away at the blockage from that side. But nothing seemed to work. The explosions threw rocks and earth in the air only for them to seemingly land back in the same place. But then the river came to the rescue. After heavy rain the Chagres began to flood. In three days the level of the lake had risen by 3 feet. A small trench was dug out by hand over the Cucaracha blockage, and a trickle of water began to flow across it. As the water bit into the mud, the stream increased to a torrent, further clearing rocks and mud from its path. Soon the dry section behind the blockage was full, and dredges from the Pacific side could start work on the slide from the other direction. Floodlights allowed round-the-clock digging. An old French ladder dredge made the 'pioneer cut' through the slide on 10 December 1913, to open the channel for the first time. Although there were still millions of cubic yards of slide material to be removed before an ocean-going vessel could make the transit, the waters of the two oceans were now unmistakably united.

By April 1914, the channel was wide enough for tugs and boats to pass through it, and when an American Hawaiian Steamship Line vessel showed up in the Bay of Panama wanting to move a cargo of tinned pineapples and sugar from Hawaii across the Isthmus, Goethals was able to have it taken through by lighter. Others followed, and by June the canal had taken over $7000 in tolls.

Practice continued, however, on dealing with larger ships in the locks. In June, an old Panama Railroad steamer locked up and down the Gatún tier very slowly and carefully but without mishap. Meanwhile plans were finalised for a grand celebration to mark the official opening of the Panama Canal, set for 15 August

1914. The boat given the honour of passing through the canal was the *Ancon*, another Panama Railroad steamer. This was to be followed by a fleet of international warships – symbolising global concord – sailing through from the Atlantic to the Pacific, arriving in San Francisco in time for the opening of the Panama-Pacific International Exposition. Leading the fleet would be the *Oregon*, whose famous race round the Horn during the Spanish War had done so much in the United States to build a case for a publicly funded canal. On board this vessel would be the US President himself.

There was a final practice run on 3 August, using the *Ancon*'s sister ship the *Cristobal*. Not everything went to plan. The 'mules' in the locks had great difficulty holding the ship, as one burnt out a motor and another had its cable snapped. Nevertheless, it was not these incidents that overshadowed the first interoceanic transit by a sea-going vessel. Towards the end of the trip news came through from Europe: Germany had declared war on France. Philippe Bunau-Varilla had returned to Panama for the canal's opening, and was on board the *Cristobal* with his daughter Giselle. As soon as he heard the news, he declared to the other passengers, 'Gentlemen, the two consuming ambitions of my life are fulfilled on the same day. The first, to see an ocean liner sail through the Panama Canal: the second, to see France and Germany at war.' 'French genius' had been vindicated by the opening of the canal, he wrote in a piece for the New York *Sun* the next day, as Germany invaded Belgium, bringing the British Empire into the war. And now, forty-four years after the humiliation of Sedan, wrote Bunau-Varilla, his countrymen had the chance to 'vindicate' their national dignity and power.

So just as the great civilising quest of the Panama Canal was coming to an end and the United States was taking her place as a global power, 'Old Europe' was embarking on a ruinous and bloody war. The ending of the Victorian world of de Lesseps, foreshadowed by the Panama story, was now complete. Although the Panama-Pacific Exposition went on as planned, the festivities at the canal were cancelled. The *Ancon* still made the official first trip, but it was a muted affair. There were no international dignitaries in attendance. Observing the transit from shore, Goethals followed its progress by railroad. Claude Mallet was on board, however; he was surely the only man present at both the de Lesseps inauguration of the project on a boat in the Bay of Panama in early 1880, and also at the official opening thirty-four years later. He wrote in his memoirs, 'The canal has been a life's work to me. As it progressed, I became more and more absorbingly interested in every detail of the enterprise.'

'Everything went like clockwork,' reported Winifred James, English wife of one of the *Ancon*'s guests. 'On board were all the foremost Panamanian citizens and politicians, headed by the President of the Republic and his pretty wife. The ministers and consuls of the different Powers stood in groups together talking of the war and to the ladies alternately . . . We slid slowly up to the first lock and through it; and everybody looked at his watch and wondered when it would be breakfast time.'

There was no razzmatazz, no music, only a few flags among the largely silent crowd lining the tops of the lock walls. On board, the guests were served 'mugs of cold tea and dishes of broken meats'. Occasionally someone would get a burst of energy and go around slapping people's backs, saying what a great day it was, but on the whole most remained solemn. According to Winifred James, 'Not very late in the morning the Panamanians began to inquire' if there was anything to drink on the ship. There wasn't.

Because of the war, the news of the official opening, so long anticipated, was consigned to the inside pages. 'The Panama Canal is open to the commerce of the world. Henceforth ships may pass to and fro through that great waterway,' announced the *New York Times* on page fourteen. The Philadelphia *Record*, however, commented that the 'Unostentatious dedicatory act [was] a more appropriate celebration of the triumph of the arts of peace than if it had been associated with martial pomp and an array of destroyers and battleships.' The peaceful American 'Army of Panama' had fought for its country well. 'The practical completion of this great achievement wins little attention from a world intent upon war and the news from Belgium and Alsace,' wrote another paper. 'Americans should find solemn pride in the thought that they have added much to a world from which other nations are taking so much away.' How much, of course, could not have been imagined.

The day after the opening Winifred James came across a squad of Martiniquan ex-canal workers in uniform marching down to the station to entrain to Colón and thence to fight for France in the war with 'the tricolour waving above them and the *Marseillaise* playing them on. Beside them walked their women,' she writes, 'the older ones wearing the gaudy bandana of the Martinique woman, which is the gayest of all the turbans. They were, men and women alike, smiling a little in a hypnotized way, caught up in the trance of the music and the flying colours. Probably none would come back. But that was nothing: the band was playing and they were going to fight for their country.'

POSTSCRIPT

WHOSE CANAL IS IT ANYWAY?

'[Poyah] faced Ella, piling up the goods on the counter. "I's a man, man," he said, meeting Ella's frosting eyes. "I wuz a brakesman in Palama . . . I wuz de bes' train hooper on de Isthmus!"'

– Eric Walrond, 'Panama Gold', 1926

Whatever the news agenda at the beginning of August 1914, the opening of the Panama Canal was an achievement, an epic battle won, that still thrills nearly a hundred years later. Regardless of the motivation or the means, the end result of cutting in two the eight-thousand-mile-long maritime barrier of the American continent fulfilled the dream of four hundred years. In 1914 scores of books and articles were published in the United States celebrating the achievement of the American construction leadership. And rightly so. The completion of the canal was at the time history's greatest engineering marvel and a victory of man over nature that would not be matched until the moon landings.

When one considers that this world record-breaking construction was carried out in tropical conditions, in a previously disease-ridden area with endemic political unrest, with no local workforce and two thousand miles from home base, the achievement becomes even more impressive.

That six pairs of locks were constructed, from start to finish, in approximately four years is remarkable, and would be considered a perfectly reasonable timeframe even today. The Gatún dam, also, unlike the hastily and cheaply built Jonestown dam, was a solid, professional and lasting piece of work. And by making the Chagres River an ally rather than an enemy of the canal, the Americans mastered a problem that had ruined the French effort.

So the United States engineers managed to complete in ten years a goal which had defeated the French over twenty-two. Between 1881 and 1903, the two French companies removed 73 million cubic yards of spoil. Although the Americans only took out 14 million cubic yards during the difficult early years, from 1907 onwards Goethals' 'Army of Panama' moved an astonishing 219 million cubic yards. Much credit for this excavation must go to Stevens' Railroad Era workers. Stevens built the digging machine, having learnt from the mistakes of Wallace; Goethals merely had to turn the handle and, now and then, tighten the screws.

Furthermore, the end product was vastly superior to anything conceived by the French. The fact that the canal is still, a hundred years later, a key artery of world commerce, is testimony to how the Americans 'thought big'.

The French failed, of course, because of disease and because they ran out of money. They were also unlucky, suffering floods, fires and revolution on the Isthmus, as well as the financial mugging by Trenor Park of the Panama Railroad Company. But they massively underestimated the size of the task they had taken on. De Lesseps' first calculation of the amount of excavation needed for the sea-level plan was only a quarter of that removed by the Americans for their lock-canal project. Furthermore, the sea-level plan, the culmination of centuries of wishful thinking about the Isthmus, was simply impossible – the slides in the Cut would have been even worse if the digging had gone down another 60 or 70 feet. As Goethals commented, there was simply not enough money in the world to pay for such a plan. For a private company looking for a return on their investment, it was hopeless.

The Americans, as has been seen, repeated many of the mistakes of the French, but they also had key advantages in addition to their much more reliable financial backing. Crucial to the morale of the white American workforce was the defeat of yellow fever and taming of malaria, made possible by discoveries that largely post-dated the de Lesseps effort. The presence of the US military and their firm grip on Panamanian politics meant that the Americans suffered none of the political instability, revolution and violence that the French had to work around, in spite of the long-established local antipathy towards the 'Yankees'. Advances in precision manufacturing, assembly-line production and steel technology, driven in part by the naval armaments and motor industries, meant that the US plant was far superior to that of the French. The Bucyrus shovels were capable of excavating at a rate three or four times greater than the best French machines. The Americans also had better drills and explosives and superior expertise in railroad transportation. And in Stevens and Goethals they found determined and accomplished leadership.

For all Theodore Roosevelt's bellicosity, the war against the Panama jungle and mountains would be the only battle he would fight as President. On several occasions – the choice of the Panama route, the creation of the Republic of

Panama, the backing of Gorgas and the choice of the lock and lake plan – his intervention was decisive. Certainly, Roosevelt was in no doubt where the credit should go. In 1908, as he was preparing to leave office, he wrote of the canal project to a newspaper editor in London: 'This I can say absolutely was my own work, and could not have been accomplished save by me or by some man of my temperament.' Out of office, he was even more boastful, saying to an audience at the University of California at Berkeley in 1911, 'I am interested in the Panama Canal because I started it. If I had followed traditional, conservative methods I would have submitted a dignified State paper of probably 200 pages to Congress and the debates on it would have been going on yet; but I took the Isthmus, started the canal and then left Congress not to debate the canal, but to debate me and in portions of the public press the debate still goes on as to whether or not I acted properly in getting the canal but while the debate goes on the canal does too.' What was key, Roosevelt pronounced in his autobiography, was that 'somebody [namely, himself] was prepared to act with decision'.

The speech in California, quoted and misquoted in newspapers across the United States, caused a sensation, and reignited the controversy of America's role in the 'Panama Revolution'. The whole affair, argued a contributor to the *North American Review* in 1912, had been a 'Chapter of National Dishonour'. A prominent US historian described it as 'an affront to international decency'. Alfred Mahan himself replied to these attacks, writing, 'The summary ejectment of Colombia from property which she could not improve herself, and against the improvement of which by another she raised frivolous obstacles, is precisely in line with transactions going on all over the world . . . India, Egypt, Persia, Tripoli, Tunis, Algiers, Morocco, all stand on the same general basis as Panama.' Their occupation, he went on, was part of the 'advance of the world'.

Many Americans agreed that the ends justified the means, but most were uneasy about this sort of high-flown imperialism, especially after difficulties from 'insurgents' persisted in the Philippines and Cuba. In fact, Democratic Party policy was now that Colombia had a legitimate grievance for the loss of Panama, and after the 1910 mid-term elections they had a majority in the House of Representatives. So when Roosevelt made his 'I took the Isthmus' speech, a resolution was introduced and approved calling for a fresh Congressional investigation into the whole affair. There were also renewed protests and demands for international arbitration from Colombia, and this time, with Roosevelt off the scene, they were heard sympathetically. In 1911 the Taft administration sent James du Bois to Bogotá to secure an agreement.

The envoy was shocked at the strength of feeling in Colombia. 'Confidence and trust in the justice and fairness of the United States, so long manifested, has vanished completely,' he reported back to Washington the following year. He also found the country's leadership to be nothing like Roosevelt's famous description: 'instead of "blackmailers" and "bandits" the public men of Colombia compare well with the public men of other countries in intelligence and respectability,' he wrote. 'I deplore Colonel Roosevelt's bitter and misleading attack.'

In 1914 the Wilson government offered Colombia a 'sincere apology' and an indemnity of $25 million, but such was the vehemence of Roosevelt's attack on this measure that Congress backed off ratifying the new treaty. Then, in January 1919, Roosevelt died, and a major impediment to the deal was removed. There was also a new and powerful incentive to repair relations with Bogotá – what was thought at the time to be the world's largest reserve of oil had been found under the soil of Colombia, and the Anglo-Dutch company Shell looked set to control the supply. So with Colombia's permission to remove the clause stating 'sincere regret', the treaty was ratified in early 1921 under Harding's Republican administration, and the $25 million, dubbed 'canalimony' by one wit, was paid over.

Doubtless the agreement had as much to do with oil reserves and other business opportunities in Latin America as it did with righting a past wrong. US investment in Colombia increased tenfold in the eight years after the deal. But for the New York *World* the paying of the indemnity was a vindication of their investigation into 'a most sordid and shameless conspiracy into which Theodore Roosevelt had dragged the United States Government in order to satisfy his personal ambition'. 'The most flagrant act of Prussianism in the history of the United States,' the paper concluded, 'is now definitely repudiated by the political party that ardently defended it for nearly eighteen years.'

On 24 August 1914 the *Pleiades* steamed into New York to be met by a cacophony of whistles from all the ships in the harbour. The vessel and its cargo, 5000 tons of lumber and general merchandise, was unremarkable enough. What had caused the outbreak of celebrations was that the *Pleiades* was the first ship to trade between San Francisco and New York via the new canal. In her fourteen years plying the trade routes of the world, it was her most profitable trip yet. The canal had shaved nearly eight thousand miles off the journey between the two cities, almost halving the time at sea. Now two voyages could be made in the time of one. The benefits of the canal for US trade and shipping were there for all to see.

The first years of the canal, however, saw continuing challenges. Two months after the transit of the *Pleiades*, a huge slide at East Culebra completely blocked the channel in half an hour. As before, Goethals, still in charge, ordered his dredges and shovels to dig it all out once more, but the following year saw more slides and the waterway blocked again for all but the smallest vessels, this time for seven months, and President Wilson was forced to return to a two-ocean navy, exactly what the canal had been built to avoid.

In fact, the problem of slides was never solved. All the canal maintenance teams could do was to remove the spoil and keep their fingers crossed. As recently as 1974, 250,000 cubic yards slid into the Cut, reducing it to one-way traffic and costing more than $2 million to remove.

The canal was also sporadically closed when, during the dry season, the level of Lake Gatún fell below that needed to operate the locks. So early 1935 saw the completion of a new structure at Alhajuela, the Madden Dam, which held a higher, secondary reserve of water to hold back extreme floods and to feed the larger, lower lake when necessary.

When it was found that the new aircraft carriers were too wide for the locks, the United States army engineers started work on two giant new lock basins at either end of the canal. But with America's entry into World War II, the project was shelved. By the end of the war, the United States fleet was so vast that the canal's original purpose – the avoidance of having to support a two-ocean navy – had been outgrown, although much use was made of the canal for ferrying men and materials for the Korean and Vietnam wars. Just as useful strategically were the army and air force bases in the Zone, from where US power could be (and was) projected throughout Central American and northern South America.

The canal is now Panama's after two generations of struggle against the United States to regain control of their country. The bases are gone, and the canal has returned to the peaceful purpose always intended by idealists like Humboldt and de Lesseps. The checkpoints that prevented Panamanians from driving into the Zone are still there but abandoned and dilapidated, and the neat rows of identical houses for the US administrators and military are now home to Panamanians, who, with their 'dislike of uniformity', have been busy personalising them, adding scruffy lean-tos or extending verandas in a higgledy-piggledy fashion.

Opponents of the 1999 handover argued that the Panamanians would be unable to run the canal efficiently, but they have been proved wrong. Canal improvements have continued steadily and threats to the water supply, so vital for the huge locks, have been addressed. Now the ACP, the Autoridad del Canal de Panamá, has proposed an ambitious plan to buld two giant new lock systems at either end of the canal to cope with the increasing number of post-Panamax container ships and the currently vast traffic from manufacturing centres in East Asia to the US's East Coast. Work is due to begin in 2007.

While wishing the project well, it is impossible to avoid hearing echoes of the canal's long history. There is a Technical Commission of international worthies (who even considered a sea-level scheme); the work will be done, it is planned, by machines rather than men; by 2025 the enlarged capacity will be contributing eight times the canal's current $500 million annual payment to the Panama treasury, or, as Paterson said three hundred years ago, 'trade will increase trade, and money will beget money.'

Between 1904 and 1914 the US Government paid out about $400 million for their canal. It was not until the 1950s, however, that the venture started showing a profit, far longer than private capital would have required. There were, of course, other costs as well. According to the official figures, just over six thousand employees died in ICC hospitals during the American construction period, of whom about three hundred were from the United States. As we have seen, this overall figure is likely to be an underestimate. Those who suffered most were the humble 'silvermen'.

At the end of the construction period, some of the canal workers had made good, some had not. A number did return home with a large enough nest egg to buy some land, or set up a small business, or just to impress their friends. 'The returned Panama Canal labourer is an uncommonly vain fellow,' commented one observer in Barbados. 'He struts along in all the glory of a gay tweed suit, a cylindrical collar and a flaring necktie.' Like 'Colón Man' in Jamaica a generation before, returnees to Barbados brought with them a less subservient attitude and a new cosmopolitanism. They would be at the forefront of the social upheavals of the 1920s and 1930s that eventually led to political decolonisation. Also, support for Marcus Garvey's Universal Negro Improvement Association would be strongest among those who had worked in Central America. Garvey had worked as a newspaper editor in Colón during the last years of the construction period, and his approach was characterised by internationalism, accentuating the identity of interests among blacks all over the world in place of narrow national loyalties. He was also a materialist – 'Wealth is strength, wealth is power,' he wrote – and saw the Panama money earned by the blacks as a liberating force.

But not everyone returned home with their pockets rattling with coins. Jamaican Z. McKenzie remembered: 'The completion of the waterway Brought great Desolation on the W.I. [West Indian] employees . . . The wage during the Canal Construction was so small that we could not put by any Savings in the Bank. Hence the majority of us left empty handed, to live or die.' A popular song of the time in Barbados goes: 'Look de Panama man come home from sea/As skinny as a Church rat,/An' all he had in he grip fo' me/Was a wide-brimmed Panama hat.' In fact many were like Albert Bannister, who wrote, 'I would be glad to go home but I can't go home empty handed.' Some 45,000 Barbadians went to Panama, and of these only about half returned home at the end of the construction period. Barbados only had a population of 200,000, so the effect on the island was dramatic. For one thing, the planters no longer had the pool of cheap labour that had sustained their inefficient practices. And even more than in Jamaica a generation before, the demographics of the country were radically altered. As late as 1921 there were less than four hundred males per thousand females on the island.

Of those who did not return, some enlisted in the British or French armies and were shipped to Europe. Others took jobs on the United Fruit or the sugar plantations in Guatemala, Cuba and South America. A large number also stayed in Panama, of whom about 7000 were kept on in the employ of the canal, often at lower wages than they had been paid during the construction period. Their treatment by the canal authorities was pretty much consistent with what had gone before, with the Gold/Silver Roll system as firm as ever and strikers or activists ruthlessly deported. If anything, some of them comment, it was worse, as the *esprit de corps* of the construction period did not last. They were also targeted by nationalist Panamanians, who from the 1930s and 1940s onwards tried to purge their country of non-Hispanic elements.

Saddest of all, perhaps, was the treatment of the old-timers who had worked during the construction and then stayed with the canal for the rest of their working lives. Initially, there was no pension at all, then in the 1930s the canal authorities offered the men 'disability relief' of one dollar a month per year worked, up to a maximum of $25. Inspectors would come to their houses and if they had possessions of any value, or if another family member worked, this sum was reduced.

One man who had worked for thirty-eight years for the canal, without a single week's holiday and having suffered numerous injuries in the course of his work, was told that he was to be retired and was given fifteen days' notice to quit his Canal Zone quarters. He eventually found a small apartment in Colón for $25 a month, which was the sum total of his pension. His son described in 1946 how his father was 'receiving not enough to live a comfortable life for his remaining days. Broken in body and spirit, he calmly smokes his faithful old pipe waiting for the call of his Maker.'

'Who dug the canal?' asks Jules LeCurrieux. 'Who suffered most even until now? Who died most? Who but the West Indian negroes.' LeCurrieux worked from 1906 to 1938, in the course of which he was blinded in one eye while building the relocated railroad. When he was 'retired', he received $17.50 a month, 'which was too small to live on'.

Many of the letters to the 'Competition for the Best True Stories . . .' held in 1963, contain pathetic pleas for help with subsistence. A doctor who treated a lot of the old-timers in their last days tells how the majority had chronic health problems not caused by the difficult conditions in which they had worked, but, shockingly, from malnutrition.

Most of the West Indians signed off their accounts for the competition with mixed feelings. Unlike the thousands who worked and died on the French canal, at least there was something to show for their efforts. As one digger put it, 'I am glad to see that all my sweat, tears, and all those deaths were not in vain.' Having been part of the great achievement of the canal was a source of great pride. 'It is a job well done,' wrote one Jamaican, 'and a help to mankind.' 'I got to be a man,' says another.

Harrigan Austin, who had arrived in 1905 hungry enough to attack bags of sugar on the wharf, writes about the 'untold benefits to the world at large' that the canal brought. ''Tis reasonable in any big war or any such projects as this, something will happen,' he goes on. 'Some must suffer for the good and welfare of the others for where there is no Cross there may be no Crown . . . Thank God, the canal has been finished and has become a blessing to the world at large. A great accomplishment, the work of a Great Nation – May God Bless America.'

George Martin, who looked back fondly on the days he could afford ham and ice cream, concluded: 'The work of the construction days was a hard and rough struggle, but it was done cheerfully, and faithfully; thus giving the American people their hearts' desire.'

Other old-timers were less gracious about their treatment. Benjamin Jordan, who back in 1905 had lied about his age to get selected for a Karner contract, testified in 1984 that although he had not let the 'discrimination take hold' of him, there were now, at the end of his life, feelings that he could no longer 'put in a corner'. 'For my years with the Panama Canal,' he said, 'there is a feeling that I have not been treated as I should. I still enjoy life, like some of the others who survived. The fact still remains: much blood was split, and no one cared about it. But I'm still alive, under God's care and will always remember: the good that you do lives with you.'

SOURCE NOTES

FO Public Record Office, Kew, London, Foreign Office Records
RG Record Group, National Archives, Suitland, Maryland, USA

Preface: The Battle to Build the Canal (pp. xv–xix)

p. xvi Bahamas-born Albert Peters . . .: Isthmian Historical Society, 'Competition for the Best True Stories of Life and Work on the Isthmus of Panama During the Construction of the Panama Canal'

p. xvi Alfred Dottin . . .: 'Competition for the Best True Stories . . .'

p. xvii Constantine Parkinson . . .: *ibid.*

p. xvii 'Some of the costs of the canal are here': Franck, *Zone Policeman 88*, p. 85

p. xvii 'some sort of semi-slavery': Harrigan Austin, 'Competition for the Best True Stories . . .'

p. xvii 'Many times I met death at the door', J. T. Hughes, 'Competition for the Best True Stories . . .'

p. xvii 'We worked in rain, sun, fire': Prince Green, *ibid.*

p. xviii 'greatest liberty ever taken with nature': James Bryce, quoted in LaFeber, *The Panama Canal: The Crisis in Historical Perspective*, p. 4

Chapter 1: 'The Keys to the Universe' (pp. 1–12)

p. 1 'I am not capable by writing . . .' National Library of Scotland: Adv. MS 83.7.3, f.22

p. 3 'no mountain range at all': quoted by Lionel Wafer in Duval, *From Cadiz to Cathay*, p. 10

p. 4 'Do but open these doors': National Library of Scotland, Adv. MS 83.7.3, f. 44v

p. 4 'talks too much and raises people's expectations': National Archives of Scotland: GD26/13/43/27

p. 4 'very free and not at all shy . . .': Burton (ed.), *The Darien Papers*, p. 62

p. 6 'here is food hunting and fowling': National Archives of Scotland: GD406/1/6489
p. 6 'We have had but little Trade as yet': National Archives of Scotland: GD150/3375
p. 7 'In our passage from Calidonia hither': National Library of Scotland, Adv. MS 83.7.3, f.22
p. 7 'nothing left but a howling wilderness': Letter from Alexander Shields, 2 February 1700, Registrar General for Scotland: OPR453/9, p. 139
p. 9 'shewing them the great maine sea': Peter Martyr: *de orbe novo*. In Hakluyt's *Collection of the early voyages, travels, and discoveries*, Vol. 5, p. 253
p. 10 'If there are mountains there are also hands': de Vedia (ed.), *Historiadores primitivos de Indias*, Vol. 1, p. 222
p. 10 'would open the door to the Portuguese': quoted in Anguizola, *Philippe Bunau-Varilla*, pp. 3–4
p. 11 'I intend . . . to possess the Lake of Nicaragua': Duval, *From Cadiz to Cathay*, p. 15

Chapter Two: Rivalry and Stalemate (pp. 13–19)

p. 13 'I am assured . . . a canal appeared very practicable': Jefferson: *Writings*, Vol. 1, p. 518
p. 13 'The American continents . . . by the free and independent condition': quoted in Siegfreid, *Suez and Panama*, p. 224
p. 13 'would immortalise a government occupied with the interests of humanity': Humboldt, *Political essay on the kingdom of New Spain*, p. 77
p. 15 Goethe, who . . . quoted Collin, *Theodore Roosevelt's Caribbean*, p. 129
p. 15 'veritable capital of the world': quoted in Siegfreid, *Suez and Panama*, p. 223
p. 16 'present a human barrier of such formidable power': Lloyd, 'On the facilities for a ship canal communication . . . through the Isthmus of Panama', *Inst. of Civil Engineers, Minutes of Proceedings*, Vol. 9, 1850, p. 242
p. 16 'superstitious . . . Billiards, cockpits, gambling and smoking': Lloyd, 'On the facilities for a ship canal communication . . . through the Isthmus of Panama', *ibid.*, p. 59
p. 16 A visitor from Bogotá in the 1830s . . .: Castillero, *Historia de Panama*, p. 87
p. 17 'an absurdity': 'A letter of Alexander von Humboldt, 1845', *American Historical Review*, Vol. 7, No. 4 (July 1902), p. 705, quoted in Mack, *The Land Divided*, p. 128

Chapter Three: Goldrush (pp. 20–24)

p. 20 'low, miserable town, of thirty thatched huts': Hotchkiss, *On the Ebb*, p. 84
p. 20 'The house are only hovels': 'Across the Isthmus in 1850: The Journey of Daniel A. Horn', Bayard Taylor, *Eldorado or Adventure in the Path of Empire*, New York, 1856, quoted in Perez-Penero *Before the Five Frontiers*, p. 82
p. 21 'Half are full-blooded negroes': quoted in Perez-Penero, *ibid.*, p. 84
p. 21 'Early in the morning went ashore in a skiff': Richards (ed.), *California Gold Rush Merchant*, p. 7

p. 21 'the birthplace of a malignant fever': Marryat, *Mountains and Molehills*, pp. 1–3

p. 21 'as disconcerting as hell': John Easter Minter, *The Chagres: River of Westward Passage*, p. 238, quoted in Perez-Penero, *Before the Five Frontiers*, p. 85

p. 21 'The eye does not become wearied': Marryat, *Mountains and Molehills*, pp. 1–3

p. 21 Some of the men carried as much as 300 pounds: Borthwick *Three Years in California*, p. 34, quoted in Brands, *The Age of Gold*, p. 81

p. 21 'so like a nightmare': Frémont, *A Year of American Travel*, p. 34

p. 22 'the main street is composed almost entirely of hotels': Marryat, *Mountains and Molehills*, p. 5

p. 22 'disposition to roam': Seacole, *Wonderful Adventures*, p. 8
'Those bound for the gold country': Seacole, *ibid.*, p. 19

p. 23 'mustard emetics, warm fomentations': Seacole, *ibid.*, p. 25

p. 23 'I knew nothing of the great risk in travelling alone': Julius H. Pratt, 'To California by Panama in '49', *Century Illustrated Monthly Magazine*, Vol. 41, No. 6, April 1891, pp. 915–17

p. 23 a 'little prejudice': Seacole, *ibid.*, p. 14

p. 24 'is it surprising that I should be somewhat impatient': Seacole, *ibid.*, p. 41

p. 24 'I never travelled with a nigger yet': Seacole, *ibid.*, p. 57

Chapter Four: The Panama Railroad (pp. 25–35)

p. 26 'No imposing ceremony inaugurated the "breaking ground"': Otis, *History of the Panama Railroad*, p. 26

p. 26 'It was a virgin swamp': Otis, *ibid.*, p. 26

p. 26 'carried his noonday luncheon in his hat': Otis, *ibid.*, p. 12

p. 26 'wore the pale hue of ghosts': Seacole, *Wonderful Adventures*, p. 64

p. 27 'intended to be, to a certain extent prohibitory': Otis, *History of the Panama Railroad*, p. 24

p. 28 'British consul's precarious corrugated iron dwelling': Otis, *ibid.*, p. 5

p. 28 'Three sides of the place were a mere swamp': Seacole, *Wonderful Adventures*, p. 64

p. 28 'as filthy and odorous as any slavers': Schott, *Rails across Panama*, p. 177

p. 29 'We could name many persons who were walking the streets of the City': quoted in Senior, 'The Panama Railway', *Jamaica Journal*, June 1980, p. 68

p. 29 all but two of the fifty American technicians then on site . . .: McCullough: *The Path between the Seas*, p. 38

p. 29 One railroad historian . . .: Schott, *Rails across Panama*, pp. 192–3

p. 29 'Workers who toppled over in the jungle': Schott, *ibid.*, p. 139

p. 29 a grim trade sprung up in cadavers . . .: Schott, *ibid.*, p. 68

p. 30 'a bare-footed, coatless, harum-scarum looking set': Tomes, *Panama in 1885*, p. 123

p. 30 in the process publicly flogging the Panamanian official . . .: Conniff, *Panama and the United States*, p. 28

p. 31 A 40-foot deep cut was dug near Paraíso . . .: Nelson, *Five Years at Panama*, p. 148

p. 31 'Iron and steam, twin giants subdued to man's will': Seacole, *Wonderful Adventures*, p. 63

p. 31 'no one work . . . has accomplished so much': Otis, *History of the Panama Railroad*, p. 15

p. 32 estimates are as high as $7–9 million, or $170,000 per mile . . .: New York *Tribune*, 13 March 1855

p. 32 standing at $295 a share . . .: McCullough, *The Path between the Seas*, p. 35

p. 33 'better class of shop-keepers are Mulattoes': Panama *Star and Herald*, 27 December 1856

Chapter Five: The Competing Routes (pp. 36–39)

p. 36 'The Department has entrusted to you a duty': McCullough, *The Path between the Seas*, p. 20

p. 36 'It is proved beyond all doubt . . . that Dr. Cullen never was in the interior': Mack, *The Land Divided*, p. 255

p. 37 'mystical and imaginative': Ripley, *The Capitalists and Colombia*, p. 45

p. 38 'it [would] one day be covered with sails from every clime': *H. Misc. Doc. No. 113*, 42 Cong., 3 Sess., p. 41, quoted in Mack, *The Land Divided*, p. 243

p. 39 'the deep cut would probably be subject to land-slides': *Sen. Ex. Doc. No. 15*, 46 Cong., 1 Sess., p. 5

Chapter Six: '*Le Grand Français*' (pp. 41–48)

p. 41 'We are, gentlemen, soldiers under fire': Bunau-Varilla, *Panama*, p. 52

p. 44 'Astonish[ing] the world by the great deeds': quoted in Siegfreid, *Suez and Panama*, p. 236

p. 45 'man eminent for originality': *The Times*, 2 July 1870

p. 45 'You exercise a charm', de Lesseps, *Recollections of Forty Years*, Vol. 2, p. 294

p. 45 'will help to wed the whole universe': quoted in Siegfreid, *Suez and Panama*, p. 60

p. 45 'Is it not a glorious thing for us to be able to carry out': de Lesseps, *Recollections of Forty Years*, p. 172

p. 46 'like brothers': Marshall, *Demanding the Impossible*, p. 340

p. 47 'Doubt is no longer possible' Congrès des Sciences Géographique, Cosmographique et Commerciales Antwerp, 1871: *Compte-rendu*, Vol. 1, p. 361

p. 47 'to the attention of the great maritime powers': Congrès des Sciences Géographique, Cosmographique et Commerciales Antwerp, 1871: *Compte-rendu*, Vol. 1, p. 362

Chapter Seven: The French Explorers (pp. 49–57)

p. 49 *'oeuvre grandiose'*: Reclus, *Panama & Darien*, p. 18

p. 50 'veritably the Stock Exchange of Panama': Reclus, *ibid.*, p. 75

p. 50 'Perfidious friends, more dangerous than the climate': Reclus, *ibid.*, p. 83

p. 51 'Our hope is to fill these waters with all the ships': Reclus, *ibid.*, p. 184

p. 52 'You should start preparations immediately': Wyse to Reclus, 2 February 1878, quoted in Fauconnier, *Panama: Armand Reclus*, p. 105

p. 55 Reclus, in particular, was concerned about the dimensions of the massive tunnel . . .: Fauconnier, *ibid.*, p. 139

p. 55 Three were in Darién . . .: Wyse, Reclus and Sosa, *Rapport sur les Etudes de la Commission Internationale d'exploration de L'Isthme Americain*, p. 48

p. 56 'no more glorious name': Reclus, *Panama & Darien*, p. 417

p. 56 Chesnaye, was, he wrote to Reclus, 'an inconvenience': Wyse to Reclus 24 August 1878, quoted in Fauconnier, *Panama*: *Armand Reclus*, p. 142

Chapter Eight: The Fatal Decision (pp. 58–67)

p. 58 'who for some time past': Hugh Mallet to Foreign Office, 5 January 1879, FO55/269

p. 58 'Tragic scenes': New York *Tribune*, 5 May 1879

p. 58 'Business has been paralysed': Hugh Mallet to Foreign Office, 17 June 1879, FO55/269

p. 62 'involved so much uncertainty': Menocal, 'Intrigues at the Paris Canal Conference', *North American Review*, September 1879, p. 16

p. 62 'provide for the whole drainage': *Instructions to Rear Admiral Daniel Ammen . . .* , p. 11

p. 62 'threw off the mantle of indifference': quoted in McCullough, *The Path between the Seas*, p. 78

p. 63 De Lépinay himself, although favouring Panama over Nicaragua . . .: Congrès International D'Études du Canal Interocéanique, *Compte Rendu Des Séances*, pp. 293–9, 355

p. 64 he claims that he investigated its potential . . .: Wyse, *Le Canal de Panama*, p. 191

p. 64 He also argued that the new lake . . .: *Instructions to Rear Admiral Daniel Ammen . . .* , p. 19

p. 65 'the resolution is indefinite': *ibid.*, p. 20

p. 65 'May that illustrious man': *Le Matin*, 30 May 1879, quoted in Anguizola, *Philippe Bunau-Varilla*, p. 38

p. 66 'A careful examination of the names of the French delegates': Menocal, 'Intrigues at the Paris Canal Conference', *North American Review*, September 1879

p. 66 'relative consideration of natural advantages': *Instructions to Rear Admiral Daniel Ammen . . .* , p. 10

p. 66 'a comedy of most deplorable kind': Johnston, 'Interoceanic ship canal discussion', *Journal of the American Geographical Society*, Vol. 11, 1879, pp. 172–80

p. 66 Menocal, for his part, was sure that the whole Panama project was doomed . . .: *Instructions to Rear Admiral Daniel Ammen . . .* , p. 21

p. 67 'prefigures for us an era of complications and difficulties': New York *World*, 1 January 1880

p. 67 'The financial organs were hostile': quoted in McCullough, *The Path between the Seas*, p. 102

Chapter Nine: The Riches of France (pp. 68–80)

p. 69 'Mr Lesseps' enterprise': Panama *Star and Herald*, 1 January 1880
p. 69 'wearing the diplomatic smile for which he was noted': Robinson, *Fifty Years at Panama*, p. 139
p. 69 'The Canal will be made': Robinson, *ibid.*, p. 140
p. 70 'every one of the [city's] 14,000 inhabitants': New York *World*, 22 January 1880
p. 70 'Such an air of neatness and real cleanliness': New York *Tribune*, 22 January 1880
p. 70 'gave éclat to the occasion': Robinson, *Fifty Years at Panama*, p. 143
p. 71 'unanimous in their expressions of gratification': Panama *Star and Herald*, 8 January 1880
p. 71 'His mind is unalterably made up on one point': New York *Tribune*, 22 January 1880
p. 71 'Mr. Lesseps is an accomplished horseman': Panama *Star and Herald*, 5 January 1880
p. 71 'as dark as Arabs and as wild': Robinson, *Fifty Years at Panama*, p. 144
p. 71 'bright with myriad lights': Panama Star and Herald, 7 February 1880
p. 72 dance 'all night like a boy': Robinson, Fifty Years at Panama, p. 146
p. 72 'The engineering difficulties . . . are being solved one by one': Panama *Star and Herald*, 16 January 1880
p. 72 'the moral, if not material support', *ibid.*, 14 February 1880
p. 73 'vastly to the regret of the people': *ibid.*, 14 February 1880
p. 73 'Blanchet does nothing . . . the tachometers are broken': Wyse to Reclus, 24 January 1880, quoted in Fauconnier, *Panama: Armand Reclus et le Canal des Deuz Océans*, p. 170
p. 73 'sham and charlatanism': Rodrigues, *The Panama Canal*, p. 66
p. 73 'from first to last he was perfectly conscientious and honest': Robinson, *Fifty Years at Panama*, p. 141
p. 74 'proper predominance over the sea': New York *World*, 2 January 1880
p. 74 *New York Times* thought that the sea-level canal . . .: *New York Times*, 25 January 1880
p. 74 'All this looks like business': New York *Tribune*, 23 January 1880
p. 74 'Now is the time for the Government to make up its mind': *ibid.*, 10 February 1880
p. 75 'where one is used to working for the civilisation of the world': Fauconnier, *Panama: Armand Reclus et le Canal des Deuz Océans*, p. 167
p. 75 'the enterprise of M. Lesseps is of an entirely private nature': Panama *Star and Herald*, 17 February 1880
p. 75 'I have offered America 300,000 of the shares': *ibid.*, 21 April 1880
p. 76 'The policy of this country is a canal under American control': *Sen. Exec. Doc. No. 112*, 46 Cong., 2 Sess., pp. 1–2
p. 76 'When M. de Lesseps gets ready to leave Washington to-morrow': New York *Tribune*, 9 March 1880

p. 76 'President's message assured the political stability of the canal': *Bulletin du Canal Interocéanique*, 15 March 1880

p. 76 'In these provincial tours': Johnston, 'Interoceanic ship canal discussion', *Journal of the American Geographical Society*, Vol. 11, 1879, pp. 172–80

p. 77 'few miles of oozy quagmire and jungle': *The Times*, 3 May 1880

p. 77 'It is a region': London *Standard*, 8 December 1880

p. 77 500-franc shares in Suez were now worth . . .: McCullough, *The Path between the Seas*, p. 125

p. 78 'Capital and science have never had such an opportunity': quoted in *Bulletin du Canal Interocéanique*, 1 December 1880

p. 78 'At that time they realized the poetry of capitalism': quoted in Siegfreid, *Suez and Panama*, p. 240

p. 79 'The company now has a legal existence and a name': Panama *Star and Herald*, 12 March 1881

p. 79 'The worry is that it will weaken the United States strategically': New York *Tribune*, 16 December 1880

p. 80 'insist on acquiring from Colombia the territory': New York *Tribune*, 6 January 1881

Chapter Ten: '*Travail Commencé*' (pp. 81–95)

p. 81 '*Travail Commencé*': quoted in *Bulletin du Canal Interocéanique*, 15 February 1881

p. 82 '*Mais, bah!*': Cermoise, *Deux Ans à Panama*, pp. 1–2

p. 82 'A great thoughtfulness took hold of us all': Cermoise, *ibid.*, p. 25

p. 83 'horrifying tangle of trees': Cermoise, *ibid.*, p. 42

p. 84 'implacable enemy of our great enterprise': Cermoise, *ibid.*, p. 67

p. 84 'passed her days *à vagabonder* barefoot and bare-headed': Cermoise, *ibid.*, p. 73

p. 84 'thoroughly daunted': Cermoise, *ibid.*, p. 80

p. 85 'a steam bath': Cermoise, *ibid.*, p. 107

p. 85 'I am a free man': Cermoise, *ibid.*, p. 81

p. 85 'In spite of the faults, the Colombians': Cermoise, *ibid.*, p. 83

p. 86 'smoking cigarettes when the din became too loud': Cermoise, *ibid.*, p. 77

p. 86 'Flamed with millions of fireflies': Cermoise, *ibid.*, p. 86

p. 86 The first criticism that Reclus makes is levelled at the choice of men . . .: Reclus to Charles de Lesseps, 30 April 1881, quoted in Fauconnier, *Panama: Armand Reclus et le Canal des Deuz Océans*, pp. 186–8

p. 88 Ernst Dichman, did everything in his power . . .: Bennett to Foreign Office, 11 May 1880, FO55/274, also *Star and Herald*, 8 May 1880

p. 88 'all alliances with the United States': Arthur O'Leary in Bogotá to Granville, 5 April 1881 FO420/36

p. 88 considered denouncing the 1846 treaty . . .: letter from Carlos Holguin, 9 July 1881, FO420/36

p. 88 'an alliance against the United States': Secretary of State Blaine to James P. Lowell, 24 June 24 1881, 'Correspondence respecting the Projected Panama Canal Presented to both houses of Parliament 1882' FO420/36

p. 88 'would be glad to see England and France take joint measures': letter from Mr Langley in Madrid to Granville, 23 September 1881, FO420/37

p. 89 'enable the United States to keep military possession of the canal': Sackville-West to Granville, FO420/37, 12 January 1882

p. 89 it would be 'manifestly unjust' for the Americans to request abrogation: Blaine to Mr Lowell, 19 November 1881, FO420/37

p. 89 'Mr Blaine had overshot the mark': New York *Herald*, 20 January 1882

p. 90 *'Viva el canal', 'Viva los franceses'*: Cermoise, *Deux Ans à Panama*, pp. 144–5

p. 90 *'gros bonnets'*: Fauconnier, *Panama: Armand Reclus et le Canal des Deuz Océans*, p. 229

p. 90 'He works a lot': Fauconnier, *ibid.*, p. 201

p. 91 'Everyone had his own room!': Cermoise, *Deux Ans à Panama* p. 109

p. 91 'We said goodbye with a certain sadness': Cermoise, *ibid.*, p. 128

p. 92 Only one in ten newly arrived labourers . . .: McCullough, *The Path between the Seas*, p. 133

p. 92 'Mr. de Lesseps contemplates making up what is short in the labor': Panama *Star and Herald*, 12 March 1881

p. 93 'learn a foreign language' or 'seek adventure': quoted in Newton, *The Silver Men*, p. 7

p. 94 'A trip to Colón?': Barbados *Herald*, 6 August 1885

p. 94 'pretentious, and always complaining': quoted in Siegfreid, *Suez and Panama*, p. 252

p. 94 'They were excellent workers': Cermoise, *Deux Ans à Panama*, p. 247

p. 95 'are left when ill to die in the streets of Colón': 'Governor's Report on the blue book 1881–2', p. 32, quoted in Petras, *Jamaican Labor Migration*, p. 112

Chapter Eleven: Fever (pp. 96–105)

p. 96 'He got up, dressed, and went to the captain's cabin': *Bulletin du Canal Interocéanique*, 1 September 1881

p. 99 'There was a dismal period for the administration': *Deux Ans à Panama*, p. 129

p. 99 'At the moment, the state of health conditions': Verbrugghe to Reclus, 5 October 1881, quoted in Fauconnier, *Panama: Armand Reclus et le Canal des Deuz Océans*, p. 208–9

p. 99 'No epidemic of maladie had manifested itself': Panama *Star and Herald*, 1 October 1881

p. 99 'Mr Blanchet's death is an irreparable loss': Bennett to Granville, 10 November 1881, FO420/36

p. 99 The best estimate is that about fifty men died . . .: Panama *Star and Herald*, 22 February 1882, and Newton, *The Silver Men*, p. 126

p. 100 'the hospital rooms are so vast and well-ventilated': Jos, *Guadeloupéens et Martiniquais au Canal de Panamá*, p. 46

p. 100 'She is one of those rare women whose personal zeal is contagious': New York *Herald*, 22 August 1881

p. 101 There's only one certain way to diagnose fever': Cermoise, *Deux Ans à Panama*, p. 148

p. 101 '*la section de la grande tranchée*': Chamberlaine report, 18 May 1882, FO420/37

p. 102 'invaded by a persistent tiredness': Cermoise, *Deux Ans à Panama*, p. 230

p. 103 'It all but produced an earthquake': Nelson, *Five Years at Panama*, p. 178

p. 104 'I have spent two of the best years of my youth': Cermoise, *Deux Ans à Panama*, p. 301

p. 104 'After two years' work . . . we are much farther advanced': *Bulletin du Canal Interocéanique*, 8 December 1882

p. 104 'incompetent': New York *Herald*, 17 January 1882

p. 104 'exaggerated statements' Panama *Star and Herald*, 3 February 1882

p. 104 'applications for shares showering him from all quarters of France': New York *Tribune*, 28 September 1882

p. 104 'The truth is that during the trial period': Chambre des Députés, *5e Législature, Session de 1893, Rapport Général . . .* , Vol. 1, p. 451

p. 105 'aged a good deal': New York *Tribune*, quoted in Panama *Star and Herald*, 23 June 1883

Chapter Twelve: Jules Dingler (pp. 106–123)

p. 106 'busy . . . and bright with hope': Nelson, *Five Years at Panama*, p. 230

p. 106 'The work moves steadily on': Panama *Star and Herald*, 21 November 1884

p. 107 'I intend to show the world that only the drunk and the dissipated': Haskin, *The Panama Canal*, p. 194

p. 107 'This result has surprised all': Panama *Star and Herald*, 9 November 1883

p. 107 'determined to have a finger in the canal pie': Robinson, *Fifty Years at Panama*, p. 151

p. 108 an American naval officer . . .: Rodgers, *Progress of Work on Panama Ship-Canal* 27 January 1884 (*Sen. Doc. No. 123*, 48 Cong., 1 Sess.)

p. 108 'So the visitor to Gatún': Panama *Star and Herald*, 14 October 1884

p. 109 'From morning till night': Rodgers, *Progress of Work on Panama Ship-Canal*, 27 January 1884 (*Sen. Doc. No. 123*, 48, 1 Sess.)

p. 110 'A stampede took place which is hardly possible to describe': quoted in Panama *Star and Herald* 26 January 1884

p. 110 'Now and again you see a great swell with a watch and gold chain': Senior, 'The Colon People', *Jamaica Journal*, March 1978, p. 70

p. 110 'The infatuation to go seems to have taken hold': *Daily Gleaner*, 7 May 1883, quoted in Senior, 'The Colon People', *Jamaica Journal*, March 1978, p. 64

p. 110 'a flag of liberation': Senior article, 'West Indian Participation in the Construction of the Panama Canal', p. 40

p. 110 'We are not of those who think it a calamity': quoted in Panama *Star and Herald*, 14 February 1884

p. 111 'the great outflow from the Colony of labourers': quoted in Panama *Star and Herald*, 18 May 1883

p. 111 Kill my partner/Kill my partner: 'West Indian Work Songs', Manuscript Collection of the Canal Zone Library-Museum, Library of Congress, Washington, D.C., Box 33

p. 112 'They have a way of shifting for themselves': Panama *Star and Herald*, 29 January 1884

p. 112 'There have been many serious complaints': Mallet to Foreign Office, 19 February 1884, FO55/304

p. 113 'The practice in this consulate': Mallet to Foreign Office, 23 August 1886, FO55/325

p. 113 'I perceive that these men are partial in their protection': letter of 29 March 1883, FO55/297

p. 113 'a ruinously expensive method': New York *World*, 8 February 1885

p. 113 'Damage amounting to thousands of dollars': Montreal *Gazette*, 24 August 1884

p. 114 'of a character and complexity to defy description': Sibert and Stevens, *The Construction of the Panama Canal*, p. 79

p. 114 the Company lost some 10 per cent of the work it paid for . . . J. Bigelow private diary 2 March

p. 114 'There was no system or organisation': Jeremiah Waisome, 'Competition for the Best True Stories . . .'

p. 115 'The rainy season, at last set in': Panama *Star and Herald*, 25 May 1883

p. 115 'The heavy downpours of late': *ibid.*, 26 May 1884

p. 115 'the water will have to be hung up on the sides of the mountains': paper read before the Franklin Institute, 22 October 1884, by Charles Colné

p. 115 'Fresh engineering difficulties present themselves': Admiral Lyon to Foreign Office, 29 November 1883, FO420/50

p. 116 'A day of reckoning is coming': Montreal *Gazette*, 24 August 1884

p. 116 'It is probable the present company will go into bankruptcy' New York *Herald*, 1 November 1884

p. 116 'It is generally believed here that the present Company': Mallet to Foreign Office, 5 July 1884, FO55/304

p. 117 'It would be a pity that a work such as this should be left partially completed': Panama *Star and Herald*, 5 July 1884

p. 117 'Under such circumstances, there is something amounting to heroism': New York *Herald*, 2 December 1883

p. 117 'Death becomes a grim joke, burial a travesty': New York *Tribune*, 8 August 1886

p. 118 'Probably if the French had been trying to propagate Yellow Fever': Gorgas, *Sanitation*, p. 232

p. 118 'thousands dying with yellow fever': Charles Wilson, unpublished memoir, p. 74, Box 11, Manuscript Collection of the Canal Zone Library-Museum, Library of Congress, Washington, D.C.

p. 118 'As for the men': Montreal *Gazette*, 24 August 1884

p. 118 'burials averaged from thirty to forty per day': Nelson, *Five Years at Panama*, p. 7

p. 118 'a Jamaican arrived in Panama by the afternoon train in a dying condition': Panama *Star and Herald*, 17 September 1883

p. 119 'found lying before the door of her Majesty's consulate in a dying state': Mallet to Foreign Office, 12 December 1884, FO55/304

p. 119 'My poor husband is in a despair that is painful to see': Edgar-Bonnet, *Ferdinand de Lesseps*, p. 184

p. 119 'Seldom, if ever, during many years residence': Panama *Star and Herald*, 24 January 1884

p. 119 'Mr Dingler was but 20 years of age': *ibid.*, 25 February 1884

p. 120 'exalted the energy of those who were filled with a sincere love': Bunau-Varilla, *Panama*, p. 44

p. 120 cases among steamer crews . . .: Nelson, *Five Years at Panama*, p. 129

p. 120 In October, a British bark . . .: Panama *Star and Herald*, 14 October 1884

p. 120 'abominable neglect of all sanitary measures': *ibid.*, 28 January 1884

p. 121 'host of idle loafers, who infect the town': *ibid.*, 25 October 1884

p. 121 'thanks to abstemious habits': Nelson, *Five Years at Panama*, p. 17

p. 121 'Woe to the feeble person who doesn't know how to quench his thirst!': Reclus, *Panama & Darien*, p. 130

p. 121 'designed for nothing but hasty drinking': Cermoise, *Deux Ans á Panama*, pp. 52–3

p. 121 'a veritable sink of iniquity': Bishop, *The Panama Gateway*, p. 88

p. 121 'the hardest drinking and the most immoral place I have ever known': Mallet, *Pioneer Diplomat*, p. 26

p. 121 'spirit of venality and corruption': Robinson, *Fifty Years at Panama*, p. 61

p. 121 'There is a general belief held by many intelligent people': Nelson, *Five Years at Panama*, p. 18

p. 122 'assembly rooms, provided with books, periodicals, and various indoor games': *Bulletin du Canal Interocéanique*, 18 July 1883

p. 122 'A great enormous hall with a stone floor was the bar-room': Cermoise, *Deux Ans á Panama*, pp. 52–3

p. 122 'passions run high owing to the constant proximity of death': Dansette, *Les Affaires de Panama*, pp. 24–5

p. 122 'the Sword of Damocles hangs over everyone': Mimande, *Souvenirs d'un Echappe de Panama*, pp. 60–61

p. 122 'death and *la fête* are perpetually hand in hand': Cermoise, *Deux Ans á Panama*, p. 145

p. 122 'foreign men of dubious reputation': Perez-Penero, *Before the Five Frontiers*, p. 113

p. 122 'which as usual was occasioned by the vile rum': Panama *Star and Herald*, 22 May 1883

p. 123 'an agglomeration of all nations': *ibid.*, 18 September 1883

p. 123 billed as a hole of 33 metres paid at 50 piastres per cubic metre . . . Zévaès, *Le Scandale du Panama*, p. 24

Chapter Thirteen: *Annus Horribilis* (pp. 124–141)

p. 124 'not unlikely': *New York Times*, quoted in Panama *Star and Herald*, 3 February 1882

p. 124 'the British recently took possession of Egypt and the Suez canal': New York *Sun*, 6 November 1884

p. 125 'be so fortified as to become a second Gibraltar': Lord Lyons in Paris to Granville, 22 December 1884, FO420/50

p. 125 muttering darkly to the British ambassador . . .: Sackville-West to Granville, 26 December 1884, FO420/50

p. 126 Britain should also have one, built at Tehuantepec: Sackville-West to Granville, 28 December 1884, FO420/50

p. 126 'free to colonize in Central and South America': Sackville-West to Granville, 26 December 1884, FO420/50

p. 126 'more serious than diplomatic correspondence?': Sackville-West to Granville, 26 December 1884, FO420/50

p. 127 'described every Canal functionary': F. R. St John, British Minister in Bogotá to Granville, 2 March 1885, FO420/51

p. 127 Minister shared his grave concerns with the British ambassador . . .: Sackville-West to Granville, 27 January, 1885 FO420/51

p. 128 'At 2 a.m. on the 16th': Panama *Star and Herald*, 19 March 1885

p. 129 'rebel bullets and cannon balls': Leay note to Mallet, 3 April 1885, FO55/313

p. 129 'The firing was hot and reckless in the extreme': Mallet to Foreign Office, 4 April 1885, FO55/313

p. 130 'The entry of the American marines into the city': Mallet to Foreign Office, 27 April 1885, FO55/313

p. 130 'The presence is only temporary': US Navy Department, *Report of Commander Bowman H. McCalla upon the naval expedition to the Isthmus of Panama, April 1885*

p. 131 'The State will never be free from such revolutionary nonsense': Panama *Star and Herald*, 20 March 1885

p. 132 'were ignorant of Isthmian affairs': Mallet to Foreign Office, 10 August 1885, FO55/313

p. 133 Arthur Webb, a Jamaican . . .: eyewitness reports submitted to vice-consul Leay, sent to London by Mallet, FO55/313

p. 133 'I have never before witnessed anything so horribly sickening': Statement to Leay by C. H. Burns 10 May 1885

p. 134 'In all these fights between Jamaicans and Colombians': Panama *Star and Herald*, 9 May 1885

p. 134 'It must also be borne in mind': Mallet to Governor of Jamaica, 6 July 1885, FO55/313

p. 135 'the enterprise will be ready for the world's commerce': Mallet to Foreign Office, 4 March 1885, FO55/313

p. 135 'The Panama canal is in such a state that its ultimate completion is beyond question': New York *Tribune*, 8 May 1885

p. 136 'Mr Varilla's tremendous mental capacity': letter from Wulsin, 28 May 1906, in Bigelow Papers, New York Public Library, Box 24

p. 136 'His versatility was fantastic': quoted in Anguizola, *Philippe Bunau-Varilla*, p. 68

p. 136 Menocal visited in August . . .: *New York Times*, 18 August 1885

p. 136 'Is Monsieur de Lesseps a Canal Digger, or a Grave Digger?': *Harper's Weekly*, 3 September 1881

p. 137 'like a human bunch of grapes': Bunau-Varilla, *Panama*, p. 54

p. 137 'Rain poured in torrents': Panama *Star and Herald*, 4 December 1885

p. 138 'I saw they were covered with the most enormous and deadly spiders': Bunau-Varilla, *Panama*, p. 55

p. 138 Kimball, who toured the works soon after . . .: Kimball, *Special Intelligence Report on the Progress of the Work during the year 1885*

p. 140 'The contagion of my confidence in our success': Bunau-Varilla, *Panama* p. 53

p. 140 'It is an impressive fact that there is money value in the prestige of M. de Lesseps': New York *Tribune*, 16 August 1885

p. 140 'As for human life that is always cheap': Kimball, *Special Intelligence Report on the Progress of the Work during the year 1885*, p. 32

p. 140 'under no circumstances would negroes be admitted': J. B. Poylo to Mallet, 8 August 1885

p. 141 'Many a man of them had been happy to enlist': Bunau-Varilla, *Panama*, p. 44

p. 141 'I had reached a state of semi-coma': Mallet, *Pioneer Diplomat*, p. 31

p. 141 'should the works cease': Consul Sadler to Marquis of Salisbury, 27 January 1886, FO 55/325

Chapter Fourteen: Collapse (pp. 142–157)

p. 142 'scientific order, the active discipline that prevailed everywhere': Bunau-Varilla, *Panama*, p. 58

p. 142 'that it was probably a scheme to use my name as an ex-Minister to France': Bigelow private diary, 31 January 1886

p. 143 'The success of the Panama business does not (need I even mention) only interest the investors in the Company': de Molinari, *A Panama*, p. 25

p. 143 'I was pleased to find that Grace and the old Baron got on admirably together': Bigelow private diary, 18 February 1886

p. 144 'When we left they gave us repeated cheers': Bigelow private diary, 19 February 1886

p. 144 'take a share in its management': *ibid.*, 22 February 1886

p. 144 'His stay is one continued fête': Sadler to Foreign office, 26 February 1886, FO55/325

p. 145 'Even if the piercing of the Isthmus presents enormous difficulties': de Molinari, *A Panama*, p. 68
p. 145 'often conflicting and rarely more than approximative': Bigelow, *The Panama Canal. Report of the Hon. John Bigelow*, p. 4
p. 145 'insalubrity': Bigelow, *The Panama Canal. Report of the Hon. John Bigelow*, p. 12
p. 145 'people of small means': Bigelow, *ibid.*, p. 22
p. 145 'secure to the United States': Bigelow, *ibid.*, p. 24
p. 147 'a violent vibration of my bed': Bunau-Varilla, *Panama*, p. 61
p. 148 'They are trying to shelve me': *The Times*, 13 July 1886
p. 148 'It indeed looks as though "the beginning of the end" is in sight': New York *Tribune*, 15 July 1886
p. 148 'In a year hence the machinery accumulated by the company': *ibid.*, 30 July 1886
p. 148 'M. de Lesseps in his reports': *ibid.*, 31 July 1886
p. 148 'Nothing is ever done by the canal company without a great amount of pomp': *ibid.*, 8 August 1886
p. 149 Of thirty Italians who had arrived in a shipment twelve months previously . . .: *ibid.*, 22 August 1866
p. 149 'Every month or two I would lose a man, perhaps two men': *Hearings 3110, Sen. Doc. No. 253*, 57 Cong., 1 Sess., p. 100
p. 149 some men, having expired from fever . . .: New York *Tribune*, 29 August 1866
p. 149 'On one side of the room is a long bar': *ibid.*, 8 August 1866
p. 150 'a dismal failure': Bunau-Varilla, *Panama*, p. 68
p. 151 'It is magnificent, but it is not business': *The Times*, 15 October 1886
p. 151 'From all I could learn by conversation with the officials': Rogers, *Intelligence Report of the Panama Canal*, p. 34
p. 151 'as to Culebra, I leave you to form your own conclusions': Rogers, *ibid.*, p. 56
p. 151 'In the present uncertainty': Rogers, *ibid.*, p. 34
p. 153 'During my stay in Colón': Sweetman, *Paul Gaugin*, p. 169
p. 154 'the perfect, the final, project': Bunau-Varilla, *Panama*, p. 80
p. 156 'All France is joined in the completion of the Panama Canal': Chambre des Députés, *5e Législature: Session de 1893, Rapport Général . . .* , pp. 101–2
p. 156 'I appeal to all Frenchmen': *Bulletin du Canal Interocéanique*, 2 December 1888
p. 156 journalist Emily Crawford . . . New York *Tribune*, 13 and 14 December 1888

Chapter Fifteen: Scandal (pp. 158–171)

p. 158 'like a stroke of paralysis': Robinson, *Fifty Years at Panama*, p. 147
p. 159 'a few labourers . . . retained': Captain Rolfe to Commodore Hana, 3 March 1889, FO420/88
p. 159 'There are hundreds absolutely starving': Panama *Star and Herald*, 6 April 1889
p. 159 7244 were repatriated . . .: Governor's Report on the Blue Book, 1888–9
p. 159 But stern warnings from the new US President Benjamin Harrison . . .: inauguration speech, 4 March 1889
p. 160 'champagne had flowed in torrents': Drumont, *La Dernière Battaille*, p. 362

p. 160 'He had apparently recovered all his strength': Smith, *The Life and Enterprises of Ferdinand De Lesseps*, p. 310

p. 160 'dissipating' funds 'in a manner . . . more consistent with the personal views and interests of the administrators': Chambres des Députés: *5e Législature, Session de 1893, No. 2921*, Vol. 3, pp. 217–18

p. 162 'excessive optimism': Cour d'Appel de Paris, Ire Chambre: *Plaidoirie de Me. Henri Barboux pour MM. Ferdinand et Charles de Lesseps*, p. 185

p. 163 'With extraordinary imprudence, the press published the most deceptive articles': Chambre des Députés: *5e Législature, Session de 1893, No. 2921*, Vol. 1, p. 378

p. 164 'face was drawn and his skin yellowed': Courau, *Ferdinand de Lesseps*, p. 237

p. 167 'one of the enterprises which made some of the most scandalously excessive profits': quoted in Skinner, *France and Panama: The Unknown Years*, p. 75

p. 167 the bankers who had creamed some 75 million francs . . .: Skinner, *France and Panama: The Unknown* Years, p. 79

p. 169 'laid down its arms without fighting': Bunau-Varilla, *Panama*, p. 156

Chapter Sixteen: The 'Battle of the Routes' (pp. 173–187)

p. 173 'The construction of such a maritime highway': Annual Message, 5 December 1898

p. 174 'The Nicaragua canal is a purely national affair': New York *Herald*, 30 January 1902

p. 175 'He is one of the readiest talkers in town': New York *World*, 4 October 1908

p. 175 'involved almost every branch of professional activity': quoted in Harding, *The Untold Story of Panama*, p. 3

p. 176 'There was a hole in Panama into which a lot of French money had been sunk': United States Congress House of Representatives, *The Story of Panama: hearings on the Rainey resolution*, p. 6

p. 176 'the lawyer Cromwell': Bunau-Varilla, *Panama*, p. 10

p. 176 'Once you have touched Panama, you never lose the infection': Harding, *The Untold Story of Panama*, p. 88

p. 177 'the revolutionist who promoted and made possible the revolution': United States Congress House of Representatives, *The Story of Panama: hearings on the Rainey resolution* p. 61

p. 177 'masterful mind, whetted on the grindstone of corporation cunning': quoted in United States Congress House of Representatives, *ibid.*, p. 94

p. 177 'no one in the United States doubted that the Panama Canal in itself was an impossibility': Harding, *The Untold Story of Panama*, p. 5

p. 177 'We must make our plans with Napoleonic strategy': Harding, *ibid.*, p. 5

p. 177 'ubiquitous and ever present': McCullough, *The Path between the Seas*, p. 273

p. 178 'mysterious influences': Harding, *The Untold Story of Panama*, p. 2

p. 178 'under the control, management and ownership of the United States': quoted in Skinner, *France and Panama: The Unknown Years*, p. 138

p. 178 'for hours on the most profound study of the technical sides of the question':

United States Congress House of Representatives, *The Story of Panama: hearings on the Rainey resolution*, p. 224

p. 178 'the scales had fallen from their eyes': Bunau-Varilla, *Panama*, p. 166

p. 179 'all the difficulties of obtaining the necessary rights . . . on the Panama route': *Sen. Doc. No. 5*, 56 Cong., 2 Sess. pp. 42–3, quoted in Mack *The Land Divided*, p. 425

p. 180 So heated had his effort become . . .: Birmingham, *Our Crowd*, p. 280

p. 180 'Towards midnight . . . as I was about to go out for a breath of fresh air': Bunau-Varilla, *Panama*, p. 184ff

p. 181 'almost demented state of the mind': Bunau-Varilla, *ibid.*, p. 187

p. 183 'No single great material work which remains to be undertaken': quoted in McCullough, *The Path between the Seas*, p. 249

p. 183 'If that canal is open to the war ships of an enemy': Thayer, *Theodore Roosevelt: An Intimate Biography*, Vol. 2, p. 339

p. 183 'strengthen our position enormously': Beale, *Theodore Roosevelt and the Rise of American Power*, pp. 147–8

p. 184 'ruptured the dikes placed against so-called American imperialism' . . .: Miner, *The Fight for the Panama Route*, p. 232

p. 185 'legitimate means': United States Congress House of Representatives, *The Story of Panama: hearings on the Rainey resolution*, p. 169

p. 185 'The Colombians . . . have negro blood enough to make them lazy': *Harper's Weekly*, 23 August 1902

p. 185 'Talk about buying a lawsuit': New York *World*, quoted in Panama *Star and Herald*, 22 February 1902

p. 186 'poison the minds of people': *Sen. debate*, 57 Cong., 1 Sess.

p. 186 'It is the certainty of moral defilement': Anguizola, *Philippe Bunau-Varilla*, p. 215

p. 187 'If the vote were to be taken under this impression': Bunau-Varilla, *Panama*, p. 247

Chapter Seventeen: 'I Took the Isthmus' (pp. 188–212)

p. 189 'a crowd of French jail-birds, cleverly advised by a New York railroad wrecker': Collin, *Theodore Roosevelt's Caribbean*, p. 218

p. 189 'Few of the members who will assemble in Bogotá': Panama *Star and Herald*, 14 March 1903

p. 189 'What a difference there is in our nature': Mallet private letter, 30 January 1904

p. 189 'a great deal of illness': Mallet private letter, 1 June 1903

p. 190 'for the Isthmus of Panama it is a question of life or death': Miner, *The Fight for the Panama Route*, p. 290

p. 190 'give rise to unpatriotic feelings': Miner, *ibid.*, p. 292

p. 190 'Not an atom of our sovereignty nor a stone of our territory': *El Correo Naçional*, 3 February 1903

p. 191 'our nation has insisted': John Bigelow Papers, New York Public Library, Box 24

p. 191 'It is the conviction of his irresistible superiority and vigour that makes the Yankee': quoted in Miner, *The Fight for the Panama Route*, p. 265

p. 191 'from approbation to suspicion and from suspicion to decided opposition': Skinner, *France and Panama: The Unknown Years*, p. 208

p. 191 'History will say of me': Miner, *The Fight for the Panama Route*, p. 233

p. 192 'We pointed out': United States Congress House of Representatives, *The Story of Panama: hearings on the Rainey resolution*, pp. 278–9

p. 192 'greedy little anthropoids': Anguizola, *Philippe Bunau-Varilla*, p. 226

p. 192 'If Colombia should now reject the treaty': Miner, *The Fight for the Panama Route*, p. 285

p. 192 'The only party': Anguizola, *Philippe Bunau-Varilla*, p. 223

p. 193 'President Roosevelt is determined': New York *World*, 13 June 1903

p. 193 'Any amendment whatever or unnecessary delay in the ratification': Miner, *The Fight for the Panama Route*, p. 308

p. 193 'Make it as strong as you can': Pringle, *Theodore Roosevelt: A Biography*, p. 311

p. 194 'the foolish and homicidal corruptionists in Bogota': Bishop, *Theodore Roosevelt and his Time Shown in His Own Letters*, Vol. 1, p. 278

p. 194 'I know, for having heard him say so, how intensely he wants it': Jusserand, *What Me Befell*, p. 252

p. 194 'We might make another treaty, not with Colombia, but with Panama': New York *Herald*, 15 August 1903

p. 194 'It is altogether likely there will be an insurrection on the Isthmus': Bemis (ed.), *The American Secretaries of State and their Diplomacy*, Vol. IX, pp. 163–4

p. 195 'a thousand offers in the direction of assisting the revolution': United States Congress House of Representatives, *The Story of Panama: hearings on the Rainey resolution*, p. 362

p. 195 'The United States would build the Panama Canal': United States Congress House of Representatives, *ibid.*, pp. 359–60

p. 196 'Revolutionary agents of Panama [are] here': Collin, *Theodore Roosevelt's Caribbean*, p. 249

p. 196 'All is lost': Bunau-Varilla, *Panama*, p. 291

p. 197 'I have no doubt that he was able to make a very accurate guess': Miner, *The Fight for the Panama Route*, p. 359

p. 197 'I held all the threads of a revolution on the Isthmus': Bunau-Varilla, *Panama*, p. 297

p. 198 'If I succeeded in this task the Canal was saved': Bunau-Varilla, *ibid.*, pp. 329–30

p. 199 'If you will aid us': United States Congress House of Representatives, *The Story of Panama: hearings on the Rainey resolution*, p. 390

p. 200 'There was nothing that did not show the greatest cordiality': Espino, *How Wall Street Created a Nation*, p. 106

p. 201 'resumed their independence': Mack, *The Land Divided*, p. 463

p. 201 '*Viva La Republica de Panama!*': Panama *Star and Herald*, 11 November 1903

p. 202 'a second Boer War': New York *Tribune*, 24 November 1903

p. 202 'It is preferable to see the Colombian race exterminated': Anguizola, *Philippe Bunau-Varilla*, p. 276

p. 202 'revolution of the canal, by the canal, for the canal': Birmingham *Age-Herald*, 7 November 7 1903

p. 202 'It is another step in the imperial policy': Pittsburgh *Post*, 7 November 7 1903

p. 202 'hot-headed and immature': Little Rock Arkansas *Gazette*, 11 November 1903, quoted in Graham, *The 'Interests of Civilization'?*, p. 49

p. 202 'It begins to look as if nobody can touch that Panama ditch': Salt Lake *Herald*, quoted in *Literary Digest*, 27 November 1903

p. 203 'stolen property': *New York Times*, 29 December 1903

p. 203 'The thing is done, there is no way of undoing it': Houston *Post*, 12 November 1903

p. 203 'The world must move on': San Francisco *Chronicle*, 18 November 1903

p. 203 'The country ought to be ringing with the protests': Graham, *The 'Interests of Civilization'?*, pp. 127–8

p. 203 'Mr President, I am glad you did not start the rabbit to running': quoted in New York *Commercial Advertiser*, 12 November 1903

p. 203 'I did not lift a finger to incite the revolutionists': Collin, *Theodore Roosevelt's Caribbean*, p. 267

p. 204 'It is reported that we have made the revolution': Jusserand, *What Me Befell*, p. 253

p. 205 'Mr. Secretary of State, the situation harbours the same fatal germs': Bunau-Varilla, *Panama*, p. 358

p. 206 'What do you think, Mr Minister': Bunau-Varilla, *ibid.*, p. 366

p. 206 'Remember that ten days ago the Panamanians were still Colombians': Weisberger, 'The strange affair of the taking of the Panama Canal Zone', *American Heritage*, 27, p. 75

p. 206 'As for you poor old dad': McCullough, *The Path between the Seas*, p. 392

p. 207 'inflammatory, unnecessary and offensive': Collin, *Theodore Roosevelt's Caribbean*, p. 281

p. 207 'So long as the delegation has not arrived in Washington': Bunau-Varilla, *Panama*, pp. 372ff

p. 207 'very satisfactory, vastly advantageous to the United States': McCullough, *The Path between the Seas*, p. 392

p. 207 the new treaty was many times worse for Panama, as Hay later admitted . . .: Cullom, *Fifty Years of Public Service*, p. 383

p. 208 'We separated not without emotion': Bunau-Varilla, *Panama*, p. 377

p. 208 'I greeted the travellers with the happy news!': Bunau-Varilla, *ibid.*, pp. 377–8

p. 209 'What do you think of the canal treaty?': Mallet private letter, 25 November 1903

p. 209 'it sounds very much like we wrote it ourselves': Bunau-Varilla, *Panama*, p. 427

p. 209 'stolen in the most bare-faced manner from Colombia': 58 Cong., 2 Sess., p. 871, 19 January 1904

p. 209 'rising as one man', 'but the one man was in the White House': 58 Cong., 2 Sess., p. 706, 13 January 1904

p. 209 'but the beginning of systematic policy of aggression': 58 Cong., 2 Sess., p. 709, 13 January 1904

p. 209 'I fear that we have got too large to be just': 58 Cong., 2 Sess., p. 401, 19 December 1903

p. 209 'You might whip the dog, but would you throw away the rabbit?': *New York Times*, 8 January 1904

p. 210 'to our future political, commercial, and naval expansion': quoted in Graham, *The 'Interests of Civilization'?*: p. 126

p. 210 'From the point of view of world politics': Charles Henry Huberich, *The Trans-Isthmian Canal: A Study in American Diplomatic History*, Austin, Texas, 1904, p. 31, quoted in Graham *The 'Interests of Civilization'?*, p. 126

p. 210 'the great Frenchman, whose genius has consecrated the Isthmus': quoted in Anguizola, *Philippe Bunau-Varilla*, p. 289

p. 211 'somebody unexpectedly seized my hands': Bunau-Varilla, *Panama*, p. 428

p. 211 'great uneasiness caused among my friends by my action': Roosevelt to Spooner, 20 January 1904, quoted in Collin, *Theodore Roosevelt's Caribbean*, p. 306

p. 211 'The one thing for which I deserved most credit in my entire administration': Major, *Prize Possession*, p. 63

p. 211 'more to an empire than a Republic': Graham, *The 'Interests of Civilization'?*: p. 144

p. 211 'Tell our speakers to dwell more on the Panama Canal': Collin, *Theodore Roosevelt's Caribbean*, p. 306

p. 211 'We stand today, apparently in the shadow of a great defeat': Graham, *The 'Interests of Civilization'?*, p. 150

Chapter Eighteen: 'Make the Dirt Fly' (pp. 213–237)

p. 213 'In America, anything is possible': van Hardeveld, *Make Dirt Fly*, p. 5

p. 213 George Martin . . .: 'Competition for the Best True Stories . . .'

p. 213 'There is nothing in the nature of the work': New York *Tribune*, 23 January 1904

p. 213 'solid inevitability': Skinner, *France and Panama: The Unknown Years*, p. 2

p. 214 'the results be achieved': McCullough, *The Path between the Seas*, p. 408

p. 214 'Old Man of the Sea': Gorgas and Hendrick, *William Crawford Gorgas*, p. 170

p. 214 'I feel that the sanitary and hygiene problems': Roosevelt to Morison, 24 February 1904, quoted in McCullough, *The Path between the Seas*, p. 406

p. 215 At first the police were used . . .: Mallet private letter, 4 April 1904

p. 215 70 per cent had the enlarged spleen of the malaria carrier . . .: Gorgas, *Engineering Record*, May 1904

p. 215 'he was sure every member of the Commission hoped that a sea-level canal would be built': Mallet to Lord Landsdowne, 27 April 1904, FO881/8429

p. 216 'They have taken all the meat and left the bone': Mallet private letter, 16 June 1904

p. 216 'The Isthmus is swarming with Yankees already': Mallet private letter, 10 June 1904

p. 216 'Panama without mosquitoes?': Mallet private letter, 2 June 1904

p. 216 'Such a God forsaken place where the French have so finally failed!': Hibbard account, Manuscript Collection of the Canal Zone Library-Museum, Library of Congress, Washington, D.C., Box 22

p. 217 'the rainy season had just commenced': Karner, *More Recollections*, p. 11

p. 217 had reverted to jungle . . .: Patrick Kenealy: *Chagres Yearbook*, 1913, p. 188

p. 217 'There was, I realised, a stupendous piece of work before us': quoted in Dock (ed.), *A History of Nursing*, p. 300

p. 217 'A more prolific source would be hard to imagine': Le Prince and Orenstein, *Mosquito Control in Panama*, p. 20ff

p. 218 'only jungle and chaos from one end of the Isthmus to the other': McCullough, *The Path between the Seas*, p. 439

p. 218 Joseph Le Prince found several trees . . .: Le Prince and Orenstein, *Mosquito Control in Panama*, p. 17

p. 218 'records to show that they bought': Karner, *More Recollections*, p. 57

p. 218 'excellently recorded [which] proved to be of great use': Sibert and Stevens, *The Construction of the Panama Canal*, p. 11

p. 218 'a fairly good condition, and were systematically stored': Wallace 'Preliminary Work on the Panama Canal', *Engineering Magazine*, October 1905

p. 218 'Splendid workmanship was shown on these machines': Sibert and Stevens, *The Construction of the Panama Canal*, p. 13

p. 219 'considerable work had been done on the channel from La Boca to Miraflores': Sibert and Stevens, *ibid.*, p. 11

p. 219 'vastly more that the popular impression': McCullough, *The Path between the Seas*, p. 441

p. 219 The idea of a plan that would render much of the French digging . . .: James Thomas Ford, 'The Present Condition and Prospects of the Panama Canal Work', Institute of Civil Engineers, *Minutes of Proceedings*, 5 February 1901

p. 220 'afford convenient passage for vessels of the largest tonnage': quoted in Miner, *The Fight for the Panama Route*, pp. 408–9

p. 220 'A sea-level canal alone satisfied the requirements of the case': 'Correspondence on the Panama Canal Works', Institute of Civil Engineers, *Minutes of Proceedings*, p. 172

p. 220 A sea-level canal 'would be an ideal solution of the question': 'Correspondence on the Panama Canal Works', Institute of Civil Engineers, *Minutes of Proceedings*, p. 203

p. 220 'wish success to those who were proposing to complete the canal': 'Correspondence on the Panama Canal Works', Institute of Civil Engineers, *ibid.*, p. 189

p. 221 'Lizards and gaudy snakes crawled and scuttled': Waldo, 'The Panama Canal Work, and the Workers', *Engineering Magazine*, p. 327

p. 221 'indescribably filthy': Maltby, 'In at the Start at Panama', *Civil Engineering*, August 1945, p. 324

p. 221 'As the use of cheap steel had not become the practice': Sibert and Stevens, *The Construction of the Panama Canal*, pp. 86–7

p. 221 Soon after his arrival, Maltby tried to organise . . .: Maltby, 'In at the Start at Panama', *Civil Engineering*, June 1945, p. 262

p. 222 'I intend that those fellows on the hill': quoted in Gorgas and Hendrick, *William Crawford Gorgas*, p. 161

p. 222 240,000 perfectly-built hinges . . .: Bates, *Retrieval at Panama*, p. 30

p. 223 'Suitable quarters and accommodations could not be provided': Wallace 'Preliminary Work on the Panama Canal', *Engineering Magazine*, October 1905

p. 223 'before there was any way to care for them properly': Maltby, 'In at the Start at Panama', *Civil Engineering*, June 1945, p. 262

p. 223 'A heavy suitcase in each hand, no light anywhere': van Hardeveld, *Make Dirt Fly*, pp. 8–9

p. 223 'apprehension', 'homesickness': Jessie Murdoch account, *Chagres Yearbook*, 1913, p. 43ff

p. 224 'the wreck of the French companies': James Williams unpublished account Manuscript Collection of the Canal Zone Library-Museum, Library of Congress, Washington, D.C., Box 35

p. 224 'We were supposed to have furniture issued to us': 'The Early Days' by John J. Meehan, *Chagres Yearbook*, 1913, p. 137ff

p. 224 'I am thoroughly sick of this country and everything to do with the canal': quoted in Pepperman, *Who Built the Panama Canal?*, p. 61

p. 224 'There have been many errors and much wastage and pilfering of money': Boni *Panamá, Italiay los Italianos*, p. 129

p. 225 'supplies taking for ever to arrive': Wallace, 'Preliminary Work on the Panama Canal', *Engineering Magazine*, October 1905

p. 225 A request was posted for twenty-five track foremen . . .: Lewis: *The West Indian in Panama*, p. 97

p. 225 'were not examined at all': Karner, *More Recollections*, p. 25

p. 225 'I have two sons who wish to go to Panama to work on the Canal': letter of 17 February 1905, Charles M. Swinehart file Record Group 185, National Archives, Washington, D.C., National Records Centre, Suitland, Maryland

p. 225 'railroad men who were blacklisted on the American railroads': McCullough, *The Path between the Seas*, p. 444

p. 225 'Americans will put their coats on for meals', Sands, *Our Jungle Diplomacy*, p. 14

p. 226 'the people of Panama look upon Americans as noisy, grabbing bullies': Fraser, *Panama and What it Means*, p. 185

p. 226 'The average Americans has the utmost contempt for a Panaman': Bishop, *The Panama Gateway*, p. 121

p. 226 'Panama must conduct itself as a civilized nation': Mallet private letter, 10 January 1904

p. 226 'revolutionary firebrand' and 'notorious hater of foreigners': US consul Barrett to Hay, 9 August 1904, Hay Papers, Library of Congress

p. 227 'any Latin American nation who fused her destiny with that of the United States': Mellander, *The United States in Panamanian Politics*, p. 71

p. 227 'I look upon the Republic of Panama as doomed': Mallet private letter, 29 June 1904

p. 227 'Opinion here amongst the natives is spreading': Mallet private letter, 25 July 1904

p. 228 'very loud spoken', 'vulgar' and 'full of self-assurance': Mallet private letters, 25 July 1904, 9 March 1905
p. 228 Soon after, an anonymous fly-sheet was distributed . . .: Mallet private letter, 28 August 1904
p. 228 'festering with intrigue': Sands, *Our Jungle Diplomacy*, p. 14
p. 229 'made for an ambiguous and most delicate diplomatic situation': Sands, *ibid.*, p. 47
p. 229 'whither Theodore Roosevelt and his "Yankee imperialism" might be tending': Sands, *ibid.*, p. 44
p. 229 'political jungle of its own creation': Sands, *ibid.*, p. 3
p. 229 'a ready-made politico-diplomatic pattern for expansion': Sands, *ibid*,' p. 4
p. 229 'Don Santiago was aware': Sands, *ibid.*, p. 62
p. 229 started appearing in New York newspapers . . . e.g.: New York *Herald*, 4 October 1904
p. 230 'a kind of Opera Bouffe republic and nation': LaFeber, *The Panama Canal: The Crisis in Historical Perspective*, p. 40
p. 230 'the most infectious chuckle in the history of politics': McCullough, *The Path between the Seas*, p. 456
p. 230 'Though the heaviest man, in weight, in the room': Karner, *More Recollections*, p. 78
p. 231 One shipment of laborers was met by agents of the Municipal Engineering Division . . .: Lindsay-Poland, *Emperors in the Jungle*, p. 135
p. 231 'there was no surplus throughout Central or South America': Wood, in Goethals (ed.), *The Panama Canal: An Engineering Treatise*, Vol. 1, p. 191
p. 231 estimated that some eight to ten thousand workers . . .: Hains, 'The Labor Problem on the Panama Canal', *North American Review*, July 1906
p. 231 'useless to discuss the question of utilizing the white race': 1906 *Hearings*, quoted in Major, *Prize Possession*, p. 81
p. 231 'The native population is wholly unavailable': Hains, 'The Labor Problem on the Panama Canal', *North American Review*, July 1906, p. 49
p. 232 'fairly industrious; not addicted to drink': Hains, 'The Labor Problem on the Panama Canal', *North American Review*, July 1906, p. 50
p. 232 'He does loaf about a good deal': Major, *Prize Possession*, p. 82
p. 232 'natural markets for unskilled labor': Wood, in Goethals (ed.), *The Panama Canal: An Engineering Treatise*, Vol. 1, p. 194
p. 233 fell to as low as a shilling (twenty-five cents) a day . . .: Karner, *More Recollections*, p. 105
p. 233 'Chronic pauperism . . . like a chronic disease is . . . undermining the population of this island': Richardson, *Panama Money in Barbados*, p. 72
p. 233 282 per thousand live births . . .: Richardson, *ibid.*, p. 77
p. 234 his first impressions of the place . . .: Karner, *More Recollections*, pp. 96–111
p. 234 'The island has always been and still is run for the whites': Edwards, *Panama, the Canal, the Country, the People*, p. 21
p. 235 'I shipped only sixteen laborers': Karner, *More Recollections*, pp. 106–7

p. 236 'which nearly buried the shovel from sight': Karner, *ibid.*, p. 40

p. 236 'there is little probability of finding a satisfactory location': Report of the Chief Engineer Isthmian Canal Commission, June 1, 1904–February 1, 1905, Washington, D.C., 1905

p. 237 'remove the principal elements of uncertainty now existing in regard to the project as a whole': *ibid.*

p. 237 'I want you to build up an organization so complete and efficient': Maltby, 'In at the Start at Panama', *Civil Engineering*, June 1945, p. 260

Chapter Nineteen: Yellow Jack (pp. 238–252)

p. 238 'in a year or so there will be no mosquitoes there!': Le Prince and Orenstein, *Mosquito Control in Panama*, p. iv

p. 238 'He says you and the children will soon be able to live here': Mallet private letter, 29 June 1904

p. 240 'yellow fever's first enounter with one who became its implacable foe': Gorgas and Hendrick, *William Crawford Gorgas*, p. 4

p. 240 'we were rather inclined to make light of his ideas': Gorgas, *Sanitation in Panama*, pp. 14–16

p. 240 'chlorinated lime': Gorgas and Hendrick, *William Crawford Gorgas*, p. 86

p. 242 'Of all the silly and nonsensical rigmarole about yellow fever': Washington *Post*, 2 November 1900

p. 243 'When I think of the absence of yellow fever from Havana': Gorgas and Hendrick, *William Crawford Gorgas*, p. 133

p. 243 'I fear an epidemic is inevitable': Gorgas and Hendrick, *ibid.*, p. 170

p. 244 'our work in Cuba and Panama will be looked upon as the earliest demonstration': *Journal of the American Medical Association*, 19 June 1909

p. 244 'the veriest balderdash': Gorgas and Hendrick, *William Crawford Gorgas*, p. 162

p. 244 'I'm your friend, Gorgas, and I'm trying to set you right': Gorgas and Hendrick, *ibid.*, p. 164

p. 244 'The world requires at least ten years to understand a new idea': McCullough, *The Path between the Seas*, p. 405

p. 244 'clean, healthy, moral Americans': McCullough, *ibid.*, p. 451

p. 244 'Consequently,' Joseph Le Prince complains . . .: Le Prince and Orenstein, *Mosquito Control in Panama*, p. 299

p. 244 'without having to wipe the mosquitoes off every second': Mallet private letter, 3 August 1904

p. 244 'a convert to the mosquito theory': Mallet private letter, 19 January 1905

p. 245 'existed at practically every house in town' Le Prince and Orenstein: *Mosquito Control in Panama*, p. 272

p. 245 'To attempt it is a dream, an illusion': quoted in Rink, *The Land Divided, the World United*, p. 102

p. 245 'Le Prince,' Johnson said, 'you're off on the upper storey!': Gorgas and Hendrick, *William Crawford Gorgas*, p. 172

p. 246 'Some yellow fever cases exist in the San Tomas hospital': Mallet private letter, 14 January 1905

p. 246 'If you should be unwell here or if anything should happen to you': Barrett Papers, Library of Congress

p. 246 'revealed a dishpan of water standing outside the cook's headquarters': Gorgas and Hendrick, *William Crawford Gorgas*, pp. 178–9

p. 246 'the saddest incident in the history of the American colony': Davis to Wallace, 17 January 1905, General Correspondence RG 185

p. 246 'one of the saddest graveyards in the world': Sullivan, *Our Times*, Vol. 1, p. 455

p. 246 'darkened the whole Isthmus': Gorgas and Hendrick, *William Crawford Gorgas*, p. 174

p. 246 'The rush to get away': Gorgas and Hendrick, *ibid.*, p. 173

p. 247 'the place where the "ghost walks"': Galveston *Daily Gleaner*, quoted in Panama *Star and Herald*, 13 January 1905

p. 247 'Unless something is done and done quickly': Panama *Star and Herald*, 20 February 1905

p. 248 'It was like 'the ending of many a bright young man I have seen on the battle-field': Davis to Wallace, 2 May 1905, quoted in Duval, *And the Mountains will Move*, p. 176

p. 248 'Everybody here seems to be sitting on a tack': Pepperman, *Who Built the Panama Canal?*, p. 59

p. 248 stenographer 'rose from his chair and shrieked': Carr, 'The Panama Canal', *The Outlook*, 5 May 1906

p. 248 'yellow fever . . . completely filled the atmosphere': Gorgas and Hendrick, *William Crawford Gorgas*, p. 168

p. 248 One returning nurse . . .: New York *Tribune*, 6 July 1905

p. 248 'Things are looking pretty sour at Culebra': Panama *Star and Herald*, 26 June 1905

p. 249 about three-quarters of the white workforce . . .: Pepperman, *Who Built the Panama Canal?*, p. 58

p. 249 'The military regime in Panama': Waldo, 'The Panama Canal Work, and the Workers', *Engineering Magazine*, 5 January 1907, p. 323

p. 250 'gruff, domineering': Gorgas and Hendrick, *William Crawford Gorgas*, p. 194

p. 250 'Smells and filth, Mr President': Gorgas and Hendrick, *ibid.*, pp. 198–202

p. 250 'huge in all three dimensions': Sands, *Our Jungle Diplomacy*, p. 62

p. 251 'It would perhaps be difficult to find any spot on earth': Panama *Star and Herald*, 8 May 1905

p. 251 'ill-paid, over-worked, ill-housed, ill-fed': Magoon to Shonts, 3 June 1905, quoted in Duval, *Mountains*, p. 178

p. 252 'For mere lucre you change your position overnight': Pepperman, *Who Built the Panama Canal?*, p. 121

p. 252 'We felt like an army deserted by its general': Maltby, 'In at the Start at Panama', *Civil Engineering*, July 1945, p. 322

p. 252 'the effect upon the workers at the Isthmus was deplorable': Gorgas and Hendrick, *William Crawford Gorgas*, p. 174

Chapter Twenty: Restart (pp. 253–271)

p. 253 'Then I was asked to meet . . . Cromwell': Stevens, 'An engineer's recollections', *Engineering News-Record*, 5 September 1935, p. 332

p. 253 shocked at discovering that there were more canal employees . . .: Pepperman, *Who Built the Panama Canal?*, p. 130

p. 253 'The condition of affairs on the Isthmus': Stevens, 'A Momentous Hour at Panama', *Journal of the Franklin Institute*, July 1930

p. 254 'scared out of their boots, afraid of yellow fever': Stevens' statement, 16 January 1906, *Hearings No. 18*, 59 Cong., 2 Sess., Vol. 1, p. 38

p. 254 'no organization worthy of the name': Stevens, 'An engineer's recollections', *Engineering News-Record*, 5 September 1935, p. 256

p. 254 'the idiotic howl about "making the dirt fly"': Stevens, 'The Truth of History', in Bennett, *History of the Panama Canal*, p. 218

p. 254 'I believe I faced about as discouraging a proposition': Stevens, 'The Truth of History', in Bennett, *ibid.*, p. 210

p. 254 'keep his eye on the ball': Pepperman, *Who Built the Panama Canal?*, p. 135

p. 254 'There are three diseases in Panama': Bishop: *Goethals: Genius of the Panama Canal*, p. 133

p. 254 'I have had as much or more actual personal experience in manual labor': Maltby, 'In at the Start at Panama', *Civil Engineering*, July 1945, p. 324

p. 255 'Come to Panama on the first train. Stevens': Maltby, *ibid.*, p. 322

p. 255 'I cannot conceive how they did the work they did': Stevens' statement, 16 January 1906, *Hearings No. 18*, 59 Cong., 2 Sess., Vol. 1

p. 255 'I determined from the start': Stevens, 'The Truth of History', in Bennett, *History of the Panama Canal*, p. 212

p. 256 'a machine in every way superior to any in existence': Sibert and Stevens, *The Construction of the Panama Canal*, p. 88

p. 256 'The problem was simply one of transportation': Sibert and Stevens, *ibid.*, p. 76

p. 256 'two streaks of rust and a right of way': Maltby, 'In at the Start at Panama', *Civil Engineering*, July 1945, p. 324

p. 256 'thirty years behind the times': Mallet to Foreign Office, 30 August 1905, FO881/8765

p. 256 no sidings 'worthy of the name': Sibert and Stevens, *The Construction of the Panama Canal*, p. 41

p. 256 'A collision has its good points as well as its bad ones': Stevens, 'The Truth of History', in Bennett, *History of the Panama Canal*, p. 121

p. 256 'crank' for Chinese labour . . .: Karner, *More Recollections*, p. 131

p. 256 'two colored men stepped from a rowboat to the landing': Karner, *ibid.*, p. 114

p. 257 Arthur Bullard . . .: Edwards, *Panama, the Canal, the Country, the People*, p. 29ff

p. 257 Benjamin Jordan . . .: 'Diggers' documentary produced and directed by Roman Foster

p. 257 'one of the greatest engineering feats the world has ever undertaken': John Bowen in 'Diggers' documentary

p. 257 'Why you don't hit de manager in de head': Richardson, *Panama Money in Barbados*, p. 106

p. 258 'They slept higgledy-piggledy in the most cramped of postures': Fraser, *Panama and What it Means*, p. 23

p. 259 'Everything looked so strange, so different to home': Egbert Leslie 'Diggers' documentary

p. 259 the ICC was importing prostitutes at the US tax-payers' expense . . .: New York *Herald*, 18 November 1905

p. 259 'these women had been imported for a very definite social purpose': Sands, *Our Jungle Diplomacy*, p. 21

p. 259 'As a corporation it could perform many functions': Maltby, 'In at the Start at Panama', *Civil Engineering*, July 1945, p. 324

p. 260 the fare of £2 10s. ($14) . . .: Richardson, *Panama Money in Barbados*, p. 121

p. 260 Harrigan Austin . . .: 'Competition for the Best True Stories . . .'

p. 260 John Butcher . . .: *ibid.*

p. 262 'One could scarcely breathe God's free air without being clubbed and kicked': Colón *Independent*, 20 April 1906

p. 262 the 'serious disturbance': Mallet to Foreign Office, 6 May 1905, FO881/8765

p. 262 'In Jamaica a constable is a peacemaker': Slosson and Richardson, 'An Isthmian Carpenter's Story: A Jamaican Negro': *The Independent*, 19 April 1906

p. 262 'Instead of the canal bringing with it those good old times': Colón *Independent*, 6 December 1904

p. 263 On a wage of seldom more than a dollar a day . . .: Major, *Prize Possession*, p. 101

p. 263 Coffee and bread brought to the works by West Indian women . . .: Amos Clarke, 'Competition for the Best True Stories . . .'

p. 263 'Things were very different in those days': Slosson and Richardson, 'An Isthmian Carpenter's Story: A Jamaican Negro', *The Independent*, 19 April 1906

p. 263 'In their anxiety to save money', Karner, *More Recollections*, p. 142

p. 263 'I have looked into hundreds of their pots': Carr, 'The Silver Men', *The Outlook*, 19 May 1906

p. 264 in October 1905, there were twenty-six, all West Indians . . .: United States Isthmian Canal Commission, *Population and Deaths from various diseases*

p. 264 In November 1905, journalist Poultney Bigelow . . .: Bigelow, 'Our Mismanagement at Panama', *The Independent*, 4 January, 1906

p. 265 'Notwithstanding nearly six thousand new laborers were brought in': Stevens to Shonts, 14 December 1905, RG 185 2-B-1

p. 265 frail 'disposition to labor': United States Isthmian Canal Commission *Annual Report* 1906

p. 265 'The West Indian's every movement is slow and bungling': Carr, 'The Silver Men', *The Outlook*, 19 May 1906

p. 265 'They were not getting proper food in sufficient and regular amounts': Sibert and Stevens, *The Construction of the Panama Canal*, p. 133

p. 265 By February 1906, there were over fifty in operation . . .: Stevens to Shonts, 16 February 1906, RG 185 2-E-1

p. 266 'the leavings from the hotels': Chatfield, *Light on Dark Places*, p. 148
p. 266 LeCurrieux's family . . .: 'Competition for the Best True Stories . . .'
p. 266 Henry de Lisser visited one of the barracks . . .: Lisser, *Jamaicans in Colon and the Canal Zone*, quoted in Newton, *The Silver Men*, p. 149
p. 267 'The discipline maintained in the labour camps is severe': Mallet, letter to Governor of Jamaica, 30 November 1906, FO 371/300
p. 267 'This rule worked well': Sibert and Stevens, *The Construction of the Panama Canal*, p. 97
p. 267 'At midnight when everyone is asleep': Colón *Independent*, 12 September 1906
p. 268 'and it was as if with these clothes we had donned a more formal manner': van Hardeveld, *Make Dirt Fly*, p. 18ff
p. 271 'Special inducements were added': Sibert and Stevens, *The Construction of the Panama Canal*, p. 119

Chapter Twenty-One: The Railroad Era (pp. 272–282)

p. 272 'the great waste of money and the utter futility of the whole procedure': Stevens, 'A Momentous Hour at Panama', *Journal of the Franklin Institute*, July 1930
p. 273 'that there was a big, thick white cloud of smoke': Alfonso Suazo, 'Competition for the Best True Stories . . .'
p. 273 'We became so clean, orderly, and "dried out"': Maltby, 'In at the Start at Panama', *Civil Engineering*, August 1945, p. 360
p. 274 'the strong smell of decomposed fish has gone': Mallet private letter, 5 September 1906
p. 275 'the day of the good-for-nothing tropical tramp had nearly passed': Carr, 'The Panama Canal', *The Outlook*, 5 May 1906
p. 275 'weeding out the faint-hearted and incompetent': Pepperman, *Who Built the Panama Canal?*, pp. 11–12
p. 275 'The men themselves . . . have distinctive virtues': Carr, 'The Panama Canal', *The Outlook*, 5 May 5 1906
p. 275 'Normal family life is becoming established': Slosson and Richardson, 'Life on the Canal Zone', *The Independent*, 22 March 1906
p. 276 'Most of the young men on the Isthmus have absolutely no places of amusement': John Barrett, in Panama *Star and Herald*, 27 February 1905
p. 276 'positive forces for evil': quoted in Mack, *The Land Divided*, p. 549
p. 276 'use money appropriated for the construction of the Canal': Karner, *More Recollections*, p. 22
p. 277 'Stevens lives on the line': Mallet private letter, 5 September 1905
p. 277 'Stevens' sturdy, competent presence': Sands, *Our Jungle Diplomacy*, p. 40
p. 278 'I am not running things': Chatfield, *Light on Dark Places*, p. 64
p. 278 'So many men sent down here drink to excess': Chatfield, *ibid.*, p. 51
p. 278 'Like many other people here in positions of authority': Chatfield, *ibid.*, p. 67
p. 278 clerk starting work on a salary of $2500 . . .: Carr, 'The Panama Canal', *The Outlook*, 5 May 1906

p. 278 could now report more regular wages and less police harassment . . .: Mallet to Sir Edward Grey, 31 January 1906, FO881/8892

p. 278 'have returned [from Panama] with money': *Barbados Agricultural Reporter*, 3 March 1906, quoted in Richardson, *Panama Money in Barbados*, p. 116

p. 278 'innate respect for authority': Sibert and Stevens, *The Construction of the Panama Canal*, p. 125

p. 279 described conditions as 'unsatisfactory': Mallet to Foreign Office, 14 August 1905. FO881/8765

p. 279 'a capacity to develop into sub-foremen': Stevens to Shonts, 14 December 1905, RG185 2-B-1

p. 279 'No one will ever know . . . no one can realize': Stevens, 'The Truth of History', in Bennett, *History of the Panama Canal*, p. 211

p. 280 'Stock Gambler's Plan to Make Millions!': New York *World*, 17 January 1904

p. 280 'a group of canal promoters and speculators and lobbyists': *New York Times*, 29 December 1903

p. 280 'I do not think there is a place on the face of the globe': Pepperman, *Who Built the Panama Canal?*, p. 279

p. 280 'I have heard all those things and many more': Chatfield, *Light on Dark Places* p. 189

p. 281 'Many of the prominent American newspapers': Townley Report, 3 May 1906, FO881/8892

p. 282 'I had been told to build a house': McCullough, *The Path between the Seas*, p. 481

Chapter Twenty-Two: The Digging Machine (pp. 283–306)

p. 283 'Such a canal would undoubtedly be the best in the end if feasible': *Engineering Record*, 24 February 1906

p. 283 'One genius proposed to wash the entire cut': Sibert and Stevens, *The Construction of the Panama Canal*, p. 76

p. 284 'discovered an unknown way through this mysterious labyrinth': Bigelow, *The Panama Canal and the Daughters of Danaus*, p. 40

p. 284 'Mr Randolph . . . advises M. P. Buneau Varilla': Note of 7 November 1905, Bigelow Papers, New York Public Library Box 24

p. 284 'The most important document in the engineering history': *Engineering Record*, 10 January 1906, p. 211

p. 285 'Should they climb over the hill or remove it?': Report of the Board of Consulting Engineers, 1906

p. 285 'narrow gorge' would be 'tortuous': Shonts quoted in Pepperman, *Who Built the Panama Canal?*, p. 208

p. 285 'Such a waterway is far from meeting the conception': Abbot, 'The Panama Canal. Projects of The Board of consulting Engineers', *Engineering Magazine*, July 1906, p. 483

p. 287 'personal study of the conditions': Stevens, 'An engineer's recollections', *Engineering News-Record*, 5 September 1935, p. 40

p. 287 'the difficulties and dangers of navigation': Washington *Post*, 20 February 1906

p. 288 'simply preposterous piece of work': quoted in Bigelow, *The Panama Canal and the Daughters of Danaus*, p. 35

p. 288 'the greatest engineering conflict of the canal': Bates, *Crisis at Panama*, p. 32–4

p. 289 'Hard rains had set in by this time': van Hardeveld, *Make Dirt Fly*, p. 47

p. 289 'came bounding up the steps three at a time': van Hardeveld, *ibid.*, p. 44

p. 290 'Distinctive social lines were drawn on the Isthmus' [Sibert and Stevens *The Construction of the Panama Canal*

p. 290 'sounded as though some one was throwing boulders': Chatfield, *Light on Dark Places*, p. 118

p. 290 'The meat served is almost always beef, and such beef!': Chatfield, *ibid.*, p. 127

p. 291 'was *worse than usual*, which was only just possible': Chatfield, *ibid.*, p. 155

p. 291 'severely manhandled': Taft to Magoon, 4 June 1906, quoted in Major, *Prize Possession*, p. 121

p. 292 'guarantee public order and constitutional succession': Mellander, *Charles Edward Magoon*, p. 78

p. 292 'in that territory [in] which [disorder] can be prevented': Mellander, *Charles Edward Magoon*, p. 80

p. 293 'party feeling is very bitter': Magoon annual report, 1905, RG185

p. 293 *Diaro de Panama*, described the choice for the voters . . .: 21 April 1906

p. 293 'a senior Conservative declared . . .': De La Guardia, quoted in Mellander, *The United States in Panamanian Politics*, p. 88

p. 293 'suppress any insurrection in any part of the Republic.' Taft to Magoon 26 April 1906 General Correspondence (ICC) 1905–14, RG185

p. 293 'customary' for Government candidates to win elections . . .: Mallet to Foreign Office, 21 May 1906, FO371/101

p. 293 'explicit directions have been given to the police': Panama *Star and Herald*, 18 May 1906

p. 293 Mallet puts this down: Mallet to Foreign Office, 30 May 1906, FO371/101

p. 293 'Negro influence': quoted in Major, *Prize Possession*, p. 118

p. 293 'The police voted the first time in uniform': Sands, *Our Jungle Diplomacy*, p. 65

p. 294 'prevailing clannishness': Sibert and Stevens, *The Construction of the Panama Canal*, p. 115

p. 294 'There is no sense in putting so many different races together': Slosson and Richardson, 'An Isthmian Carpenter's Story: A Jamaican Negro': *The Independent*, 19 April 1906

p. 294 'some sort of hazy idea had gotten into their heads': Sibert and Stevens, *The Construction of the Panama Canal*, p. 115

p. 294 'three separate nationalities of laborers': Stevens to Shonts, 4 May 1906, Panama Canal Commission File (PCC) 2-E-1, RG185

p. 295 'The American is too proud to work with his hands!': Lewis, *The West Indian in Panama*, p. 35

p. 295 'Everybody in his area was so scared of disease': interview with Mr William Donadío, Panama City, 17 August 2004

p. 295 'and those of any colour or background were accepted': Lomonaco report, quoted in Marco, *Los Obreros espanoles*, p. 24

p. 296 'The Spaniard is certainly the more intelligent and better worker': Thompson, 'The Labour problems of the Panama Canal', *Engineering* (London) 3 May 1907

p. 296 'We knew we had with us genuine workers': van Hardeveld, *Make Dirt Fly*, p. 102

p. 296 'It did exactly what was expected in changing the self-confidence of the negroes': Sibert and Stevens, *The Construction of the Panama Canal*, p. 118

p. 296 'practically all with malaria': Carr, 'The Work of the Sanitary Force', *The Outlook*, 12 May 1906

p. 296 During the headline-grabbing yellow fever epidemic of May to August . . .: Bishop, *The Panama Gateway*, p. 243

p. 297 seventy-five people a day with the disease . . .: Chatfield, *Light on Dark Places*, p. 142

p. 297 'This rainy season has been a heavy trial on the canal builders': New York *Herald*, 1 August 1906

p. 297 the cases that came to the attention of the medical system numbered nearly twenty-two thousand . . .: Newton, *The Silver Men*, pp. 152–4

p. 297 an astonishing 80 per cent of the overall workforce . . .: Le Prince and Orenstein, *Mosquito Control in Panama*, p. 228

p. 297 West Indian Rufus Forde . . .: 'Competition for the Best True Stories . . .'

p. 298 Jamaican James Williams . . .: *ibid.*

p. 298 St Lucian Charles Thomas . . .: *ibid.*

p. 298 Clifford Hunt . . .: *ibid.*

p. 299 'The question of controlling malaria': Le Prince and Orenstein, *Mosquito Control in Panama*, p. 24

p. 299 'like fighting all the beasts of the jungle': Gorgas and Hendrick, *William Crawford Gorgas*, p. 226

p. 299 'Very patient negroes were necessary': Le Prince and Orenstein, *Mosquito Control in Panama*, p. 113

p. 300 'larvae of dragon flies and water beetles': Le Prince and Orenstein, *ibid.*, p. 185

p. 301 'the cause of many break downs in the constitution': Mallet private letter, 10 June 1906

p. 301 John Prescod . . .: 'Competition for the Best True Stories . . .'

p. 301 'The prevailing illness is malaria': Chatfield, *Light on Dark Places*, p. 103

p. 301 'I went to the Cristóbal dispensary this morning': Chatfield, *ibid.*, p. 150

p. 301 Albert Peters . . .: 'Competition for the Best True Stories . . .'

p. 301 Barbadian Amos Parks . . .: *ibid.*

p. 301 'the horrible and unfamiliar noise at night': van Hardeveld, *Make Dirt Fly*, p. 49ff

p. 302 Thus the excavation was planned to proceed . . .: Sibert and Stevens, *The Construction of the Panama Canal*, p. 77

p. 303 The US locomotives could haul four or five times the volume . . .: Pepperman, *Who Built the Panama Canal?*, p. 38

p. 303 a single rock weighing some 34 tons . . .: Sibert and Stevens, *The Construction of the Panama Canal*, p. 89

p. 305 'ran over a colored man': Chatfield, *Light on Dark Places*, p. 198

p. 305 'Oh, give me my Teddy, give me my Teddy': van Hardeveld, *Make Dirt Fly*, p. 72

Chapter Twenty-Three: Segregation (pp. 307–320)

p. 307 'There is much talk about the anticipated visit of the president': Chatfield, *Light on Dark Places*, p. 150

p. 307 'seemed obsessed with the idea that someone was trying to hide something from him': Maltby, 'In at the Start at Panama', *Civil Engineering*, September 1945, p. 422

p. 308 'When the president was at Cristóbal': Chatfield, *Light on Dark Places*, p. 210

p. 308 'A Strenuous Exhibition on the Isthmus': McCullough, *The Path between the Seas*, p. 496

p. 308 'He was intensely energetic': Maltby, 'In at the Start at Panama', *Civil Engineering*, September 1945, p. 421

p. 309 'Every man seems animated with the idea that he is doing a necessary part of the canal': Thompson, 'The Labour problems of the Panama Canal', *Engineering* (London), 3 May 1907

p. 309 At Gatún over a hundred new borings had been made on the dam site . . .: Mallet to Grey, 31 January 1907, FO881/8897

p. 309 'something which will redound immeasurably to the credit of America': Pepperman, *Who Built the Panama Canal?*, pp. 13–14

p. 309 'the heartiest contempt and indignation': Pepperman, *ibid.*, p. 292

p. 310 'Those cooking sheds with their muddy floors': McCullough, *The Path between the Seas*, pp. 502–3

p. 310 'The higher death rate is, in our opinion, due to circumstances': Colón *Independent*, 24 August 1906

p. 310 'wretched little houses rest on stilts': Chatfield, *Light on Dark Places*, p. 137

p. 311 'a striking lack of appreciation': Lindsay-Poland, *Emperors in the Jungle*, p. 35

p. 311 'racial and ethnic discrimination by the U.S. Government': Conniff, in 'Publication of the Proceedings of Symposium held at the University of the West Indies Mona, Jamaica, June 15–17, 2000', p. 43

p. 311 'Panama is below the Mason and Dixon Line': Franck, *Zone Policeman 88*, p. 65

p. 311 'If the stronger and cleverer race is free to impose its will': Woodward, *The Strange Career of Jim Crow*, p. 72

p. 311 In 1896 Louisiana had contained 130,000 black voters . . .: Woodward, *ibid.*, p. 85

p. 312 These included foremen, office clerks and teachers . . .: Petras, *Jamaican Labor Migration*, p. 150

p. 312 'the solution to troubles growing out of the intermingling of the races': Haskin, *The Panama Canal*, p. 160

p. 313 'It would, I think, be very impolitic to separate': Commissary manager to Stevens, 15 February 1907, RG185 2-C-55

p. 313 The white schools, housed in new buildings and well staffed and equipped . . .: Westerman, Conniff, Newton and Harper, 'Tracing the Course of Growth and Development in Educational Policy for the Canal Zone Colored Schools, 1905–1955' (University of Michigan School of Education 1974), N. L. Engelhardt 'Report on the Survey of the Schools of the Panama Canal Zone' (Mt Hope 1930)

p. 314 'Any northerner can say "nigger" as glibly as a Carolinian': Franck, *Zone Policeman 88*, p. 225

p. 314 'was made to feel the prejudice against her color': New York *Age*, quoted by the Colón *Independent*, 29 December 1905

p. 314 'My father read of Panama and thought it a wonderful place': Mrs Taylor, interviewed by Eunice Mason

p. 314 Jeremiah Waisome . . .: Competition for the Best True Stories . . .'

p. 315 'often seen the threat of the slave-driver in the foreman's eye': Carr, 'The Silver Men', *The Outlook*, 19 May 1906, p. 118

p. 315 'Among the white employees on the "gold roll" some times an employee would use his hands': letter of Ralph B. Irwin Manuscript Collection of the Canal Zone Library-Museum, Library of Congress, Washington, D.C., Box 35

p. 315 'the he-man type': van Hardeveld, *Make Dirt Fly*, p. 114ff

p. 315 'it cost twenty-five dollars to lick a Jamaican negro': Grier, *On the Canal Zone*, p. 71

p. 315 'straighten himself up and say to the foreman': Karner, *More Recollections*, p. 40

p. 316 'developed an excessive regard for the English': 'Publication of the Proceedings of Symposium held at the University of the West Indies Mona, Jamaica, June 15–17, 2000', p. 64

p. 316 '"Pay me," I says, "or I'll stick de British bulldog on all yo' Omericans!"': Walrond, *Tropic Death*, p. 42

p. 316 'You couldn't talk back': Constantine Parkinson in 'Diggers' documentary

p. 317 'little better than the West Indian negro': Gaillard to Goethals, 12 July 1907, RG 185

p. 317 'the efficiency of the Spaniards did not hold up': Sibert and Stevens, *The Construction of the Panama Canal*, p. 118

p. 317 nearly half of those recruited during 1906 were gone by the beginning of the following year . . .: Mallet to Grey, 31 January 1907, FO881/8897

p. 317 But clashes between Spaniards and police continued . . .: Navas, *El movimiento obrero en Panamá*, pp. 143–4

p. 317 'People are falling ill the whole time': *El Socialista*, 21 December 1906, quoted in Marco, *Los Obreros espanoles*, p. 18

p. 318 'handed over to the care of a black, who, as you'd expect, doesn't bother himself': *El Socialista*, 24 May 1907, quoted in Marco, *Los Obreros espanoles*, p. 19
'The labourers' lives are not highly valued': *El Socialista*, 14 May 1909, quoted in Marco, *Los Obreros espanoles* p. 32

p. 318 A Naples paper claimed that most of the workers had died . . .: Boni, *Panamá, Italia y los Italianos*, p. 144
p. 318 'My own private opinion': Stevens to Shonts, 18 January 1907, RG 185 2-E-1
p. 318 the deployment of no less than sixty-three Bucyrus shovels . . .: Waldo, 'The Present Status of the Panama Canal', *Engineering*, 15 March 1907
p. 319 'Stevens must get out at once': Roosevelt to Taft, 12 February 1907, Taft Papers, Taft-Roosevelt, Box 3, quoted in Duval, *And the Mountains will Move*, p. 259
p. 319 'blow up the Republican Party': McCullough, *The Path between the Seas*, p. 506
p. 319 'an immoderate amount of adulation': Mallet to Foreign Office, 9 April 1907, FO 881/9201
p. 319 'well-planned and well-built machine': Stevens, 'An engineer's recollections', *Engineering News-Record*, 5 September 1935, p. 52
'I know you pretty well now and without raising the question of your competence': Maltby, 'In at the Start at Panama', *Civil Engineering*, September 1945, p. 423
p. 320 'Don't talk, dig': Panama *Star and Herald*, 1 March 1907
p. 320 'in the charge of men who will stay on the job': Mack, *The Land Divided*, p. 501

Chapter Twenty-Four: 'The Army of Panama' (pp. 321–339)

p. 321 'Colonel Goethals here is to be chairman': Sullivan, *Our Times*, vol. 1, p. 466
p. 321 'most absolute despot in the world' . . .: Bishop, *Goethals: Genius of the Panama Canal*, p. 239
'It was asserted that the Department of Government': Goethals (ed.), *The Panama Canal: An Engineering Treatise*, Vol. 1, p. 46
p. 322 'a case of just plain straight duty': Bishop, *Goethals: Genius of the Panama Canal*, p. 149
p. 322 'I expect to be chief of the division of engineers': Panama *Star and Herald*, 19 March 1907
p. 322 'The magnitude of the work grows and grows on me': Goethals to his son, 17 March 1907, Goethals Papers, Library of Congress
p. 323 'Yellow Peril': van Hardeveld, *Make Dirt Fly*, p. 99
p. 324 'the President, in his talks, praised the men for their patriotism': Sibert and Stevens, *The Construction of the Panama Canal*, p. 268
p. 324 'was not necessary to the work of building the canal': Petras, *Jamaican Labor Migration*, p. 180
p. 324 'I have no complaint of any kind against the Isthmian Canal Commission': Panama Police Department Report, 14 May 1907, RG185 2-E-1
p. 325 'Supposed ill treatment has been the source of a great deal of trouble': US Senate 'Report of Special Commission appointed to investigate conditions of labor and housing of Government employees of the Isthmus of Panama', 8 December 1908, p. 51
p. 325 'Few or none of the gold-payroll employees bothered to learn enough of the languages': Sands, *Our Jungle Diplomacy*, p. 28
p. 325 'professional agitators': *Canal Record*, 25 March 1908

p. 325 'In Panama, Spaniards are treated as of less importance than negroes': *El Liberal*, 7 May 1907, quoted in Marco, *Los Obreros espanoles*, p. 24

p. 326 'Without doubt, the workers were rather merry': *El Socialista*, 22 November 1907, quoted in Marco, *ibid.*, p. 27

p. 326 'At the present time all of our superintendents and foremen are unanimously of the opinion': Goethals to Spanish chargé d'affaires at the legation in Panama, 17 April 1909, RG85

p. 327 'The biggest boss is King Yardage': Blythe, 'Life in Spigotty Land', *Saturday Evening Post*, Philadelphia, 21 March 1908

p. 328 'the last vestige of fear and uncertainty seemed to have left': van Hardeveld, *Make Dirt Fly*, p. 131

p. 328 'By mid-1908': Jessie Murdoch account, *Chagres Yearbook*, 1913, p. 58

p. 328 By May 1908 there were well over a thousand families on the Zone . . .: 'Report of Special Commission appointed to investigate conditions of labor and housing of Government employees of the Isthmus of Panama', 8 December 1908, p. 7

p. 328 'It is doubtful, to be sure, whether one-fourth of the "Zoners"': Franck, *Zone Policeman 88*, p. 220

p. 329 'one of the brand new cottages over the hill': van Hardeveld, *Make Dirt Fly*, p. 110ff

p. 329 'not until the business depression' . . . 'Report of Special Commission appointed to investigate conditions of labor and housing of Government employees of the Isthmus of Panama', 8 December 1908, pp. 11–12

p. 329 the turnover of skilled workers was nearly 60 per cent . . .: Haskin, *The Panama Canal*, p. 529

p. 330 'Anyone who stays here through a year of it becomes depressed': Ghent, 'Work and Welfare on the Canal', *The Independent*, 29 April 1909, p. 910

p. 330 'They fill a necessary place in the somewhat artificial life on the canal zone': 'Report of Special Commission appointed to investigate conditions of labor and housing of Government employees of the Isthmus of Panama', 8 December 1908, p. 19

p. 330 'to be furnished by Sidney Landon, character delineator': *Canal Record*, 11 September 1907

p. 331 'a really active American community': van Hardeveld, *Make Dirt Fly*, p. 130

p. 331 'He says it is not home, but on the order of a boarding school': 8 May 1907 Manuscript Collection of the Canal Zone Library-Museum, Library of Congress, Washington, D.C., Boxes 9–10

p. 331 'Every day I am better pleased that I came': 12 June 1907

p. 332 'things are not always very clean': 6 July 1907

p. 332 'I have adapted myself pretty well': 27 July 1907

p. 332 'The Tivoli is giving a reception and dance tonight': 6 July 1907

p. 332 'Pineapples are only fifteen cents': 17 July 1907

p. 332 'things have already begun to slack up': 27 June 1907

p. 332 'Now what can they tell about it?': 12 November 1907

p. 332 'the novelty has worn off': 1 December 1907

p. 332 'Empire Lady Minstrels': 22 March 1908
p. 333 'Nothing ever happens here': 6 November 1908
p. 333 'Am beginning to like Culebra better': 7 December 1908
p. 333 'broke into Culebra society': 10 January 1909
p. 333 'Pyne stupid and Saville unpleasant': 27 April 1909
p. 333 'It is pretty bad at present': 28 February 1909
p. 333 'a dry season night down here, with a moon': 28 October 1910
p. 333 'She's just about the nicest thing in the girl line there is': 17 November 1911
p. 334 'The commissary is an assured success': Ghent, 'Work and Welfare on the Canal', *The Independent*, 29 April 1909, p. 914
p. 334 'First of all, there ain't any democracy down here': Edwards, *Panama, the Canal, the Country, the People*, p. 572
p. 334 'the establishment of a an autocratic form of government': Thompson, 'The Labour problems of the Panama Canal', *Engineering* (London), 3 May 1907
p. 334 'enlightened despotism': Franck, *Zone Policeman 88*, p. 205
p. 334 'omnipresent': van Hardeveld, *Make Dirt Fly*, pp. 99–100
p. 334 'Goethals dominates over everybody and everything': Mallet's annual report 1910, FO881/9841
p. 334 'You can't realize what the Chief Engineer is': Courtney Lindsay, 3 April 1908
p. 334 'The system is one that would be very repugnant to Englishmen': Thompson 'The Labour problems of the Panama Canal', *Engineering* (London), 3 May 1907
p. 334 'judicial terrorism': Ghent, 'Work and Welfare on the Canal', *The Independent*, 29 April 1909, p. 913
p. 335 'there has grown up in Panama circles somewhat of a tendency to monopolise patriotism': Waldo, 'The Panama Canal Work, and the Workers', *Engineering Magazine*, p. 15
p. 335 'Caste lines are as sharply drawn as in India': Franck, *Zone Policeman 88*, p. 219
p. 336 'a narrow ribbon of standardized buildings': Sands, *Our Jungle Diplomacy*, p. 25
p. 337 'a chestless youth': Franck, *Zone Policeman 88*, p. 10ff
p. 337 'We have such control in Panama': Major, *Prize Possession*, p. 126
p. 337 'extremely tactful and friendly towards everybody': Mallet's annual report 1910, FO881/9841
p. 337 'a racial inability to refrain long from abuse of power': George Weitzel, quoted in Major, *Prize Possession*, p. 125
p. 337 'It is really farcial to talk of Panama as an independent state': Mallet to Grey, 22 August 1910, FO371/944
p. 337 'docile to American wishes': Mallet to Foreign Office, 8 May 1913, FO881/10293
p. 338 a story that accused Taft's influential brother Charles . . .: New York *World*, 3, 6 October 1908
p. 338 'I have never known in my lengthy experience in company matters': Harding, *The Untold Story of Panama*, p. 35
p. 338 'There are many peculiar circumstances about the Panama canal': Niemeier, *The Story of Panama*, p. 124

Chapter Twenty-Five: 'Hell's Gorge' (pp. 340–353)

p. 340 'a tropical glacier – of mud instead of ice': Smithsonian online exhibition http://www.sil.si.edu/Exhibitions/Make-the-Dirt-Fly

p. 340 'it required night and day work to save our equipment': Mallet to Foreign Office, 31 January 1908, FO 881/9201

p. 340 'the magnitude of the task is much greater than was at first thought': Mallet to Foreign Office, 31 January 1908, FO 881/9201

p. 340 'most formidable of the canal enterprise': Goethals (ed.), *The Panama Canal: An Engineering Treatise*, Vol. 1, p. 337

p. 340 'like snow off a roof': McCullough, *The Path between the Seas*, p. 551

p. 341 Antonio Sanchez . . .: interview with Mr William Donadío, Panama City, 17 August 2004

p. 341 'the old hill politely slid back again': Smithsonian online exhibition http://www.sil.si.edu/Exhibitions/Make-the-Dirt-Fly

p. 341 'Today you dig and tomorrow it slides': Albert Banister in 'Competition for the Best True Stories . . .'

p. 341 'was a land of the fantastic and the unexpected': Bishop, *The Panama Gateway*, pp. 193–4

p. 341 'The difficulties we are liable to encounter are unknown': Mallet to Foreign Office, 31 January 1908, FO 881/9201

p. 341 'This material has or will ultimately make its own design': Sibert and Stevens, *The Construction of the Panama Canal*, p. 166

p. 341 'found themselves handling hard rock one hour': Goethals (ed.), *The Panama Canal: An Engineering Treatise*, Vol. 1, p. 346

p. 342 'The Cut is a tremendous demonstration': Archer, *Through Afro-America*, p. 287

p. 343 'heroic human endeavour': James, *A Woman in the Wilderness*, p. 96

p. 343 'They are generally comfortable men and women of 50 or more': William Baxter account, *Chagres Yearbook*, 1913, p. 59ff

p. 343 Arnold Small . . .: 'Diggers' documentary

p. 343 John Prescod . . .: 'Competition for the Best True Stories . . .'

p. 343 'I had never saw so much rain in all my life': Rufus Forde, *ibid.*

p. 344 'The different levels varied from ten to twenty feet': Franck, *Zone Policeman 88*, p. 115

p. 345 Jamaican Z. McKenzie . . .: 'Competition for the Best True Stories . . .'

p. 345 Amos Clarke . . .: *ibid.*

p. 346 'Two days later, all of us who had become such close friends': van Hardeveld, *Make Dirt Fly*, pp. 136–7

p. 346 'at the end of long years of patient, exacting work': Sibert and Stevens, *The Construction of the Panama Canal*, p. 10

p. 346 George Martin . . .: 'Competition for the Best True Stories . . .'

p. 347 netted 1800 specimens in a week . . .: Le Prince and Orenstein, *Mosquito Control in Panama*, p. 210

p. 348 'fine body of disciplined and skilled workmen': Mallet to Foreign Office, 8 May 1913, FO881/10293

p. 348 the West Indian labourer had lived down his bad reputation . . .: Haskin, *The Panama Canal*, p. 154, p. 162

p. 348 'the negroes from the British West Indies': Franck, *Zone Policeman 88*, p. 123

p. 348 'What was the black culture': Victor Smythe in Russell, *The Last Buffalo*, p. 15

p. 349 'They lived chiefly in windowless, six-by-eight rooms': Franck, *Zone Policeman 88*, p. 40

p. 349 'On Sunday morning every religious community is busy': Carr, 'The Panama Canal', *The Outlook*, 5 May 1906

p. 349 'a forum for expression on many issues': Westerman, *Los immigrantes antillanos en Panamá*, p. 23

p. 349 'tame them and provide a relief valve': Navas, Luis *El movimiento obrero en Panamá*, p. 250

p. 350 'were the unquestioned leaders of glamour and glitter': Russell, *The Last Buffalo*, p. 34

p. 350 'To see people at night': Benjamin Jordan in 'Diggers' documentary

p. 350 'the elegant quadrille dances': Russell, *The Last Buffalo*, p. 37

p. 350 'the vortex of trouble on the Isthmus': Franck, *Zone Policeman 88*, p. 79

p. 351 'The majority of them were employed as team drivers': Enrique Plummer, 'Competition for the Best True Stories . . .'

p. 351 'although no such requirement is made of white employees': Goethals to Henry A. Hart, John Thomas, 18 March 1910 2-C-55, pt 1, General Correspondence, 1904–14, RG 185

p. 351 no more US blacks were given Gold Roll contracts . . .: Gaillard (acting chairman and Chief Engineer) to Jackson Smith, 11 February 1908, 2-C-55, pt 1, General Correspondence, 1904–14, RG 185

p. 351 by February 1909 only one such employee remained, a Henry Williams . . .: letters of 8 February, 22 May 1909, 2-C-55, pt 1, General Correspondence, 1904–14, RG 185

p. 351 'that he was colored and not eligible for employment on the Gold Roll': letter to Goethals, 10 January 1910, 2-C-55, pt 1, General Correspondence, 1904–14, RG 185

p. 352 In July 1912 there were only 69 . . .: letter to Goethals, 9 July 1912, 2-C-55, pt 1, General Correspondence, 1904–14 RG 185

p. 352 The following year, only fifteen remained . . .: letter to Goethals, 5 March 1913, 2-C-55, pt 1, General Correspondence, 1904–14, RG 185

p. 352 'the knowledge that there are ten hungry applicants': Panama *Star and Herald*, 7 September 1907

p. 352 'openly unfriendly towards all coloured West Indians': Mallet's annual report 1910, FO881/9841

p. 353 a 104-year-old resident of Colón . . .: interview with Mr Foster Burns, Colón, 19 August 2004

Chapter Twenty-Six: 'Lord How Piercing!' (pp. 354–365)

p. 355 'an ugly denuded waste of land': Lee, *The Strength to Move Mountains*, p. 172
p. 356 Barbadian Edgar Simmons . . .: 'Competition for the Best True Stories . . .'
p. 356 'Chagres River plunges': Lee, *The Strength to Move Mountains*, p. 183
p. 357 'the outbreak of yellow fever journalism': Lee, *ibid.*, p. 185
p. 357 'fire in the rear . . .': Lee, *The Strength to Move Mountains*, p. 189
p. 358 'It is not hard to realize': Franck, *Zone Policeman 88*, p. 307
p. 359 On a couple, deer were even found . . .: Sibert and Stevens, *The Construction of the Panama Canal*, p. 258
p. 359 'No one expected on returning to work in the morning': Sibert and Stevens, *ibid.*, p. 205
p. 359 American Harry Cole . . .: 'Some Episodes in connection with the Construction of the Pacific Division of the Panama Canal' 1908–1914 by Harry O. Cole, unpublished memoir 1947, Box 1 Manuscript Collection of the Canal Zone Library-Museum, Library of Congress, Washington, D.C.
p. 361 'They did not look with favour': Cameron, *The Impossible Dream*, p. 189
p. 362 'Joseph Seeley, a West Indian laborer', *Canal Record*, 28 December 1913
p. 362 'These locks are more than just tons of concrete': Lee, *The Strength to Move Mountains*, p. 263
p. 364 'Yuh gets more money for that job than working in the cut': 'West Indian Work Songs' Box 33 Manuscript Collection of the Canal Zone Library-Museum, Library of Congress, Washington, D.C.
p. 364 'I found myself racing across the narrow plank bridges': Franck, *Zone Policeman 88*, p. 181
p. 364 Eustace Tabois . . .: 'Diggers' documentary
p. 364 'hurting many and sometimes killing men instantly': James Ashby, 'Competition for the Best True Stories . . .'
p. 364 'The family of those men working on those locks': T. H. Riley, 'Competition for the Best True Stories . . .'
p. 364 Jamaican Nehemiah Douglas . . .: *ibid.*
p. 364 'poured like sand': Rufus Forde, *ibid.*

Chapter Twenty-Seven: The Land Divided, The World United (pp. 366–370)

p. 368 'There was a reverent silence': van Hardeveld, *Make Dirt Fly*, p. 142
p. 369 'Gentlemen, the two consuming ambitions of my life': Cameron, *The Impossible Dream*, p. 254
p. 369 'On board were all the foremost Panamanian': James, *A Woman in the Wilderness*, p. 94ff
p. 370 'Unostentatious dedicatory act a more appropriate celebration': Lee, *The Strength to Move Mountains*, p. 283

Postscript: Whose Canal is it Anyway? (pp. 371–378)

p. 373 'This I can say absolutely was my own work': Roosevelt, *The Letters of Theodore Roosevelt*, Vol. 6, p. 1444

p. 373 'I am interested in the Panama Canal because I started it': *New York Times*, 24 March 1911

p. 373 'somebody was prepared to act with decision': Roosevelt, *An Autobiography*, p. 553

p. 373 argued a contributor to the *North American Review* . . .: Chamberlain, Leander T., 'A Chapter of National Dishonor', *North American Review*, February 1912

p. 373 'an affront to international decency': Elmer Ellis, quoted in Graham, *The 'Interest of Civilization'?*, p. 178

p. 373 'The summary ejectment of Colombia': Mahan 'Was Panama "A chapter of national dishonor"?, *North American Review*, 196, 1912

p. 374 dubbed 'canalimony': Graham, *The 'Interests of Civilization'?*, p. 158

p. 374 'a most sordid and shameless conspiracy': New York *World*, 24 April 1921

p. 376 'The returned Panama Canal labourer is an uncommonly vain fellow': Richardson, *Panama Money in Barbados*, p. 149

p. 376 Z. Mackenzie . . .: 'Competition for the Best True Stories . . .'

p. 376 'Look de Panama man come home from sea': Frederick, *Colón Man a Come*, p. 128

p. 377 One man who had worked for thirty-eight years . . .: Petras, *Jamaican Labor Migration*, pp. 227–8

p. 377 A doctor who treated a lot of the old-timers . . .: Interview with Dr Hedley C. Lennon, Panama City, 18 August 2004

p. 378 'I am glad to see that all my sweat': Alfred E. Dottin, 'Competition for the Best True Stories . . .'

p. 378 'It is a job well done': S. Smith, 'Competition for the Best True Stories . . .'

p. 378 'I got to be a man': 'Diggers' documentary

BIBLIOGRAPHY

Abbot, Willis, J., *Panama and the Canal in Pictures and Prose*, New York and London, 1913

Acosta, José de, *Historia natural y moral de las Indias occidentales ó América*, Vol. 2, Madrid, 1787

Anguizola, Gustave, *Philippe Bunau-Varilla: The Man Behind the Panama Canal*, Chicago, 1980

Archer, William, *Through Afro-America: An English Reading of the Race Problem*, London, 1910

Arias, Harmodio, *The Panama Canal: A Study in International Law and Diplomacy*, London, 1911

Bailey, Thomas, *A Diplomatic History of the American People*, New York, 1947

Barrett, J., *The Panama Canal, What it is, What it Means . . .*, Washington, 1913

Barry, Tom, and Lindsay-Poland, John, *Inside Panama*, Albuquerque, New Mexico, 1995

Bates, Lindon Wallace, *The Crisis at Panama*, New York, 1906

—*Retrieval at Panama* New York 1907

Beale, Howard, *Theodore Roosevelt and the Rise of American Power*, Baltimore, 1956

Beach, Rex, *The Ne'er do Well*, New York, 1911

Beisanz, John and Mavis, *The People of Panama*, New York, 1955

Bemis, Samuel Flagg (ed.), *The American Secretaries of State and their Diplomacy*, New York, 1958

Bethell, Leslie (ed.), *Cambridge History of Latin America*, Vol. VII, Cambridge and New York, 1990

Bennett, Ira E., *History of the Panama Canal*, Washington, 1915

Beveridge, Albert, *The Meaning of Times and Other Speeches*, Indianapolis, 1908

Bidwell, Charles, *The Isthmus of Panama*, London, 1865

Bigelow, John, *The Panama Canal. Report of the Hon. John Bigelow, delegated by the Chamber of Commerce of New York to assist at the inspection of the Panama canal in February 1886*, New York, 1886

—*The Panama Canal and the Daughters of Danaus*, New York, 1908

Bigelow, Poultney, 'Our Mismanagement at Panama', *The Independent*, 4 January 1906

Birmingham, Stephen, *Our Crowd: The Great Jewish Families of New York*, New York, 1967

Bishop, Joseph Bucklin, *The Panama Gateway*, New York, 1913

—*Theodore Roosevelt and his Time Shown in His Own Letters*, New York, 1920

—*Goethals: Genius of the Panama Canal: A Biography*, New York, 1930

Blythe, Samuel G., 'Life in Spigotty Land', *Saturday Evening Post*, Philadelphia, 21 March 1908

Boni, Diego dal, *Panamá, Italia y los Italianos en la época de la construcción del Canal (1880–1915)*, Panama City, 2000

Borthwick, J. D., *Three Years in California*, Edinburgh and London, 1857

Brands, H. W., *The Age of Gold*, London, 2005

Brau de Saint-Pol, Lias, *Percement de L'Isthme de Panama*, Paris, 1879

Bunau-Varilla, Philippe, *Panama: The Creation, Destruction, and Resurrection*, New York, 1914

Burnett, Robert, *The Life of Paul Gauguin*, New York, 1937

Burton, John (ed.), *The Darien Papers*, Edinburgh, 1849

Cameron, Ian, *The Impossible Dream: The Building of the Panama Canal*, London, 1971

Carr, John Foster, Various articles, *The Outlook*, 28 April, 5, 12, 19 May 1906

Castillero Pimentel, Ernesto, *Panamá y los Estados Unidos*, Panama City, 1973

Castillero Reyes, Ernesto J., *Historia de la Communication Interoceanica*, 1939, quotes from 1999 edn, Panama City

— *Historia de Panama*, Panama City, 1962

Cermoise, Henri *Deux Ans à Panama: Notes et Recits d'un Ingenieur au Canal*, Paris, 1886

Chamberlain, Leander T., 'A Chapter of National Dishonor', *North American Review*, February 1912

Chambre des Députés, *5e Législature, Session de 1893, Rapport Général . . .* , Paris, 1893

Chatfield, Mary, *Light on Dark Places*, New York, 1908

Clapp, Margaret, *Forgotten First Citizen: John Bigelow*, Boston, 1947

Cole, Harry O., *Some Episodes in connection with the Construction of the Pacific Division of the Panama Canal*, unpublished memoir, n.d., Box 1 Manuscript Collection of the Canal Zone Library-Museum, Library of Congress, Washington, D.C.

Collin, Richard, *Theodore Roosevelt's Caribbean: The Panama Canal, the Monroe Doctrine and the Latin American Context*, Baton Rouge, Louisiana, 1990

Congrès des Sciences Géographique, Cosmographique et Commerciales, Antwerp, 1871, *Compte Rendu*, Antwerp, 1872

Congrès International D'Etudes du Canal Interocéanique, *Compte Rendu Des Séances*, Paris, 1879

Conniff, Michael L., *Black Labor on a White Canal – Panama 1904–1981*, Pittsburgh, Pennsylvania, 1985

—*Panama and the United States: The Forced Alliance,* Athens, Georgia, 1992

Conrad, Joseph, *Nostromo: A Tale of the Seaboard*, London and New York, 1904

Cornish, Vaughan, *The Panama Canal and its Makers*, London, 1909

Courau, Robert, *Ferdinand de Lesseps*, Paris, 1932

Cullen, Edward et al., *Over Darien by a ship canal*, London, 1856
Cullom, Shelby Moore, *Fifty Years of Public Service*, Chicago, 1911
Dansette, Adrien, *Les Affaires de Panama*, Paris, 1934
Davis, Raymond A., 'West Indian Workers on the Panama Canal: A Split Labor Market Interpretation', Ph.D. Diss., Stanford, Connecticut, 1981
Dean, Arthur H., *William Nelson Cromwell*, New York, 1957
Dock, Lavinia. L. (ed.), *A History of Nursing*, 4 vols, New York, 1912
Donadío, William Daniel, *The Thorns of the Rose*, Colón, 1999
Drumont, Edouard, *La Dernière Battaille*, Paris, 1890
Duval, Miles, *From Cadiz to Cathay*, Stanford, Connecticut, 1940
—*And the Mountains will Move* Stanford, Connecticut, 1947
Edgar-Bonnet, G., *Ferdinand de Lesseps*, Paris, 1951
Edwards, Albert (pseudonym of Bullard, Arthur), *Panama, the Canal, the Country, the People*, New York, 1911
Espino, Oviedo Diaz, *How Wall Street Created a Nation*, New York, 2001
Fauconnier, Gérard, *Panama: Armand Reclus et le Canal des Deuz Océans*, Paris, 2004
Franck, Harry A., *Zone Policeman 88*, New York, 1913
Fraser, John Foster, *Panama and What it Means*, London, 1913
Frederick, Rhonda D., *'Colón Man a Come': Mythographies of Panamá Canal Migration*, Lanham, Maryland, 2005
Frémont, Jesse Benton, *A Year of American Travel: Narrative of Personal Experience*, California, 1878
Ghent, W. J. 'Labor and the Commissary at Panama' and 'Work and Welfare on the Canal', *The Independent* 66: pt 2 (April–June 1909)
Goethals, George W., *Government of the Canal Zone*, Princeton, New Jersey, 1915
—(ed.), *The Panama Canal: An Engineering Treatise*, 2 vols, New York, 1916
Gorgas, Marie D., and Hendrick, Burton J., *William Crawford Gorgas: His Life and Work*, New York, 1915
Gorgas, William Crawford, *Sanitation in Panama*, New York, 1915
Graham, Terence, *The 'Interests of Civilization'?: Reaction in the United States Against the 'Seizure' of the Panama Canal Zone, 1903–1904*, Sweden, 1983
Great Britain, Foreign Office, *Correspondence respecting the projected Panama Canal . . . 1882*, London, 1883
Grier, Thomas Graham, *On the Canal Zone*, Chicago, 1908
Guerrero, Luis C., and Lorente, Jose G., *Los Obreros Espanoles en Panama*, Havana, 1908
Hains, Brig. Gen. Peter, 'The Labor Problem on the Panama Canal', *North American Review*, July 1906
Hardeveld, Rose van, *Make Dirt Fly*, Hollywood, California, 1956
Harding, Earl, *The Untold Story of Panama*, New York, 1959
Haskin, Frederic J., *The Panama Canal*, New York, 1913
Healy, David, *US Expansionism: The Imperialist Urge in the 1890s*, Madison, Wisconsin, 1970
Healy, Phyllis Foster, *Mary Eugenie Hibbard: Nurse, gentlewoman and patriot*, unpublished thesis, University of Texas at Austin, 1990

Heckadon-Moreno, Stanley, *Naturalists on the Isthmus of Panama*, Panama City, 2004
Helms, Mary W., *Ancient Panama: Chiefs in Search of Power*, Austin, Texas, 1976
Hill, Ralph Nading, *The Doctors who Conquered Yellow Fever*, New York, 1957
Hill, Robert A. (ed.), *Marcus Garvey and the Universal Negro Association Papers*, Berkeley, California, 1983
Hotchkiss, Charles F., *On the Ebb*, New Haven, 1878
Howarth, David, *The Golden Isthmus*, London, 1966
Humboldt, Alexander von, *Political essay on the kingdom of New Spain*, London, 1811
Instructions to Rear Admiral Daniel Ammen and Civil Engineer A. G. Menocal, U.S. Navy, Delegates on the Part of the United States to the Interoceanic Canal Congress, Held at Paris May, 1879, and Reports of the Proceedings of the Congress, Washington, D.C., 1879
Isthmian Historical Society, 'Competition for the Best True Stories of Life and Work on the Isthmus of Panama During the Construction of the Panama Canal', Balboa, 1963, Box 25 Manuscript Collection of the Canal Zone Library-Museum, Library of Congress, Washington, D.C.
Jaén Suárez, Omar, *La población del Istmo de Panamá*, Madrid, 1998
James, Winifred, *A Woman in the Wilderness*, London, 1915
Jefferson, Thomas, *Writings*, New York, 1859
Jessup, Philip C., *Elihu Root*, New York, 1938
Johnson, W. F., *Four Centuries of the Panama Canal*, New York, 1906
Johnson, Willis Fletcher (ed.), *Addresses and Papers of Theodore Roosevelt*, New York, 1909
Johnston, Dr William E., 'Interoceanic ship canal discussion', *Journal of the American Geographical Society*, Vol. 11, 1879
Jos, Joseph, *Guadeloupéens et Martiniquais au Canal de Panamá*, Paris, 2004
Jusserand, J. J., *What Me Befell*, Boston, 1933
Karner, William J., *More Recollections*, New York, 1921
Kemble, John Haskell, *The Panama Route, 1848–1869*, Berkeley, California, 1943
Kimball, William Wirt, *Special Intelligence Report on the Progress of the Work during the year 1885*, 49th Cong., 1st Sess., House of Representatives Misc. Doc. No. 395 Washington, D.C., 1886
LaFeber, W., *The Panama Canal: The Crisis in Historical Perspective*, 1978, quotes from updated edition, New York, 1989
Lamming, George, *In the Castle of my Skin*, London, 1953
Landes, David, *The Unbound Prometheus*, Cambridge, 1969
LaRosa, Michael, and Mejia, German R. (eds), *The United States Discovers Panama: The Writings of Soldiers, Scholars, Scientists and Scoundrels, 1850–1905*, Lanham, Maryland, 2004
Lee, W. S., *The Strength to Move a Mountain*, New York, 1958
Lemaitre, Eduardo *Panamá y su separción de Colombia*, Bogotá, 1971
Le Prince, Joseph A., and Orenstein, A. J., *Mosquito Control in Panama*, New York and London, 1916
Lesseps, Ferdinand de, *Recollections of Forty Years*, London, 1887
Lewis, Lancelot S., *The West Indian in Panama: Black Labor in Panama 1850–1914*, Washington, D.C., 1980

Lindsay-Poland, John, *Emperors in the Jungle: The Hidden History of the U.S. in Panama*, Durham, North Carolina, and London, 2003

Lisser, Henry de, *Jamaicans in Colon and the Canal Zone*, Kingston, 1906

Lloyd, John A., 'On the facilities for a ship canal communication . . . through the Isthmus of Panama', *Inst. of Civil Engineers, Minutes of Proceedings*, Vol. 9, 1850

Lull, Edward, *Reports of explorations and surveys for the location of interoceanic ship canal through the Isthmus of Panama*, Washington, D.C., 1879

Mack, Gerstle, *The Land Divided*, New York, 1944

Mahan, Captain A. T., *The Influence of Sea Power on History*, Boston, 1890

—*The Interest of America in Sea Power Present and Future*, Boston, 1898

—'Was Panama "A chapter of national dishonor"?', *North American Review*, October 1912

Maltby, Frank, 'In at the Start at Panama', *Civil Engineering,* June–September 1945

Major, John, *Prize Possession: The United States and the Panama Canal, 1903–1979* Cambridge, 1993

Mallet, Sir Claude Coventry, *Pioneer Diplomat*, unpublished memoir

Marco Serra, Yolanda, *Los Obreros espanoles en la construccion del Canal de Panama*, Panama City, 1997

Maréchal, Henri, *Voyage d'un Actionnaire à Panamá*, Paris, 1885

Marryat, Frank, *Mountains and Molehills, or, Recollections of a Burnt Journal*, 1855. Reprinted Philadelphia, Pennsylvania, 1962

Marshall, Peter, *Demanding the Impossible: A History of Anarchism*, London, 1992

McCain, William, *United States and the Republic of Panama*, Durham, North Carolina, 1937

McCarty, Mary Louise Allen, *Glimpses of Panama and the Canal*, Kansas City, Missouri, 1913

McKay, Claude, *Banana Bottom*, San Diego, California, 1933

McCullough, David, *The Path between the Seas*, New York, 1977

Mellander, Gustavo A., *Charles Edward Magoon: The Panama Years*, Panama City, 1999

—*The United States in Panamanian Politics: The Intriguing Formative Years*, Danville, Illinois, 1971

Menocal, A. G., 'Intrigues at the Paris Canal Conference', *North American Review*, September 1879

Mimande, Paul, *Souvenirs d'un Echappe de Panama*, Paris, 1893

Miner, Dwight Carol, *The Fight for the Panama Route: The Story of the Spooner Act and the Hay–Herran Treaty*, New York, 1940, quotes from 1966 edn, Octogon Books Inc., New York

Molinari, M. G. de, *A Panama: L'Isthme de Panama, La Martinque, Haiti*, Paris, 1887

Mollier, Jean-Yves, *Le Scandale de Panama*, Paris, 1991

Navas, Luis, *El movimiento obrero en Panamá 1880–1914*, Panama City, 1974

Nelms, Brena, *The Third Republic and the Centennial of 1789*, New York, 1987

Nelson, Wolfred, *Five Years at Panama*, New York, 1889

Newton, Velma, *The Silver Men: West Indian Labour Migration to Panama, 1850–1914*, Kingston, 1984

Niemeier, Jean Gilbreath, *The Story of Panama*, Portland, Oregon, 1968

Otis, Fessenden N., *History of the Panama Railroad*, New York, 1867
Parks, E. Taylor, *Colombia and the United States, 1765–1934*, Durham, North Carolina, 1935
Patterson, John, 'Latin American reactions to the Panamanian revolution of 1903', *Hispanic American Historical Review*, 24 (1944)
Pearcy, Thomas L., *We Answer Only to God: Politics and the Military in Panama, 1903–1947*, Albuquerque, New Mexico, 1998
Pepperman, Leon W., *Who Built the Panama Canal?*, New York, 1915
Perez-Penero, Alex, *Before the Five Frontiers*, New York, 1978
Petras, Elizabeth McClean, *Jamaican Labor Migration: White Capital and Black Labor, 1850–1930*, Boulder, Colorado, and London, 1988
Price, Roger, *The Economic Modernisation of France*, London, 1975
Pringle, Henry F., *Theodore Roosevelt: A Biography*, New York, 1931
Publication of the Proceedings of Symposium held at the University of the West Indies Mona, Jamaica, 15–17 June, 2000, 'West Indian Participation in the Construction of the Panama Canal'
Puri, Shalini (ed.), *Marginal Migrations: The Circulation of Cultures within the Caribbean*, Oxford, 2003
Reclus, Armand, *Panama & Darien*, Paris, 1884
Reyes, Raphael, *The Two Americas*, London, 1914
Richards, Benjamin. B. (ed.), *California Gold Rush Merchant: The Journal of Stephen Chapin Davis*, San Marino, California, 1956
Richardson, Bonham C., *Panama Money in Barbados 1900–1920*, Knoxville, Tennessee, 1985
Rink, Paul, *The Land Divided, the World United*, New York, 1963
Ripley, J. Fred, *The Capitalists and Colombia*, New York 1931
Robinson, Tracy, *Fifty Years at Panama*, New York, 1911
Rogers, Lt Charles C., *Intelligence Report of the Panama Canal*, Washington, 1889 (House Misc. Do. 599, 50th Cong., 1st Sess.)
Rodgers, Lt Raymond P., *Progress of Work on Panama Ship-Canal* (Sen. Doc. 123, 48th Cong., 1st Sess.), Washington, 1884
Rodrigues, José Carlos, *The Panama Canal*, New York, 1885
Roosevelt, Theodore, *The Letters of Theodore Roosevelt* (ed. Elting E. Morison et al.) Harvard, Connecticut, 1951–54, 8 vols
—*An Autobiography*, New York, 1913
Russell, Carlos E., *An Old Woman Remembers*, New York, 1995
—*The Last Buffalo*, Charlotte, North Carolina, 2003
Sands, William Franklin, *Our Jungle Diplomacy*, Chapel Hill, North Carolina, 1944
Schott, Joseph L., *Rails across Panama* New York, 1867
Scott, William Rufus, *The Americans in Panama*, New York, 1912
Seacole, Mary, *Wonderful Adventures of Mrs Mary Seacole in Many Lands*, Oxford, 1988 edn
Senior, Olive, 'The Colon People', *Jamaica Journal*, 2 parts, March and September 1978, 'The Panama Railway', *Jamaica Journal*, June 1980

Sibert, W. L. and Stevens, J. F. *The Construction of the Panama Canal*, New York, 1915
Siegfried, André, *Suez and Panama*, New York, 1940
Simon, Maron J., *The Panama Affair*, New York, 1971
Sisnett Cano, Octavio, Manuel (ed.), *Pedro Prestan: cien años después*, Panama City, 1985
Skinner, James M., *France and Panama: The Unknown Years, 1894–1908*, New York, 1988
Slosson, Edwin, and Richardson, Gardner, 'Life on the Canal Zone', *The Independent*, 22 March 1906
—'An Isthmian Carpenter's Story: A Jamaican Negro', *The Independent*, 19 April 1906
Smith, G. Barnett, *The Life and Enterprises of Ferdinand De Lesseps*, London, 1895
Stevens, John F., 'An engineer's recollections', *Engineering News-Record*, Vol. 115, 5 September 1935
—'A Momentous Hour at Panama', *Journal of the Franklin Institute*, July 1930
Stiner, William F. Klaus de Albuquerque, and Bryce-Laporte, Roy S. (eds), *Return Migration and Remittances: Developing a Caribbean Perspective*, Washington, D.C., 1982
Sullivan, Lt John T., *Report of Historical and Technical Information . . .*, Washington, D.C., 1883 (House Exec. Doc. 107, 47th Cong. 2nd Sess.)
Sullivan, Mark, *Our Times*, New York, 1928
Sweetman, David, *Paul Gauguin*, London, 1995
Thayer, William R., *Theodore Roosevelt: An Intimate Biography*, New York, 1919
Thompson, A. Beeby, 'The Labour problems of the Panama Canal', *Engineering*, (London) 3 May 1907
Tomes, Robert, *Panama in 1885*, New York, 1855
Trautwine, John C., 'Rough notes of an exploration for an inter-oceanic canal route by way of the Rivers Atranto and San Juan', *Franklin Institute Journal*, Vol. 57, No. 5, May 1854
Tuchman, Barbara, *The Proud Tower*, New York, 1966
United States Congress, House of Representatives, *The Story of Panama: hearings on the Rainey resolution* 62nd Cong., 2nd Sess., Washington, D.C., 1913
United States Congress, *Report of the Isthmian Canal Commission 1899–1901*, Washington, D.C., 1904
United States Congress, *Report of Special Commission appointed to investigate conditions of labor and housing of Government employees of the Isthmus of Panama*, 60th Cong., 2 Sess., Doc. No. 539, Washington, D.C., 1908
United States Isthmian Canal Commission, *Annual Reports*, Washington, D.C., 1906 etc.
—*Annual Reports of the Department of Sanitation*, Washington, D.C., 1908 etc.
—*Population and Deaths from various diseases . . .*, Washington, D.C., 1907
—*Report of the Isthmian Canal Commission, 1899–1901*, Washington, D.C., 1904
United States of America Navy Department, *Report of Commander Bowman H. McCalla upon the naval expedition to the Isthmus of Panama, April 1885*
—*Reports of Explorations and Surveys*, Washington, D.C., 1879
—*Papers on Naval Operations during the year ending July 1885*, Washington, D.C., 1885
United States Senate, *Report of Special Commission appointed to investigate conditions of labor and housing of Government employees of the Isthmus of Panama*, 60th Cong., 2nd Sess., Doc. No. 539, Washington, D.C., 1908

Vedia, Enrique de (ed.), *Historiadores primitivos de Indias*, Madrid, 1852

Vivian, James, 'The "Taking" of the Canal Zone: Myth and reality', *Diplomatic History* 4, 1980

Waldo, Fullerton L., 'An Engineer's Life in the Field on the Isthmus', *Engineering* December 1905

—'The Panama Canal Work, and the Workers', *Engineering Magazine*, 5 January 1907

—'The Present Status of the Panama Canal', *Engineering*, 15 March 1907

Wallace, John F., 'Preliminary Work on the Panama Canal', *Engineering Magazine*, October 1905

Walrond, Eric, *Tropic Death*, New York, 1926 quotes from Collier Books edn, 1972

Ward, Christopher, *Imperial Panama: Commerce and Conflict in Isthmian America 1550–1800*, Albuquerque, New Mexico, 1993

Watts, David, *The West Indies*, London, 1987

Weir, Hugh C., *The Conquest of the Isthmus*, New York, 1909

Weisberger, Bernard, 'The strange affair of the taking of the Panama Canal Zone', *American Heritage* 27, October 1976

Westerman, George, 'Gold and Silver Men', *Crisis*, 54: 12, December 1947

—'Historical notes on West Indians on the Isthmus of Panama', *Phylon*, 22: 4, Winter 1961

—*Los immigrantes antillanos en Panamá*, Panama City, 1980

Woodward, C. Vann, *The Strange Career of Jim Crow*, New York, 1955, quotes from 2002 paperback edn

Wyse, Lucien N-B., *Rapport sur les Etudes de la Commission Internationale d'exploration de L'Isthme du Darien*, Paris, 1877

—*Le Canal de Panama*, Paris, 1886

Wyse, Lucien N-B., Reclus, Armand, and Sosa P., *Rapport sur les Etudes de la Commission Internationale d'exploration de L'Isthme Americain*, Paris, 1879

Zévaès, Alexander, *Le Scandale du Panama*, Paris, 1931

Zimmerman, Warren, *First Great Triumph*, New York, 2002

ACKNOWLEDGEMENTS

I am very grateful to the K. Blundell Trust, administered by the Society of Authors, for their generous contribution to the research expenses of this book. Many people in a number of countries have helped me. For my research trips to the United States I am indebted to the staff of the New York Public Library, the Library of Congress and the National Archives, in particular Jackie Cohen, who went far beyond the call of duty in guiding me through the enormous amount of material – particularly the French records – held at NARA. I was also well looked after in the United States, by Peter Sollis and his family in Washington and by Ben and Louise Edwards in New York. I have also been lucky enough to receive help and encouragement from Roman Foster in New York and Eunice Mason in California as well as Barbara Adamson, Lesley Hendricks, Dr Philip C. Breunle, Donald Paine and Professor Hugh Brogan.

For help in researching the Jantje Milliery story I am indebted to Danielle Susijn and Patrick van Griethuysen in Holland. In the United Kingdom I was lucky enough to have the enthusiastic and invaluable support of the Panamanian consul and ambassador, Liliana Fernandez, who was able to open many doors for me in Panama. I must also acknowledge my debt to the staff of the British Library (where much of this book was written), the London Library, the Public Records Office, and to Carol Morgan at the Institute of Civil Engineers. Mrs Primose Mallet-Harris has been kind enough to let me quote from her grandfather's letters and offered hospitality at her home in Somerset. James Spence has provided much assistance on the engineering side of the story, and I am grateful to Dr Mary Allwood for her checking of the medical material. In France, I was greatly helped by Jane Martens, Jean-Yves Mollier and Gérard Fauconnier.

The time spent in Panama on several visits was a great pleasure in the researching of this book. If nothing else, I am delighted to have seen the amazing canal up close, an experience I can recommend to anyone. My greatest debts are to Judy Dixon and John Carlson, both of whom provided invaluable help with introductions and research material. I was also lucky enough to be taken under the wing of Gisela Lammerts Van Bueren at the Technical Resource Center at Balboa, Panama City, who provided enthusiastic assistance. From Panama I would also like to thank all those who gave me their time for interviews or assisted in other ways: Marc de Banville, Dr Angeles Ramos Baquero, Ned Blennerhassett, Foster Burns, Walter Clarke, Georges Colbourne, Graciela Dixon, William Donadío, Victor Echeverria, Terry Ford, Egla Gooden de Lynch, Cecil Haynes, Dr Stanley Heckadon-Moreno of the Smithsonian Tropical Research Institute, Eric Jackson, editor of the *Panama News*, Maria Esperanza Lavergne, Dr Hedley C. Lennon, Melva Lowe de Goodin, Jim Malcolm, UK ambassador, Gerardo Maloney, Professor of Sociology at the University of Panamá, Mercedes and Charles Morris, Marc Quinn, Lidia Ricardo, Carlos Russell, Enrique Sanchez and all at SAMAAP, Leonardo J. Sidnez, Henry F. Smith jr., Omar Jeán Suárez and Juan Tam.

I have been very lucky in having editors of great experience, skill and patience – Tony Whittome and James Nightingale at Hutchinson and Adam Bellow at Doubleday in New York – and I am grateful to all at their companies for their efforts on behalf of this book. I would also like to thank my agents, Julian Alexander in London and George Lucas in the US, for their encouragement and advice. Richard Collins has performed a very careful and skilful edit of the manuscript and Reg Piggott has drawn maps of real distinction and character.

Above all, I am indebted to my family – Anne and Paul Swain, who read early drafts of this book and provided invaluable advice and encouragement as well as a bolt hole to write in. My parents Sheila and David Parker read, translated and noted numerous works in French and Spanish respectively as well as commenting on early drafts. And my immediate family – Oliver, Thomas, Milly and their mother Hannah – while endlessly asking how many more pages I had to write, put up with good grace with my frequent absences on research trips and the many years of distraction that writing such a book as this entails.

INDEX

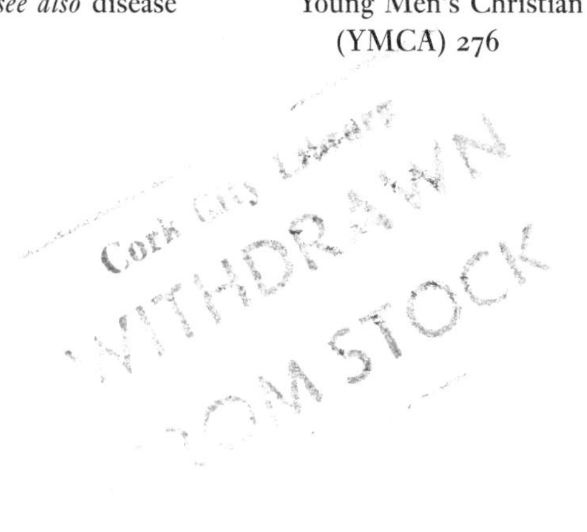